The Dean of Steam Fire Engine Builders

by Ed Hass

Typography by Helen Hegelheimer and Ed Hass.

Graphics by Ruthanne Baker.

KES-PRINT, INC.
8419 W 72 Terrace/Shawnee Mission KS 66204
913/722-0073

ISBN #0-9611166-2-5

Printed in U.S.A.

Table of Contents

About the Author	1
Acknowledgements	1
Preface	2
Chapter 1, 1848-1851	2
Chapter 2, Biographies	5
Chapter 3, 1852-1868	7
Chapter 4, 1869-1877	13
Chapter 5, 1877-1891	21
Chapter 6, 1891-1904	27
Chapter 7, 1905-1919	34
List of Cincinnati-Built Engines	38
Latta & Shawk, 1852	42
Abel Shawk, 1855-1856	44
A.B. & E. Latta, 1853-1862	46
Lane & Bodley, 1863-1868	50
C. Ahrens & Co., 1869-1877	53
Ahrens Mfg. Co., 1877-1891	74
American F.E. Co., 1891-1904	135
Ahrens F.E. Co., 1905-1910	171
Ahrens-Fox Co., 1910-1916	185
Numerical Index of Engines	198
Alphabetical Index of Engines	210
Index of Ahrens Patents	221
Ahrens Family Tree	222

About the Author

Ed Hass is a third-generation fire buff. Members of the Hass family have been chasing fire engines since 1911, when Grace Howard Hass (the author's grandmother) became a fire buff following a fire that swept Bangor, Maine.

Grace Hass passed along her interest in fire buffing to her four children. Bill Hass became an expert on "water tower" fire engines. He instilled an interest in fire apparatus in his children, one of whom is the author of this book.

Ed Hass became an Ahrens-Fox afficianado in 1966. He founded the Ahrens-Fox Fire Buffs' Association in 1970, and ever since has published their *Silver Sphere* newsletter for 400 members.

In 1979, Ed bought a 1953 Ahrens-Fox centrifugal pumper that is the last custom fire engine built in Ahrens-Fox's own factory (all subsequent Ahrens-Foxes were either built by a bus manufacturer or mounted on commercial chassis). This Ahrens-Fox is now undergoing a 100% restoration.

Ed's first book, *Ahrens-Fox: 'The Rolls-Royce of Fire Engines'*, was published in 1982.

As an outgrowth of his interest in Ahrens-Fox motor apparatus, Ed is interested in Ahrens steam fire engines. So *The Dean of Steam Fire Engine Builders* is a natural sequel to his first book.

When he is not pursuing his hobby/obsession with Ahrens-Fox fire engines, Ed earns his living as a Technical Writer in the electronics industry, in California's famed Silicon Valley. A native of New Jersey, he is a 1977 graduate of Rutgers University, with a B.A. in Journalism.

Acknowledgments

In addition to credits on individual photos (distributed throughout this book), I especially want to acknowledge the invaluable assistance of the following people and organizations in obtaining information and/or photographs for this book. Without their help, this book could not have been written.

Rick Ahrens
Larry Arnold
The late Armand Bolton
Cincinnati Fire Museum
Cincinnati Historical Society
The Clarence E. Meek Library of the Fire Department of New York
John E. Doyle
Evangelical-Lutheran Church of Osterholz, Germany
Fire Departments and Libraries in cities and towns that once used Ahrens steamers (too numerous to list individually, but I thank them all)
Firefighters Home and Museum, Hudson, N.Y.
Bill Hass
Rick Knife
Leo F. Krapp
Robert Long
The late Phil Nauman
Bruce Neal
Curt Nepper
New York Public Library
John M. Peckham
Public Library of Cincinnati and Hamilton County, Ohio
Jack Robrecht
Ross County Historical Society of Chillicothe, Ohio
Wayne Sorenson
Spring Grove Cemetery, Cincinnati, Ohio
Harold S. Walker
Dr. Lester Williams

Especial thanks to the publisher, H.H. Kesler, and to all those who believed in this book strongly enough to help finance its publication.

A photo of the Author.

Preface

Chris Ahrens was a leading figure in the steam fire engine industry from its inception (1852) to its demise (1916). He founded a family dynasty that was the fire engine industry's equivalent of the Henry Ford family.

To this day, no fire apparatus manufacturer enjoys as excellent a reputation for quality and craftsmanship as Chris Ahrens did. His fire engines held every world's record it was possible for a fire engine to achieve, and when the record was broken, it was almost always another Ahrens that broke it. Nearly every major innovation in fire apparatus for 80 years (1856-1936) was developed by members of Chris Ahrens' family, and today the company he helped start may be the oldest surviving manufacturer of self-powered vehicles in the world (133 years and counting).

It was often said that Chris Ahrens taught an entire industry how to make a quality product, and for that reason the press frequently called him "The Dean of Steam Fire Engine Builders" (hence the title of this book).

Ahrens not only made a product with an enviable reputation, he also took great pains to insure that the history of his product and of his company would survive long after he was gone. He kept meticulous records of every steam fire engine his company ever built or repaired, records which survive to this day and which form the basis for much of this book.

Until now, nobody has ever published the voluminous records that Chris Ahrens left behind, nor has anyone attempted to recount his life.

And a remarkable life it was, a shining example of the opportunities for success that America has always offered to entrepreneurs. Johann Christoph Ahrens was a German immigrant who escaped a bloody European revolution, arrived in America virtually penniless at the age of 18 (without his parents), and went on to found one of Cincinnati's most successful factories.

He became one of his city's most respected citizens, even having a street named after him. He loaned many Cincinnatians the money to start a business, which nearly cost Ahrens his life when the owner of a cigar factory he had financed shot him for refusing to make a second loan.

In writing the biography of Chris Ahrens and a history of his company, I used the premise "what if Chris Ahrens had kept as detailed a diary of his life as he did of his fire engines?" This led to my decision to write my book in the first person, as if Chris Ahrens had written it.

Unfortunately, first-person narration gives the impression that this book is fictional, which it is not. While some dialogue is invented, the facts are as accurate as it is possible to make them in recounting events of a century ago.

Chapter 1

Cincinnati, Ohio Dec. 7, 1918

Today is my 84th birthday. This morning, my old friend Fred Shepperd, publisher of *Fire & Water Engineering* magazine, dropped by my home for a visit. "Chris," he said, "I know you've had a long and successful career as a steam fire engine manufacturer. Have you ever thought of writing your autobiography?"

"As a matter of fact," I replied, "I have thought of writing a book. Not an autobiography, but the story of the steam fire engine industry, and of Cincinnati's role in that industry."

The first successful steam fire engine in the world was manufactured by Alexander Latta and Abel Shawk, right here in Cincinnati, in 1852. I became an apprentice machinist for Latta in 1854, and bought the company in 1868. I built more steamers right here in my own Cincinnati factory (named Ahrens Manufacturing Co. after myself, Chris Ahrens) than nearly any other fire apparatus manufacturer in the world. One of the last steamers ever built was also made in Cincinnati just two years ago, by my own sons and sons-in-law at Ahrens-Fox Fire Engine Company.

In that sense, I suppose this book really will be autobiographical, since my life and the history of Cincinnati's fire engine industry are heavily intertwined. For many years, I have kept a journal and a scrapbook of my activities in the steam fire engine industry. In the back of my mind, I suppose I always knew that I would some day use these records to write a book, but it was not until my discussion with Fred Shepperd this morning that I decided to actually start writing it. His promise to use his contacts in the publishing business certainly helped convince me that the time was finally right to write this book.

One of my motives in writing this book is the fact that we are now seeing the twilight of the colorful era of gleaming steam fire engines drawn by trained fire horses. No fire apparatus manufacturer is making new steam fire engines anymore, and more and more fire departments are replacing their old steamers with the new gasoline-powered fire engines. Fred and I fear that in a few years, the history of the horse-drawn fire engine might be completely forgotten, if I do not write and publish my first-hand knowledge of that history.

Before I begin this history of the steam fire engine, I will start with a little background about myself and my family, and how I came to be associated with the fire engine industry.

I was born Dec. 7, 1834, in Brettbeck, Hanover (Germany). Brettbeck is a little farming village on the edge of Osterholz, near Bremen.

My father, Johann Friedrich Ahrens von Brettbeck, was a tenant farmer -- that is, he worked the farm of the local baron for subsistence wages, and was permitted to keep a tiny portion of the crop to feed his family. His "von Brettbeck" title was more a tradition than a reality: at one time, the family had owned some land in Brettbeck, but we were no longer the land-owners that the title suggested. My mother, Anna Hartmann Ahrens, raised five children and maintained the household. I was christened Johann Christoph Ahrens on Dec. 21, 1834, but for as long as I can remember have preferred simply to be called Chris.

My parents were very religious folks. We were members of the Evangelical-Lutheran church in Osterholz. The "Johann" part of my name was for the church's patron saint, the apostle St. John. For the same reason, nearly all of my male relatives (father, brothers, uncles, etc.) were also named Johann and, like me, preferred to use their middle names. Of course, most of the middle names were *also* for saints: mine for Saint Christopher, my younger brother (Johann Peter) for Saint Peter, and so on.

Life was very hard for my parents, and one-by-one over the years, all of their 7 children (me included) sought to escape to a better way of life.

My oldest brother, Hermann, was the first to escape when he emigrated to America at age 18, in the autumn of 1847. When one of the Brettbeck villagers, Mr. Kattenhorn, emigrated to establish a small business in Cincinnati, Hermann went along as his apprentice.

Brettbeck, Hanover, Dec. 7, 1848

I turned 14 today, so why am I so unhappy? My parents did their very best to make this a happy birthday. But with the whole country in the midst of a bloody civil war, and father having to work so hard for a meager wage, happiness comes so rarely these days.

My biggest concern, however, is that now that I'm 14, I'm liable to wind-up on the battlefield myself. Both the King of Hanover and the revolutionaries are drafting all able-bodied males above the age of 14. Recruiters are everywhere, even in our peaceful little village of Brettbeck.

Because Brettbeck is a small village (about 30 people) where everyone knows everyone else, it will be virtually impossible to hide out from the recruiters if my name is picked in the draft lottery. Some neighbor or other is bound to reveal my whereabouts, thinking that he is doing something good and noble.

My oldest brother Hermann (born Feb. 22, 1829) was fortunate to have emigrated to America last autumn, before the present conflict began. When he was selected for the draft last March, he was given three choices: return to Hanover to serve in the military, pay a hefty fee for not serving, or pay someone else to serve in his place.

In October, my 17-year-old brother Frederick (born Dec., 1830) was also called. Hermann wanted Frederick to emigrate and join him in Cincinnati, but that would require our father to buy-off Frederick's military obligation. As a tenant farmer earning one thaler a week, father cannot pay the fee -- in fact, he still owes money for Hermann's emigration.

Brettbeck, March 12, 1850

My parents recently borrowed 500 thalers to pay the fees for my brothers Hermann and Frederick evading military duty (Frederick is now living with our brother Hermann in America). In the meantime, father has found it hard to pay this back because six weeks of frost and last week's heavy rains destroyed most of the crops. So Hermann has agreed to pay back this loan for us (out of his wages from Mr. Kattenhorn). He will pay it as quickly as possible, since the loan is at a very high interest rate. Mr. Kattenhorn is in Brettbeck visiting, and has agreed to help with this loan when he returns to Cincinnati.

I have become apprenticed to a blacksmith in Osterholz. It is a welcome change from the backbreaking farm labor on which I was raised in Brettbeck. So far, I have not been called for military service.

Brettbeck Sept. 8, 1851

My sister Elizabeth (born Dec. 16, 1833) missed her chance to emigrate to Cincinnati, joining our brothers Hermann and Frederick. She fears it would not be safe to make the crossing alone, so she will not be able to leave Bredbeck for another year now.

Brettbeck Nov., 1851

My brother Hermann sent newspaper clippings about a recent riot in Cincinnati. Hermann assures us that he is in no danger and we need not worry.

The first article, from September, describes what a reporter saw and heard while walking along a Cincinnati street just before the rioting began. It started with loud and distressed cries of "Fire! Fire!". Out of curiosity, the reporter followed the source of the cries, hoping to watch the brave volunteer firemen in action, and perhaps even a chance to grab a bucket and help.

The blaze turned out to be in a huge old planing mill at the corner of Augusta and John streets. Enroute to the fire, the reporter saw bankers, bakers, butchers, city councilmen, and men and boys of all ages dash out of stores and houses to their nearest fire house, grab a section of drag-rope, and haul the little hand-operated pumpers over the cobbled streets as fast as their legs would carry them.

According to another article, Washington Fire Co. #1 raced northward from their engine house on Vine between Front and Columbia, and began attaching their brass-riveted leather suction hose to the wooden fire plug at Augusta and John streets. Just then, Western Hose Co. #3 arrived after a mad dash southward from their fire hall on Fifth street between Mound and Carroll.

The foreman from Western Hose Co. loudly proclaimed that he had arrived before Washington Co., and tried to prevent Washington from using the hydrant. The Washington foreman indignantly replied that he and his men had arrived ahead of Western, and therefore had the right to use the hydrant. Apparently, neither foreman noticed that the hydrant had two connections, so that the two companies could share it.

The argument over use of the hydrant soon escalated from shouts to fists and even to hurling bricks and bottles at each other. While all of this was going on, of course, the fire in the planing mill was being completely ignored.

As the spectators watched, the flames spread from one end of the mill to the other, leaping defiantly in the air, and they felt the intense heat on their faces. The firemen, absorbed as they were in their bloody brawl, seemed to neither see the glowing flames nor feel the rising temperature. The crowd hesitated to attempt fighting the fire themselves, fearing the firemen more than the fire. Indeed, as the violence escalated, many found themselves retreating ever further to maintain a safe distance from the brawl.

In his letter, Hermann tried to explain, but I really do not understand, the intense rivalries that exist between volunteer fire companies in America. Apparently, it is a matter of utmost pride as to which fire company arrives on the scene first, and which is the first to throw a stream of water on the fire. To come in second evidently calls one's masculinity seriously into question. Long and bitter rivalries often owe their origins to really trivial disputes, then other fire companies form permanent alliances by siding with one or the other party in such a feud.

Since the planing mill fire was not being fought, two more volunteer companies were dispatched to the scene. One had a long-standing alliance with Washington, the other with Western. Soon both companies also joined the fistfight, escalating it from a brawl into a full-scale riot. The unattended fire, meanwhile, marched relentlessly on, spreading from floor to floor. Additional fire companies were sent for, and soon ten Cincinnati fire companies -- a total of over 300 firemen -- were rioting, and not one drop of water had yet touched the fire.

By sunset, the smoke was visible clear across the Ohio river in nearby Covington, Ky. Covington Fire Co. #1 ferried its hand-pumper across the river in what started as a neighborly gesture of assistance. But when Covington learned that their friends at Washington #1 were being beaten-upon by several rival fire companies, they also joined the riot rather than fighting the fire.

Cincinnati's mayor, Mark P. Taylor, attempted to end the riot, but his efforts had no effect. The riot did not end until dawn, when the firemen were too exhausted and injured to fight. By then, the planing mill had burned to its foundation, making the dispute over the hydrant a moot point.

The next clipping that Hermann sent was a week after this firemen's riot at Augusta and John streets. It seems there is now a growing movement among Cincinnatians to *do* something about it; not only to punish the firemen, but to make sure that it never happens again.

Some are saying that the problem is that Cincinnati has too many fire companies, more than they really need. There may be some truth to this, as the article says that the city has the following 14 engine companies:

Washington #1, Vine betw. Front and Columbia.
Relief #2, George betw. Plum and Western Row.
Independence #3, 4th between Walnut and Vine.
Franklin #4, 6th and Sycamore
Invincible #5, Vine between Court and Canal.
Eastern #6, Front and Parsons
Northern #7, Webster betw. Main and Sycamore.
Marion #8, Cutter between Hopkins and Laurel.
Union #9, Race between 3rd and 4th.
Deluge #10, 3rd and Lawrence.
Eagle #11.
Mohawk #12, Hamilton Rd. opposite Vine.
Brighton #13.
Western #14, 5th between Smith and Mound.

Each is equipped with a hand-operated pumper and a hand-drawn hose reel. Companies 10 through 14 were all formed earlier this year.

In addition to these 14 engine companies, Cincinnati also has three hose companies:

Independent #1, 4th betw. Broadway & Sycamore.
Deluge #2, Ludlow and Symmes.
Western #3, 5th between Mound and Carroll.

Each of these three companies is equipped with a hand-drawn hose reel or hose carriage. Also, each has recently been supplied with a hand pumper. The idea is that each engine company and each hose company now possesses the full apparatus to extinguish fires without depending on any other fire company.

There is also one hook-and-ladder company in the city, at the corner of Race and Longworth.

The article claims that the riot last week proved that all of the extra fire engines and extra firemen Cincinnati has been adding is making their fire companies *less*, not *more*, efficient.

Another article, about a month later, describes a special meeting of the Cincinnati city council that was called to discuss what to do about fire department rowdyism.

The recent riot was on everyone's mind. But many have not forgotten that only a few months ago, virtually every fire company in the city was involved in an even-larger riot, this one protesting the visit of a Catholic archbishop to the predominantly Protestant city of Cincinnati. While our family is also Protestant (Lutheran), none of us can comprehend why anyone would be so violent in their reaction to Catholics.

As a result of these two riots, several city councilmen and some of Cincinnati's prominent citizens clamored for fire department reforms to end the rowdiness of the volunteers. That is why the special council meeting was called.

Prior to the two riots this year, two Cincinnati city councilmen had been pushing for fire department reform for several years. Their names are Jacob Wykoff Piatt and James H. Walker.

Piatt, a Kentucky native, recounted how most of the older and public-spirited firemen have retired over the past few years, to be replaced by young ruffians who consider a fire merely an occasion for starting a riot. In his opinion, the engine houses have simply become a place for gangs of political parasites to loaf, or as councilman Miles Greenwood put it, a "nursery where the youth of this city are trained in vice, vulgarity, and debauchery".

The problem, as Piatt explained it, is that the firemen serve voluntarily, and therefore cannot be discharged for misconduct. But, he reasoned, place the firemen on a paid basis, with a good measure of military-type discipline thrown in, plus the threat of firing for misconduct, and the rowdiness will cease.

Piatt recounted that the first time he and Walker had proposed replacing the volunteer fire department with a paid system, the council chambers were packed with firemen who noisily demonstrated their opposition to reform.

Outside of Piatt and Walker, the rest of the council had voted against reforms that day. America's volunteer firemen have amassed such tremendous political clout, Piatt said, that his fellow councilmen feared for their re-election, and in some cases for their lives.

He and Walker had not given up, and had periodically re-introduced their fire department reform bill to the council. Each time, it was vetoed, but each time by an ever-decreasing majority. The firemen became so bitter about the reform efforts that, for his own protection, Piatt had to bring a guard of his Irish constituents to every council meeting. Once, an angry mob had even burned Piatt's effigy on his own front lawn, amidst hisses, howls, and yells.

How ironic, Piatt gloated, that it had taken two terribly destructive riots before the city council joined him in seeking reforms.

At this meeting, the vote of the city council swung the other way: for the first time the majority of councilmen supported Piatt and Walker's reform bill. Two vocal supporters were Miles Greenwood and Joseph S. Ross.

Greenwood remarked that paying the existing firemen was not enough. The reason the firemen had succeeded in terrorizing Cincinnati and other cities for so many years was their sheer numbers: Cincinnati has 18 fire companies, each with 30 to 50 members. Greenwood proposed that, in conjunction with establishing a paid fire department, Cincinnati needs to extinguish fires "without the attendance of the large numbers of men required to operate the manual pumping engines."

The article mentions that Greenwood is renting space in his foundry to one Abel Shawk, who is experimenting with building a steam fire engine that would require fewer men to operate than the hand pumpers do.

Miles Greenwood's suggestion at the 1851 council meeting led to the beginnings of the steam fire engine industry, with which I was to be associated for most of my life.

In Chapter 2, I will take a momentary diversion to sketch brief biographies of the key players who founded the steam fire engine industry: Abel Shawk, Alexander Latta, Miles Greenwood, Richard G. Bray, and Joseph S. Ross.

Then in Chapter 3, I will resume the excerpts from my journal and scrapbook, describing how those five gentlemen started the steam fire engine industry, how their efforts made Cincinnati the center of that industry, and how I emigrated to Cincinnati and began my career.

Chapter 2

Abel Shawk has always been an enigma to me. He did more than any other person in history to develop the world's first successful steam fire engine, yet history has virtually forgotten him. I met Shawk only a few times myself, and never really knew much about him.

Shawk's first steam fire engine partly owes its origins to Dr. Joseph Buchanan of Louisville, Ky. In 1819, Buchanan had invented the Buchanan Coil system for generating steam, but his invention had not met with much enthusiasm.

Buchanan's revolutionary idea was to heat a series of coiled (or spiraling) copper tubes from the outside, then inject cold water into the *inside* of the pre-heated tubes. The contact of the cold water with the interior of the heated tubes would produce steam almost instantly. Buchanan's boiler thus raised steam much more rapidly than any previous type of boiler. The Buchanan Coil method was often called the "gunpowder" boiler because, like gunpowder, it fired rapidly. Unfortunately, the name "gunpowder" conotes the danger of explosion; actually, this type of boiler tended to be safer than other types, because excessive pressure usually caused the coils to start leaking, relieving the stress that could cause a boiler to explode (although a few of the early Buchanan boilers did, indeed, explode).

Buchanan died in 1829, and his invention might have died with him were it not for two inventors in England.

In 1829, George Braithwaite and John Ericsson built the world's first steam fire engine, named *Novelty*. This engine was not considered a success: it was capable of throwing only about 200 gallons per minute, which was not very effective in extinguishing fires. After building only five steamers over a period of four years, Braithwaite and Ericsson gave up their unsuccessful fire engine enterprise in 1833.

While Braithwaite and Ericsson never used the Buchanan coil, their invention eventually led Abel Shawk to build the world's first successful steam fire engine, which *did* use a Buchanan boiler.

John Ericsson moved to America in 1839, and tried to introduce the steam fire engine here. He hoped for a better reception than his invention had received in England. In 1840, Ericsson persuaded the Mechanics Institute in New York to sponsor a contest for the best steam fire engine design. He hoped that the winning design would be adopted by one of New York's volunteer fire companies and placed in service.

But as I mentioned in Chapter 1, America's volunteer firemen were resistant to any change that threatened their fun and their political power. So Ericsson's dream of putting a steam fire engine into service in New York was not realized for another quarter-century.

At least three inventors submitted designs for steam fire engines in the 1840 Mechanics Institute contest. One was, of course, John Ericsson. His design won the gold ribbon but was never built.

Another contestant was Paul Rapsey Hodge, who actually built a steamer later that year. It was briefly placed in service at New York's Pearl Hose Co. #28 in March of 1841, but met with so much resistance from the volunteers that about two months later it was sold to a packing-box factory for use as a stationary engine.

The third entry in the 1840 steamer contest was Abel Shawk of Cincinnati. Shawk realized that every minute it took to generate steam gave a blaze time to cause more destruction. Apparently, Shawk had somehow obtained (or helped himself to) the rights to Buchanan's coil boiler, and his design included this quick-steaming boiler.

Although Shawk did not win the 1840 contest, he persevered. In 1841, he acquired a Buchanan boiler from a mill in Johnstown, Indiana. He began renting space in Miles Greenwood's Eagle Iron Works, including the use of a smith's forge. There, he dismantled this boiler and began a series of experiments to make the boiler lighter, safer, and faster, so that it would be more practical to use it on a steam fire engine. He completed his first prototype of an improved Buchanan boiler about 1849 or 1850, using coiled gas-pipe arranged in concentric layers. Shawk was granted Patent #10041 for this Improved Buchanan boiler on Sept. 20, 1853. But like Buchanan 30 years earlier, Shawk found a less-than-enthusiastic reception to this boiler.

After the 1851 firemen's riot at Augusta and John streets, Cincinnati's city council became *very* receptive to Shawk's idea of building a steam fire engine. The councilmen realized that one steamer could replace several hand pumpers and eliminate the need for hundreds of rowdy pump operators. So the council appropriated $1000 to Abel Shawk to aid him in making practical tests of a prototype steam fire engine.

Shawk canabilized a retired D.L. Farnham hand engine from one of Cincinnati's volunteer fire companies. He used its horizontal double-acting piston pump and air chamber, and added the improved Buchanan Coil with a sheet-iron furnace. All that Shawk lacked was the steam machinery to turn the steam generated in his boiler into the driving power for the pump. For that, Shawk turned to Alexander Bonner Latta, one of Cincinnati's most prominent inventors and manufacturers of steam machinery.

A.B. Latta was born on a farm in Ross County, Ohio, near Chillicothe, on June 11, 1821, the youngest of six children of John and Rebecca (Bonner) Latta. When Latta was only five years old, his father was killed in an accident, leaving Latta's mother a penniless widow. As a result, Latta attended school for only a few years, and at ten began work in a cotton mill to support the family. Later, he became an apprentice machinist at the Navy Yard in Washington, D.C.

Latta met Anthony Harkness, owner of a Cincinnati machine shop, in Washington in 1841. Latta's skills as a machinist so impressed Harkness that Harkness asked Latta to settle in Cincinnati and become superintendent of his foundry.

Latta's first assignment at Harkness was to build a monstrous lathe with a $7^1/_2$' swing and 35' ways. Latta also built a planing machine weighing 30 tons, able to handle iron 6' wide, 5' high, and 24' long.

In 1845, the Little Miami Railroad Company of Cincinnati asked Harkness to build them a steam locomotive. Latta, as foreman, supervised construction of this locomotive. He was engineer on its test run from Cincinnati to Columbus and back. The *Bull of the Woods* was the first locomotive built west of the Allegheny mountains.

Latta married Elizabeth Ann Pawson in 1847, and they had two children. About 1848, Latta built a unique locomotive for Boston & Maine Railroad. This machine had an extra pair of

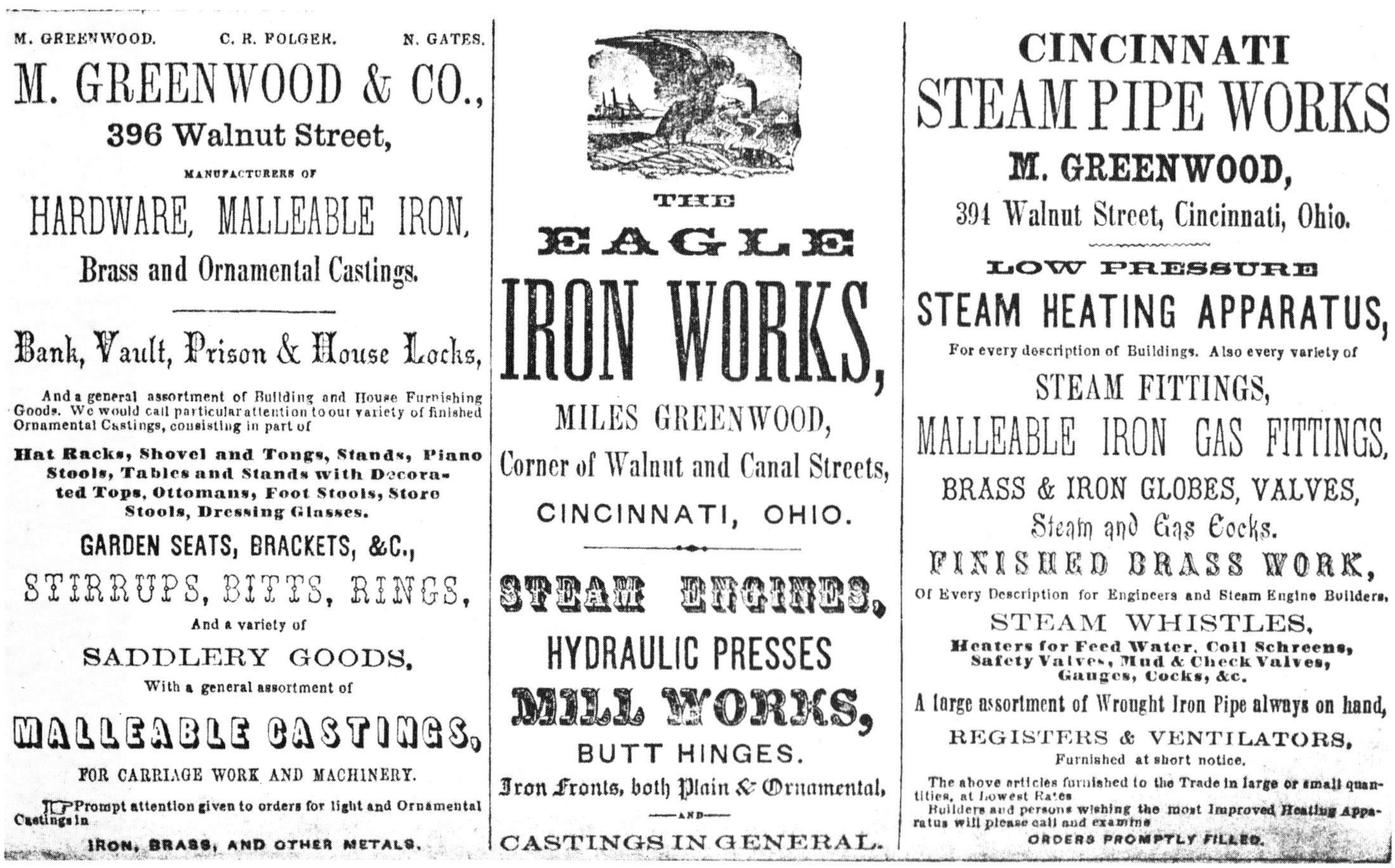

Miles Greenwood's "Eagle Iron Works" placed this ad in the Cincinnati City Directory in 1853, shortly after Alexander Latta and Abel Shawk built the world's first successful steam fire engine in Greenwood's foundry. Note the wide variety of iron products that Greenwood manufactured.
(Courtesy Public Library of Cincinnati & Hamilton County, Ohio)

steam cylinders under the water tank, and its steam pipes were fitted with ball joints.

By 1851, A.B. Latta had formed a small manufacturing business in partnership with his two older brothers, Edmundson and Finley Latta. This company was called A.B. & E. Latta, and operated in the Buckeye Works at 179 Race Street. Their main products were machine tools and small stationary steam engines. The Latta brothers provided the steam cylinders, steam chests, etc., to drive the pump on Shawk's 1851 fire engine.

Shawk built a massive wooden frame around the improved Buchanan boiler, Farnham pump, and Latta steam engine. Then Latta, Shawk, Greenwood, and Bray mounted the entire contraption on the carriage of one of the city's retired hand-drawn ladder wagons. Assembly of all of these components was performed at Miles Greenwood's Eagle Iron Works, 396 Walnut street, during the winter of 1851-1852.

Miles Greenwood was born in Jersey City, N.J., on March 19, 1807, the son of Miles Greenwood of Salem, Mass., and the former Miss Demarest of Jersey City. In 1808, the family moved to New York, and in 1817 to Cincinnati. During the next eight years (1817-1825), Greenwood struggled to support himself and his invalid father by working as a bootblack and bill-poster, cutting cord-wood, and running a small merchandise store. At the same time, he was trying to educate himself.

In 1825, Greenwood and his father moved to New Harmony, Ind., a commune started by English radical Robert Owen. The Greenwoods lived there for the next four years. After a brief training period at a Pittsburgh iron foundry in 1827, young Greenwood took charge of a factory in New Harmony. When the experimental commune failed in 1829 and Robert Owen returned to England, Greenwood returned to Cincinnati. There he was employed in the iron foundry of John and Thomas Bevan.

In 1832, Miles Greenwood and Joseph Webb established the Eagle Iron Works at 396 Walnut street, on the Miami Canal. Although it began with only 10 employees, Eagle Iron Works quickly grew to become the west's largest iron foundry. By the time Shawk's prototype steam fire engine was assembled there in 1851, Eagle Iron Works had 350 employees and annually manufactured $360,000 worth of iron and brass goods ranging from locks and hat racks to steam engines, hydraulic presses, and iron fronts for buildings.

Greenwood became a Cincinnati city councilman in 1840. He was also a volunteer fireman. His interest in the steamer, and his influence in both the city council and the fire department, were instrumental in successfully introducing the steam fire engine in Cincinnati.

Two other proponents of the steam fire engine who were instrumental in its introduction were Richard G. Bray and Joseph S. Ross.

Bray was the city fire engineer, which meant he was theoretically in charge of all of the city's volunteer fire companies. From his office at 364 West 5th St., Bray tried in vain to enforce discipline over the rowdy volunteer firemen of the city. The 1851 riot at the planing mill was the last straw for Bray, and like the members of the city council, he too began clamoring for fire department reforms. After Shawk successfully demonstrated his prototype steamer, Bray donated (from his own pocket) half of the price that Shawk and Latta quoted to build a steamer.

Joseph S. Ross is the man who persuaded the Cincinnati city council to appropriate the other half of the cost of that first steamer. Ross was born in New Brunswick, N.J., on March 5, 1803, but lived in Cincinnati from the age of three. As a young man, he worked for Jonathan W. Lyon as a carpenter, then became an independent carpenter. In 1832, he opened a feed store at 7th St. and Central, and soon after became a councilman.

About 1839, Ross built a steamboat named *Relief*, and engaged in the steamboat trade on the Red River. He lost his boat, with all of his possessions aboard, in an 1844 accident and soon after returned to Cincinnati. He became a partner in the Higdon & Ross livery stable at the southwest corner of 7th and Central, and was also re-elected to the city council.

Shawk, Latta, Greenwood, Bray, and Ross forever changed the way that fires were fought in America and, indeed, all over the world. Their efforts replaced hand fire engines manned by rowdy volunteer firemen, with mechanized fire engines manned by professional firefighters.

Chapter 3

Cincinnati, March 2, 1852 (newspaper)

Abel Shawk demonstrated his prototype steam fire engine to the City Council and about 3000 citizens this morning. Since the volunteer firemen of this city are bitterly opposed to his invention, Shawk recruited assistance from sympathetic citizens before the demonstration began.

This prototype consists of an improved Buchanan boiler that Shawk built single-handedly in Miles Greenwood's foundry, coupled to a modified Latta stationary steam engine, and a D.L. Farnham pump from one of Cincinnati's retired hand engines. Shawk built a massive wooden frame around these components. Just last week, with the assistance of Miles Greenwood, Richard G. Bray, and Alexander Latta, Shawk had lowered the entire assembly onto the chassis of an old fire department ladder wagon.

"Contraption" is the kindest word that comes to mind when one looks at this jumbled hodgepodge of components. Yet this contraption put on quite an impressive demonstration this morning, generating steam in only 4 minutes and 10 seconds from the time the match was applied to the firebox. Many in the crowd remarked that they had never seen such a quick boiler, even on the newest locomotives and river-boats.

350 feet of hose was stretched from the steamer in a single line (7 lengths). Just 50 seconds after a head of steam had been raised, water was flowing from the $^3/_4$-inch nozzle at the end of this hose, projecting a horizontal distance of 130 feet. Members of the City Council, many of them former volunteer firemen, remarked that this

performance equalled or exceeded the best they had ever seen from a hand pumper.

Cincinnati, April, 1852 (newspaper)

The City Council has just about unanimously agreed that last month's demonstration of Shawk's steam fire engine was a success. Nobody doubts that if Cincinnati had one, it would open up opportunities for much-needed fire department reform. But that is the very reason Cincinnati is not buying one: the firemen have threatened violence against any councilman who votes in favor of appropriating the $10,000 that Shawk is asking for building a new steamer.

Shawk has already dismantled his prototype. He sold the boiler to the city, to power the machinery of the Cincinnati City Shops.

Shawk has agreed with the city council that his next steam fire engine, if he builds one, will have to be all custom-made. In other words, every component will be built specifically for a steam fire engine, rather than assembling pieces of other machines as Shawk did for his first engine. That is one reason why his price is so high.

Since Alexander Latta did such a good job making the main steam engine for Shawk's prototype, Shawk has asked Latta to assist him in designing and building the city's first steam fire engine. The two inventors have formed a partnershp of sorts, and have been putting as much pressure on the city council *for* the fire engine appropriation as the firemen have *against* it.

Cincinnati, June, 1852 (newspaper)

The deadlock over financing for the proposed Latta & Shawk steam fire engine has been broken.

The city fire engineer, Richard G. Bray, has donated half of the steamer's $10,000 price out of his own pocket. As a result of his generosity, Councilman Joseph Ross (affectionately known as "Uncle Joe" to his constituents) has persuaded, or more accurately shamed, his fellow councilmen into showing at least as much public-spiritedness by appropriating the other $5000.

It looks like the city's first steam fire engine will be built after all.

Brettbeck, Sept., 1852

An anonymous letter in a month-old Cincinnati newspaper that my brother Hermann sent, describes the writer's visit to John McGowan's shop on Ohio Avenue in Cincinnati. Abel Shawk and Alexander Latta are building Cincinnati's steam fire engine there, as McGowan generously offered the use of his shop. Abel Shawk, former lock manufacturer, is a machinist for McGowan.

Cincinnati's volunteer firemen have been threatening the lives of everyone associated with the steam fire engine project, especially Latta, Shawk, and Greenwood. Miles Greenwood felt that his huge foundry was too easily-accessible for the two inventors to work with any degree of security. McGowan's little shop, on the other hand, has but one entrance, and few employees, making it much more difficult for saboteurs and assassins to sneak in undetected. And, of course, nobody goes into Latta and Shawk's workshop without permission from Miles Greenwood.

Abel Shawk was not in the workshop when the letter-writer visited, but Alexander Latta was. Drawings of the steam fire engine lined the east wall of the long, high room where the steamer was under construction. Beneath the drawings ran a long workbench, and at that workbench stood Mr. Latta himself. He was in the midst of measuring something, and the writer's entrance caused him to look up, startled.

"Did Mr. Greenwood give you permission to come in here?" he challenged.

"He did," the visitor replied. "He told me to come and see how the engine was getting on." Mr. Latta seemed to believe the stranger, probably because his mannerisms were not as threatening as one would expect from a volunteer fireman who was there to harm Mr. Latta. The truth is, the visitor really *had* just come from Miles Greenwood's Eagle Iron Works, and really *had* obtained permission to visit the McGowan workshop, but Mr. Latta had only his word for that.

"Ah! Very well," Latta said, satisfied that the visitor meant him no harm, "Very well! My name is Latta, Moses Latta, and as you come from Mr. Greenwood, you shall see what few see. Can you in any way or to any extent understand the drawings on the wall?"

The stranger confessed that he could not.

"It is very simple," Latta continued, "let me explain. The engine is intended to throw, at any time, eight streams of water; four from each side, whenever the water can be obtained in sufficient quantity for that number. It is intended to be taken to a fire under its own steam power, with four large horses guiding its direction. Its eight large streams, with the pressure we shall be able to command, will drown any fire. Even four streams, well directed, will be of wonderful value.

"But there is no certainty that this, or any other steam fire engine, will ever be allowed to work at a fire. You are, probably, not aware how bitter is the feeling of the volunteer firemen against this engine. They say it shall never throw a stream of water on a fire in this city; and I sometimes fear that I shall never live to see this grand idea brought into the service of the world.

"The recent riots here show what a mob can do in our city. My steps are dogged. Spies are continually on my track. I am worried with all sorts of anonymous communications, threatening me with all sorts of ills and evils unless I drop work on this engine, and pronounce it a failure."

Latta's eyes flashed as he proclaimed, "I'll never give up! I'll build it, and there are men enough in this city to see that it has a fair trial; and it shall have it. When it is finished, it will be heard from at the first fire, and woe to those who stand in its way!"

Cincinnati, Dec. 15, 1852

Construction of Latta & Shawk's steam fire engine was completed today, and Latta, Shawk, Greenwood, and Bray shoved the 10-ton monster through the door of John McGowan's shop and out onto Ohio Avenue for the first time.

Mr. Shawk piloted the new engine over the city streets for the first time this afternoon. The heavy steamer belched smoke as its rumbled over the limestones under its own steam power (it is guided by horses, as it lacks a steering mechanism). After driving up and down a few hills, and around the block, Mr. Shawk stopped the self-propellor. He then activated two massive jacks on the back of the engine, which lifted the rear wheels off the ground, so that they could act as flywheels for the pump.

With Mr. Latta at the pump controls, the steamer had no trouble drafting water out of a cistern and throwing an effective fire stream. Both Mr. Latta and Mr. Shawk are confident that the engine will perform well in its first public demonstration next week.

Cincinnati, Dec. 22, 1852

The first public trial of the new Latta & Shawk steam fire engine was held at Front Street and Broadway today. It took seven minutes to raise steam and throw a $1^1/2$" stream 225 feet.

Before the city accepts this steamer, they want to give it a more thorough trial on January 1st. This is advantageous for Messrs. Latta and Shawk, as it gives them time to make a few last-minute adjustments and modifications to insure that the steamer gives its best possible performance.

Cincinnati, Jan. 2, 1853

The new steam fire engine received its official trial yesterday. It raised steam in five minutes, and four horses began hauling it through the downtown streets. In another five minutes, there was sufficient steam pressure to activate the self-propellor, and the steamer proceeded under its own power, with the horses steering it.

The engine was driven down Cincinnati's steepest hill, and many spectators voiced concern that the momentum of the engine would send the 10-ton steamer careening out of control and cause it to crash. But Abel Shawk simply reversed the steam engine, which effectively counteracted the gathering momentum. The long descent of the hill was made without incident. Still not satisfied, some in the crowd began wagering that the steamer did not have sufficient power to climb the hill on its return trip.

When the steamer arrived at the cistern at Broadway and Second Street, only 15 minutes had elapsed since the boiler was lighted. It then took only another four minutes to make all hose connections and throw two powerful 1" streams.

The volunteer firemen brought out their most powerful hand engine to challenge the steamer. Even with their strongest firemen working the pump, the hand engine could not throw a stream as far as the steamer could throw two streams. The volunteers quickly withdrew in disgrace.

Miles Greenwood proclaimed "Now that we have it all our own way, let us show what this engine can do." Six hoses were connected to the steamer, and laid out in parallel. Six men stood side-by-side along a line marked in the street, each holding one nozzle. The distance from the tips of the six nozzles to the point where the water hit the limestone pavement was 175 feet.

Next, these six hose lines were siamesed into a single 50-foot length of hose with a $1^1/2$" nozzle, and a single stream was thrown 224 feet. Three of the six hoses were uncoupled from the steamer; the three remaining hoses sent three good streams over the top of the four-story Broadway Hotel.

Right on the spot, Miles Greenwood proclaimed the demonstration a success and recommended that the city immediately purchase the engine. He also suggested that this engine should be named *Uncle Joe Ross*, after the councilman who, despite the firemen's threats against him, bravely persuaded the City Council to appropriate $5000 toward the steamer's purchase. Some in the crowd felt that they should have named the steamer after Richard G. Bray, whose donation of $5000 *from his own pocket* was a far braver and nobler action than Ross's allocation of city funds.

There still remained one test for the *Uncle Joe Ross*: climbing the steep hill it had so successfully descended enroute to the demonstration. The steamer made the climb without so much as a whimper, and many citizens thus lost their wagers about its hill-climbing abilities.

Cincinnati, Feb. 20, 1853

Not willing to entrust the steam fire engine to any of the volunteer firemen under his command, City Fire Engineer R.G. Bray resigned his post on the first of this month, and organized a paid fire company, with himself as Captain. He felt that this was the only way the steamer would ever get a fair trial at an actual fire.

The *Uncle Joe Ross* got its chance last week, when Bray and his men were summoned to a fully-ablaze warehouse at Third and Main.

The steamer almost didn't make it to this fire. The engine was out on the street, steam up, four horses hitched to it, and nobody to drive the horses. It seems that none of Bray's men had ever worked with horses before. The jeering volunteer firemen who were watching weren't about to help. Most of the spectators feared the restless horses almost as much as the violent firemen. Greenwood was mad as the devil about the situation.

Two men who were standing a few feet from Greenwood, both dressed in their Sunday best, were discussing Greenwood's predicament. One was a Cincinnati resident, while the other was his visiting brother, a teamster from Vermont.

Said the Cincinnatian to his Vermont brother, "Larry, you can drive those horses, I know." Greenwood, overhearing this, said "If you can drive them, I wish you would. I'll pay you for it." So just as he was, best suit and all, this teamster jumped on the back of one of the horses, grabbed the reins, and rumbled off toward the fire. Smoke streamed from its stack and fire flashed from its fire-box as the steamer faded over the horizon. All the way down Main Street, spellbound people rushed out of their houses and followed the steamer.

When the *Uncle Joe Ross* arrived at Third and Main, four $2^1/2$" hoses were stretched from the steamer into the blazing warehouse. As the steamer was starting to control the fire, the cry "the hose is cut" rang out. The firemen/saboteurs who had thus disabled the steamer, next tried to disable its operators. But the crowd contained more citizens who were weary of fire department rowdyism than it did firemen, and the volunteers were quickly overpowered. New hoses were soon connected to the steamer, which drowned out the fire and was taken home.

Cincinnati, April 2, 1853

The volunteer firemen apparently decided not to let the presence of the steamer and Richard Bray's company of paid firemen spoil their fun. Last month, members of Marion Fire Co. #8 set fire to an old frame house in the neighborhood of their arch-rivals of Franklin Fire Co. #4. When the Franklin volunteers arrived, the Marion firemen were waiting in ambush, and one of the city's bloodiest brawls ensued.

The city council's reaction was swift and merciless. On March 16, the council abolished the volunteer fire department. Yesterday, a new fire department was established, with Miles Greenwood as Chief Engineer. Greenwood hand-picked the volunteer firemen whom he trusted, and carefully trained them for the paid fire department.

All of the hand pumpers, hose carriages, ladder wagons, and fire houses of the volunteer fire companies are now the property of the paid fire department, as is the steamer *Uncle Joe Ross*. R.G. Bray continues to be the Captain in charge of the steamer, but like everyone else in the new department, takes orders from Chief Greenwood.

The cost of running this paid fire department for one year is estimated at $30,000, of which $10,000 was appropriated by the city council. Another $5000 was contributed by the local insurance companies, who stand to save a tidy sum in property-damage settlements by the introduction of better firefighting methods. Miles Greenwood himself donated the other $15,000. Cincinnati is the first city to entirely abolish its volunteer fire department and replace it with paid, trained professional firefighters.

In this new fire department, each firefighter is to be paid $60 per year; each lieutenant $100; captains $150; pipemen and drivers, $365; assistant engineers, $300; and engineers, $1000. Miles Greenwood has volunteered to serve as Chief Engineer without salary.

Atlantic Ocean, May 15, 1853

As I have feared for the past 3½ years, I was called for military service. Seeking to avoid this, I heard that the S.S. Johann in Bremen was looking for young men to serve as cabin boys. With the arrival of spring weather, this has become their busy season, with countless trans-Atlantic steamship passengers between Bremen and New York. So I signed up last week.

My younger brother Peter (age 12) wanted to come with me to America, but has to complete two more years of school before he can emigrate.

It will be interesting to see America. I have heard much about the opportunities which that country provides for immigrants. Some of the Osterholz villagers who emigrated there (including my older brother Hermann) have written to my parents about how well they are doing in their new country. The opportunity America offers to me, of course, is to put an ocean between myself and Hanover's civil war.

The work of a cabin boy is much like farming: hard, menial, and exhausting. The hours are long, and the pay poor, but I get free meals and a place to sleep (although my quarters are cramped). As long as I am working on a German ship, I am temporarily exempt from military duty. If any of the passengers turns out to be a recruiter, he cannot stop my emigration.

It is difficult to work on a ship when all of your fellow crewmen are strangers. Captain Chon keeps us all too busy to socialize. However, I *have* befriended one member of the crew, who works in the engine room. He showed me the engine room and explained the workings of the ship's steam machinery. I must admit that his fascination with steam power is starting to rub-off on me.

New York, Aug. 28, 1853

After 16 weeks at sea, we finally put in port at New York. I resigned my post and began my westward journey to join my two older brothers in Cincinnati. My oldest brother Hermann has written that most of the city consists of German immigrants, so I should feel quite at home there.

Cincinnati, July 28, 1854

A few months ago, the citizens of Cincinnati collected $13,400 in donations to purchase a second Latta & Shawk steam fire engine. That fire engine, appropriately named *Citizen's Gift*, was demonstrated for the first time yesterday.

During this demonstration, the *Citizen's Gift* raised steam in 3 minutes 10 seconds, the fastest time that a Latta engine has achieved. It also set a distance record, throwing a 1½" stream 297'.

Immediately after the test, the new steamer was taken to Fire Co. #3, at 117 West 6th St. (between Vine and Race). This is where it will be in service. Finley Latta, brother of the steamer's builder (Alexander Latta), is to be the engineer in charge of the *Citizen's Gift*.

Abel Shawk left his partnership with Alexander Latta before this steamer was completed. They disagreed about the design of this new steamer, and Mr. Shawk is now attempting to start his own steam fire engine factory, to build steam fire engines the way he feels they should be built.

Latta's new steamer was built in the Buckeye Works, a factory at 179 Race St. that is owned and operated by Alexander Latta and his brothers Edmundson and Finley Latta. They have been building railroad locomotives and stationary steam engines there for several years.

Cincinnati, August, 1854

Because of the success of the *Uncle Joe Ross* and the *Citizen's Gift*, the Latta brothers have decided to concentrate on building steam fire engines for any city that will buy one.

To that end, they have hired me to a seven-year apprenticeship at Buckeye Works (179 Race St.), learning how to build steam fire engines. I bring some experience to this apprenticeship, having apprenticed as a blacksmith in Osterholz and learned the rudiments of steam power when I was a cabin boy. But there is still much to learn.

Cincinnati, Ohio, Nov. 16, 1854

I have been granted special dispensation from German military service. If I should ever return to Germany while still retaining my German citizenship, I will now be considered a foreign visitor, not a German citizen who is obligated for military service.

Cincinnati, Dec. 16, 1854

We have completed another steam fire engine, named the *Miles Greenwood* after the fire chief of Cincinnati's new paid fire department. It is to go on a demonstration tour of several eastern cities, starting in Washington, D.C., then moving north to Baltimore, Philadelphia, New York, and Boston. It is Mr. Latta's hope that each of these cities will buy at least one steam fire engine.

New York, N.Y., Feb. 9, 1855

Yesterday, we demonstrated the *Miles Greenwood* at City Hall Park (by "we" I mean the three Latta brothers and I). The citizens of New York seem impressed by its performance, but as in Cincinnati, the volunteer firemen are opposed to the steamer. A steamer was briefly used here 14 years ago, but even then the firemen were so hostile to it that it was quickly abandoned.

Empire Engine Co. No. 42 brought out their huge "hayrack"-style hand pumper, which was built by Henry Waterman. So much muscle is required to operate its pump that this engine has been nicknamed "Mankiller".

Some of New York's strongest firemen were working the "brakes" of the pump, and they had water faster than our *Miles Greenwood* engine did. But after a few minutes, *Mankiller's* operators dropped of exhaustion, while the *Greenwood* just kept pumping for hours. This demonstration proved the superiority of the steamer far more eloquently than mere words ever could.

Boston, March, 1855

We demonstrated the *Miles Greenwood* here in Boston, and the insurance men immediately saw its value in reducing loss of lives and property. So they bought the steamer on the spot for $12,000, and presented it to the city.

Unlike Cincinnati, Boston has given the steamer to the volunteer firemen. Firemen here are as opposed to any reforms as firemen in Cincinnati and New York. They have vowed never to let the steamer fight a fire in this city.

Cincinnati, May 12, 1855

From Boston comes word that they had a $1 million fire on April 29, and that the damage might have been considerably less had the volunteers not stubbornly refused to bring out the *Miles Greenwood* and use it in quelling the blaze. The volunteer firemen are supposed to be public-spirited citizens who have volunteered to protect the lives and property of their neighbors, so it is inexcusable not to take advantage of any new invention that assists in that effort. By such tactics, Boston's volunteer firemen will eventually seal their fates just as surely as the now-abolished volunteer fire companies of Cincinnati did.

Cincinnati, May 30, 1855

Abel Shawk and his brother George recently started a steam fire engine factory of their own, called Young America Works. Their factory is at 11 East 7th St., between Main and Sycamore.

From Philadelphia comes news that Abel Shawk successfully demonstrated his new steam fire engine there last week. This engine, named *Young America*, is mounted on four wheels, and is smaller and considerably lighter than the ones which Latta is making. Like our Latta engines, it uses the boiler Shawk developed, so it steams as rapidly as ours do (three minutes). But it is much slower to pump water (over 11 minutes), and at its best it only discharges about half as much water as the *Uncle Joe Ross* or *Citizen's Gift* can.

I have moved into a rooming-house at the northwest corner of Race and 14th streets.

Cincinnati, March 11, 1856

I have renounced my allegiance to the King of Hanover and declared my intention to become an American citizen.

Cincinnati, Sept., 1856

Mr. Latta recently sold a fire engine boiler to Lawson & Pierce of Louisville, Ky. They used it in a fire engine built for the city of Louisville, but Louisville did not buy their engine.

John Kenrick Fisher of New York recently developed a completely self-propelled steam fire engine. His engine not only moves under its own steam power, as Latta engines do, but it also has a steering mechanism that eliminates the need for horses to guide it. Lee & Larned of New York are manufacturing Fisher's engines.

Louisville, March 2, 1857

Finley Latta and I delivered a new Latta steamer here, and they put it in service yesterday. They are following the example of Cincinnati, and have used this steamer as the basis for organizing a paid fire department. The paid department also began operating yesterday, and the equipment of the volunteer fire companies has been turned over to the paid department. The new fire department consists entirely of former volunteer firemen, and as in Cincinnati, the rowdier elements have been excluded. Louisville and Cincinnati are so far the only cities to establish a paid fire department.

Cincinnati, Sept. 21, 1857

Last week, St. Louis, Mo., joined the still-slim ranks of cities operating paid fire departments. They have four Lattas, plus two built by Abel Shawk's Young America Works. Cincinnati's paid department has seven Latta steamers in service, and Louisville has one Latta steamer.

The Latta brothers claim that eventually, every major city in America will replace their volunteer fire companies and hand engines with a paid department and steam engines, as Cincinnati, Louisville, and St. Louis have done. And, of course, the Lattas hope to be the ones to sell steamers to all of those cities. As their apprentice machinist, I will be the one building all of these steamers for at least the next four years.

But the Lattas already have some competition. Not only is Abel Shawk also building steamers here in Cincinnati, but last year, Horace Silsby (of Silsby, Mynderse & Co., Seneca Falls, N.Y.) also began building steamers. Another firm, Lee & Larned, began building steamers in New York last year. Most of the sales of those two steam engine builders seem to be in the northeast, so we don't have much competition from them here in the west. Besides, we have the superior features of the quick-steaming coil boiler and the more-maneuverable three-wheel chassis, which neither Silsby nor Lee & Larned can use without infringing Alexander Latta's various patents.

Jan. 1, 1858

Yet another competitor to our Latta steamers has come along. On Christmas day, the firm of Reanie & Neafie in Philadelphia demonstrated their first steamer. It was designed by Joseph Parry of that city, and appears to be modeled on Philadelphia's 1855 Shawk steamer *Young America*.

March 10, 1859

I became an American citizen today.

August, 1859

On July 4th, Mr. N.S. Bean of Manchester, N.H., tested a new steam fire engine that he designed. The Amoskeag Manufacturing Co. of Manchester, noted builder of railroad locomotives, will begin manufacturing Bean's steamers soon. Mr. Latta is concerned, as the reputation of the Amoskeag firm will likely result in tough competition for our engines. One of the first Amoskeag steamers will go to Boston, which recently organized a paid fire department.

On the other hand, we recently lost our only competitor here in the west. Abel Shawk has given up the manufacture of steam fire engines, as it has been three years since he sold one.

I have moved to 15th and Bremen streets.

Sept. 3, 1859

My sister, Elizabeth Ahrens Willman, died yesterday. She was born in Bredbeck Dec. 16, 1833, and emigrated in 1852.

Feb., 1860

Indianapolis recently joined the growing ranks of cities with paid fire departments and steam fire engines. However, there are still quite a few cities and towns that cling to their volunteer fire companies and hand engines. So business for Mr. Latta is not as brisk as he might hope.

To increase business, Alexander Latta has published a pamphlet about the experiences of Cincinnati, Louisville, and St. Louis with Latta engines, and of other cities with other steamers.

Commenting generally about Horace Silsby, Amoskeag, and other competitors, Mr. Latta has stated "It is said there are many ways to do things, and verily if there is any other way to make a steam fire engine, besides the way we make them, it certainly will be found out, as there seems a determination to make a steam fire engine on some other plan than that now used. And why should there be such a desire to accomplish the end by some other means than the only successful plan, as experience shows? Is not the present engine good enough? It is very reliable, easily kept in order, and has been known to play water through 500 feet of hose, in 4 minutes and 15 seconds, all cold. How much quicker than this could be asked for?"

After enumerating all the advantages of our engine, Mr. Latta further comments, "It seems to us somewhat strange, that with all these peculiar merits, such a desire should exist to devise some *other means* to accomplish the ends; however, if any thing *better* comes out, we will likely adopt it. We have offered *$10,000* for a better machine."

That the competition's machines are *not* better is evidenced in the examples cited from Chicago and Boston. Both are using steamers of various makes other than Latta, and in both cities fire losses have actually *increased* since the steamer's introduction. The reason is that it takes their steamers 15 to 20 minutes to generate steam and deliver water, which is considerably slower than a hand engine. A Latta steamer can play water on a fire in less than five minutes, starting from a cold boiler. The extra ten-to-fifteen minute delay with other engines, gives a fire time enough to completely engulf a building.

April 13, 1861

Political tensions between north and south have been brewing ever since abolitionist John Brown and his 17 men tried to free all of the south's slaves at Harper's Ferry, Va., two years ago. The support Brown had expected there never arrived, and in "punishing" Brown the south unintentionally made him a martyr. It is appropriate that Brown chose a volunteer fire hall -- the traditional haven for poilitical extremists -- to make his stand against United States troops.

Yesterday, the north-south tensions that John Brown helped to ignite, finally erupted when southern secessionists fired upon Fort Sumter in Charleston, S.C., and in response the north declared war on the south.

Like all wars, the pretext for this one is both moral (slavery) and political (states rights), while the real issue that nobody admits to is economics. The industrial north uses economies of scale to undercut prices of manufactured goods from the south's few factories. For their part, the south's slave-labor has allowed them to sell agricultural goods much cheaper than the north could.

The war will greatly change this economic picture. The south will lose the biggest market for its agricultural products (the north), and its main source of supply for the manufactured weapons it needs to fight this war (again, the north). For their part, the factories of the north will lose a significant percentage, but not a majority, of its customers (in the south). In short, the northern economy is stronger, and less dependent on the south than the south is on the north. That fact alone virtually guarantees the outcome of this conflict.

Cincinnati, August, 1861

After seven years of apprenticeship at the Buckeye Works, I have now become a master machinist, and Mr. Latta has made me foreman in charge of fire engine production.

The outbreak of Civil War has slowed our production somewhat. Some of our business has traditionally been in southern cities, but now there is suspicion about buying from "Yankees"; in that regard Ohio falls under less suspicion than New York (Silsby) or New Hampshire (Amoskeag). Being so close to Kentucky, we are almost considered southerners; the sympathies of many Cincinnatians are with their Confederate neighbors in Kentucky (a sympathy which I, an immigrant to whom America has given much, do not share).

Even if a southern city does order a steamer from us, it is difficult to deliver it without crossing through the middle of a battle. For example, we had no end of trouble delivering an engine to Memphis recently. The roads are too poor and the soldiers of both sides too numerous to attempt overland delivery; shipment via riverboat is risky as the rivers are frequently blockaded; and both armies seize the railroads without warning whenever it suits their needs.

Still, when this war ends, our business should pick up again. Many war-ravaged cities will need new fire engines, and I shall supervise construction of all of these new engines.

Cincinnati, Jan., 1862

My younger brother Peter (born April 14, 1841) recently emigrated, and is now a grocer in my older brother Hermann's grocery and feed store at Sycamore and 7th. Peter and I have moved to a rented house at 13 Wade street.

Cincinnati, Sept. 3, 1862

I joined the Ohio volunteer militia yesterday. I am a sergeant in Company G, 11th regiment.

Cincinnati, Oct. 3, 1862

Yesterday, my regiment of the Ohio militia became part of the Union Army.

Cincinnati, Dec. 12, 1862

Yesterday, I married Mary Louise Seidenburg. She was born in Cincinnati on Sept. 28, 1843 to George and Charlotte (Bandermann) Seidenburg.

When we return from our honeymoon, we will move into a rented house at 125 West 2nd street.

Cincinnati, Sept. 1, 1863

I am a proud father. My son, George Frederick Ahrens, was born today. His first name is for my father-in-law, George Seidenburg; his middle name is for my father, Friedrich Ahrens.

The Latta brothers recently sold their fire engine business to Lane & Bodley, a local machine shop. The Lattas formed the Aerated Bread Co., operating out of the Buckeye Works, 179 Race St.

Lane & Bodley has hired me as Superintendent of fire engine construction at their works, located at John and Water streets. The pay is better than what the Lattas gave me, which will certainly help now that I have three mouths to feed.

Lane & Bodley was founded in Cincinnati in 1850 by Colonel Philander P. Lane and Mr. Joseph T. Bodley. Lane grew up on a farm but was an experienced machinist. Bodley had apprenticed at Miles Greenwood's Eagle Iron Works. In the 13 years since, theirs has become Cincinnati's largest manufacturer of woodworking machinery.

Cincinnati, Oct., 1863

With my new job and new addition to our family, Mary and I decided to stop renting a few rooms in someone else's house, and bought a house of our own at 86 Bremen Street. We have given my brother Peter a room in this house.

Cincinnati, July 24, 1865

My daughter Lettie was born today.

Cincinnati, Jan. 1, 1866

Last year was a year for endings. The Civil War ended after four bloody years. President Lincoln's term ended with his assassination. The life of my former employer and mentor, Alexander Latta, ended on April 28, 1865. And the volunteer firemen's tenacious resistance to reform also ended, when one of their last strongholds -- New York -- organized a paid fire department and installed several steamers last year.

1866

My mother, Anna Hartmann Ahrens, died in Bredbeck, Hanover, at age 62.

Cincinnati, April 18, 1868

My brother Johann Frederick Ahrens died yesterday. He was born in Brettbeck in Dec., 1830, and emigrated to Cincinnati in 1849.

Cincinnati, August, 1868

Colonel Philander P. Lane died recently. His partner, Joseph T. Bodley, will discontinue fire engines and build only woodworking machinery. The late Colonel's son, Henry M. Lane, is Bodley's new partner, and concurs with that decision.

I purchased the Latta patent rights from Mr. Bodley, along with the fire-engine production equipment, and will manufacture Latta steam fire engines myself. I have some ideas that should bring me more success than Lane & Bodley had.

To help finance this enterprise, I have brought in Henry Lachtrop of 468 Main Street as my business partner. He owns a saloon at the corner of Freeman and Liberty, which is doing a brisk business. He has been looking for a small new venture in which to invest his profits, and was quite interested in my new fire engine company.

We are in the process of purchasing a three-story building at 24 Webster St., and will equip it for fire engine production and hire a suitable staff (former co-workers from Lane & Bodley's now-defunct fire engine division). The new company is called C. Ahrens & Co. We expect to finish our first Latta engine by year's end.

In addition to having a new factory, I also have a new home. Mary and I sold the house at 86 Bremen and moved to 17 Woodward Street.

Chapter 4

Cincinnati, Jan. 1, 1869

The first fire engine from my new *C. Ahrens & Co.* factory on Webster St. went in service at Fire Co. No. 10 in this city today. It is built on the same plan as the Latta and Lane & Bodley steamers, that is, it has three wheels, a square boiler, and a one-cylinder horizontal piston pump. It is a 2nd Class engine, with a $5^1/_2$" pump, 10" steam cylinder, and 14" stroke. The city has named it *A.B. Latta*, in honor of its inventor.

I have bought the rights to all of Mr. Latta's steam fire engine patents, which are as follows:

#11025 Steam Generator 6-6-1854
#12682 Steam Generator 4-10-1855
#12912 Fire Engine Chassis 5-22-1855
#14963 Safety Valve 5-27-1856
#15297 Iron Spoked Wheels 7-8-1856
#18460 Steam Generator 10-20-1857
#23589 Steam Generator 4-12-1859

These patents apply to the *A.B. Latta* engine that I have just built, and to the engine I am now building for Hamilton, Ohio. Mr. Latta had sold these patent rights to Lane & Bodley, who in turn assigned them to me.

I am calling these Latta engines "Click", which is the name that Lane & Bodley gave them. As well as describing a sound, the word "click" also has the obscure definition of "pivoting motion". The name is appropriate, as the three-wheel Latta has a single front wheel that pivots 180 degrees.

Cincinnati, August, 1869

From Bredbeck comes word of the death of my father, Friedrich Ahrens von Brettbeck. He was 69 years old. My sister, Anna Ahrens Albers, is the only member of the family remaining in Bredbeck; she has been a widow for 13 years.

Cincinnati, March, 1870

Henry Lachtrop and I tested the first steam fire engine of my own design last week. It is considerably different from the four Click engines we built last year. For one thing, at 6110 pounds it is 2900 pounds lighter than a Click engine of equivalent pump capacity.

The most striking difference in this new engine is that it is on four wooden wheels rather than three iron wheels. I have found the Latta-patent iron wheels to be somewhat flimsy, and quite expensive to manufacture when only a dozen wheels were made in the entire year. It is much less expensive to buy standard wooden carriage wheels from an established wheel manufacturer, and those wheels are much sturdier.

I have also redesigned the frame, from the cross-hatched Latta frame, which is only as strong as the welds holding all the frame-pieces together. The new frame consists of two iron beams (one on each side), so it is structurally more sound.

The boiler is also new. By grouping the coils around each other, rather than in rows, the coils radiate heat to one another. Steam is thus generated more quickly. A round, rather than square jacket, seems to help the steam circulation, too.

I have also redesigned the pump, placing it vertically rather than horizontally. In this way, steam does not have to travel as far to move the pump piston, so less steam pressure is lost to friction. Due to the pump's upright orientation, I have named this new style of engine "Upright". In fact, Henry Lachtrop and I have dubbed our little Webster Street factory the "Upright Works", as we intend to produce mostly Upright engines from now on. We will continue to build Click engines for those cities which prefer them.

Indianapolis, April 1, 1870

Henry Lachtrop and I demonstrated our new Upright engine here today, after a demonstration in Louisville last week. Unknown to us, several members of the Fair Play Fire Co. #1 of Madison, Ind., were among the spectators. They were so impressed with today's demonstration that they bought our new engine on the spot.

Attitudes have certainly changed in the past 15 years! When I became Mr. Latta's apprentice in 1854, Mr. Latta had to fight for every steamer sale, because the volunteer firemen violently resisted anything that threatened their rowdyism and political clout. Yet today, with no salesmanship on our part, the volunteer firemen from Madison voluntarily bought our prototype Upright steamer, without any inducement or pressure to do so! They bought it simply because they recognized that our engine would help them fight fires more effectively, and I think this would have pleased Mr. Latta no end had he lived to see it (unfortunately, he died about five years ago, when firemen's attitudes were still hardened against his invention).

Cincinnati, June, 1870

Henry Lachtrop and I have expanded our Upright Works factory. We acquired the building at 26 Webster St., and added it to the existing factory at 24 Webster. With this additional space, we will be able to produce the new Upright engines as well as continuing to produce Click engines.

July 7, 1870

My daughter Lillie was born today.

Dec., 1870

My business partner, Henry Lachtrop, recently resigned so that he can devote all of his attention to his new saloon at 600 Main Street. My new partner is Frank Kamman of 560 Main St. (near the corner of Webster St.). Frank is an experienced machinist and therefore should be a more knowledgeable partner in our manufacturing business than a saloon-keeper like Henry Lachtrop is.

Feb. 10, 1871

On Jan. 31, my new partner Frank Kamman and I were granted U.S. Patent Number 111,298 for a new steam generator. This boiler is a vast improvement in both speed and safety over the Latta "gunpowder" boiler. I built the first new boiler last year for an engine I demonstrated in Indianapolis and later sold to Madison, Ind.

I recently hired Joe Terbeck of 1646 East Front St., as a pipefitter. He will install coils in the boilers of our steam fire engines. Joe is an experienced blacksmith and machinist, and I feel confident that he will be a valuable addition to our company. Joe and I have similar backgrounds: I was learning the blacksmith's trade in Osterholz before war forced me to flee, and of course I worked for Mr. Latta as a machinist before Henry Lachtrop and I bought the company.

Cincinnati, May 8, 1871

We are the first steam fire engine manufacturer to break the four-minute barrier in pumping water from draft. Today, our engine Registered No. 12 raised steam and threw water in only 3 minutes 57 seconds, starting from a cold boiler.

Cincinnati, Oct. 9, 1871

This morning, I had a surprize visit at my office in the Webster Street factory, from our city's former fire chief, Miles Greenwood. It seems that the present fire chief, Enoch G. Megrue, received a telegram this morning from Fire Marshal Robert A. Williams of Chicago, Ill.

Apparently, a fire started about 8:00 last night in a barn on Chicago's DeKoven Street, and the winds whipped the flames so furiously that by this morning the entire city of Chicago was in danger of burning to the ground. Fire Marshal Williams asked Chief Megrue to round up all of the men and equipment that he could spare and dispatch them as quickly as possible to Chicago.

On hearing this news, Miles Greenwood volunteered to take command of the Cincinnati forces in Chicago. This would leave Chief Megrue free to direct the Cincinnati firemen at home, in case any conflagrations should strike our own city.

Mr. Greenwood is presently a director of Cincinnati Southern Railway, so he donated the use of a locomotive, flatcars to carry the fire engines, a passenger-coach for the firemen, and a cattle-car for the fire horses. He also wired ahead for all rail-traffic between Cincinnati and Chicago to give the fire train right-of-way.

Chief Megrue felt that he could only spare two engines for this trip, belonging to Fire Companies 10 and 13. Coincidentally, both of these engines are of our own Ahrens make.

The fact that both engines being sent to Chicago were of Ahrens make, prompted Greenwood's visit to our factory this morning. "Do you," he asked, "have any completed engines on the shop floor, not yet delivered, that you could spare to help fight the Chicago inferno?"

"It so happens," I replied, "that I have recently completed an experimental engine with iron wheels, and have it on display at the Cincinnati Exposition Building. I have not yet sold this engine to any fire department, so you may certainly take it with you on the Chicago fire train. In fact," I offered, "since this is an experimental engine that the Cincinnati firemen might not know how to operate, I will go along and operate it at the fire if you want me to."

So Mr. Greenwood and I rushed off to the Exposition Building, pushed my experimental Upright engine off the exhibit floor and out onto the street, hitched-up two horses, and hauled it off to the railroad yards. I rounded-up my most experienced employees to assist me in operating this engine at the fire, and we joined the Cincinnati firemen aboard the fire-train.

It is now about 11:00 a.m., and we are just leaving the Cincinnati city limits as I write this. We are to stop in Dayton to pick up an old Latta engine and its crew from Dayton's Fire Co. No. 3, then travel non-stop to Chicago.

Chicago, Oct. 9, 1871

We arrived in the city early this evening, and by 5:30 p.m. had unloaded the two Cincinnati engines, the Dayton engine, and my Upright engine. We then checked-in with fire marshal Robert Williams.

It seems that on the afternoon of Oct. 7, some coal-storage yards on Chicago's west side caught fire, and they have been burning ever since. The fire that now threatens to devour the city started independently, about 28 hours after that. Marshal Williams fears that as the coal fire spreads east and the newer fire spreads west, the two will meet-up and end all hope of saving the city.

Firemen and fire engines from all over Illinois, Indiana, Ohio, Michigan, and Wisconsin are already hard at work on the fire that started on DeKoven street. The spirit of cooperation between the dozens of fire departments coming to Chicago's aid, contrasts with the fierce rivalries that caused the Cincinnati firemen's riots 20 years ago. In fact, 20 years ago firemen had such a reputation for violence that no fire apparatus builder would have had the courage to operate one of his engines side-by-side with the firemen, as I am doing today.

With the main fire being attacked on many fronts, what Chief Williams asked us to do was to dig-in for some long-term pumping on the smaller fire (the burning coal piles). Not the most glamorous task, perhaps, but certainly a necessary one. So the four steamers, the Cincinnati and Dayton firemen, Mr. Greenwood, and I headed for the burning coal-piles. It is now 10:00 p.m., and I have been working the experimental Upright engine for four hours now. As I write this, I have just turned over the controls so I could rest a bit and gather my thoughts.

Chicago has suffered scorching-hot weather but little rainfall this year. Most of the buildings are of wood, even the sidewalks are of wood, and everything has become so dry that a fire like this was practically inevitable. Fire Marshal Williams told us that the blaze began when a cow, upset about being milked at a late hour, kicked-over a lantern into a pile of hay. The fire spread quickly because the first fire companies were accidentally sent to a location several miles from the fire. A strong wind fanned the flames, and as the fire spread over the city, it generated its own wind that hastened the fire's spread.

By the time we arrived from Cincinnati, several steamers, and hundreds of feet of hose, had already been engulfed in the flames, making the firemen's task that much harder.

Chicago, Oct. 11, 1871

A light rain began to fall early this morning, aiding our efforts to extinguish what has already been dubbed the Great Chicago Fire. The name is appropriate, as it is the worst disaster in Chicago's history, destroying 17000 buildings, killing 250, and causing $88 million in property damage.

The last house to catch fire (on Fullerton St.) was completely extinguished shortly after 3:00 a.m., but the coal piles we were working on continued to smolder. The two Cincinnati engines, the Dayton engine, and my engine continued hosing-down the coal-piles for another five hours. By the time we shut-down, shortly after 8:00 this morning, all four engines had been pumping continuously for 38 hours, without a single shut-down for any reason. This performance must be something of a record, and certainly speaks well of our engines since all four were built by the company which I now own.

Fire officials from Kansas City, Mo., came to

Chicago during the fire, to offer their help. The magnitude of this disaster has them concerned lest a similar conflagration befall their city. Kansas City has similar conditions to Chicago: mostly wood houses, and the weather has been very hot and dry, with very little rainfall.

These Kansas City officials were impressed by the 38-hour non-stop pumping our four engines completed. They asked me to bring my experimental engine to Kansas City, and demonstrate its capabilities, with a view to installing one or more Ahrens engines in their city. So I and my engine will not be returning to Ohio with the other men and engines on Miles Greenwood's fire-train.

Kansas City, Oct. 21, 1871

I demonstrated our Upright engine here yesterday, and while the test was quite successful, they decided not to buy it. Kansas City has installed a typical big-city political machine, whose words do not agree with their deeds.

The Kansas City politicians *say* that they want to prevent a Chicago Fire in their city, but their actions contradict this. They have set up such an extensive system of graft that nobody can sell anything to the city without kicking-back a percentage into a politician's pocket.

I refuse to sell them a fire engine under such conditions: not only is it illegal and unethical, but the fire engine business is simply not profitable enough to give away a percentage of our profits in bribes. We make a product of unsurpassed quality, that can be sold on its own merits and reputation alone, and if we cannot make a sale honestly, we will not make the sale at all.

Cincinnati, Aug. 11, 1872

My son, John, was born today. Mary and I named him John Peter after his uncle (my brother) Johann Peter Ahrens. This brings the total to seven people living in our house at 23 Franklin St.: my wife and I, our four children (two boys and two girls), and my brother. Any more and we will need a bigger house!

Speaking of my brother Peter, I recently hired him to work for me, and am training him to be a machinist. He had been working as a clerk in an office at 104 Walnut Street for the past five years, but now that our company is starting to grow, I felt that I could use his help.

My former business partner, Frank Kamman, has resigned from C. Ahrens & Company, leaving me to run the place alone. So I can use all the help I can get, especially support from my family. I had thought about making my brother a partner in the company, but he told me that he is not interested in being more than another employee.

Cincinnati, Oct. 8, 1873

My brother Peter recently joined the Cincinnati Fire Dept. He drives the horses at Fire Co. No. 7, located at 17 Webster St. directly across the street from our factory.

I recently made George F. Hawekotte (pronounced **HAV'**-a-kot'-ta) of 370 West 5th St. my new partner in C. Ahrens & Co., after running the business by myself for over a year. "Fred", as he likes to be called, is the son of a silversmith, and began his own career as a machinist in 1862. He joined the Cincinnati Fire Department in 1867, and within a year had been promoted to engineer-of-steamer on one of the city's Latta engines.

Fred Hawekotte has the ideal background for a partner in C. Ahrens & Co.: an experienced machinist, engineer, and fireman, and a long acquaintance with the operation of our engines. I have so much confidence in him, in fact, that last week I sent him alone to deliver and conduct the acceptance test of a new Upright engine we built for Saint Louis, Mo.

Editor's Note: Fred Hawekotte stayed with the Ahrens company for several decades. Following is a list of Ahrens steamers that he delivered over the years. For more details on each engine, see the complete list in the back of this book.

10-1-1873 St. Louis, Mo. (Reg. #30)
12-4-1873 Hamilton, Ohio (#32)
1-1-1874 Louisville, Ky. (#33)
4-27-1874 Quincy, Ill. (#35)
7-8-1874 Indianapolis, Ind. (#37)
10-20-1874 Greensburg, Ind. (#39)
11-10-1874 Indianapolis, Ind. (#41)
1-22-1875 Shelbyville, Ind. (#42)
5-13-1875 Oxford, Ohio (#45)
6-14-1875 Chattanooga, Tenn. (#146)
8-20-1875 St. Louis, Mo. (#149)
10-1-1875 St. Louis, Mo. (#150)
11-8-1875 St. Louis, Mo. (#152)
12-17-1875 Leavenworth, Kansas (#157)
1-7-1876 Indianapolis, Ind. (#155)
2-17-1876 Cynthiana, Ky. (#159)
3-8-1876 Indianapolis, Ind. (#160)
5-4-1876 St. Louis, Mo. (#162)
5-19-1876 Cambridge City, Ind. (#163)
5-29-1876 Watertown, Wis. (#161)
9-11-1876 Aurora, Ind. (#166)
12-23-1876 Covington, Ky. (#169)
3-29-1877 Cincinnati, Ohio (#172)
6-22-1877 Georgetown, Ky. (#170)
7-6-1877 Franklin, Ohio (#178)
9-5-1877 Montreal, Quebec (#180/#182)
10-9-1877 Montreal, Quebec (#196)
3-16-1878 New Orleans, La. (#192)
11-26-1878 St. Louis, Mo. (#208)
3-1-1879 Lake Providence, La. (#204)
3-17-1879 New Orleans, La. (#206)
5-1879 New Orleans, La. (#214)
7-7-1879 Newark, Ohio (#216)
10-24-1879 Milwaukee, Wis. (#224)
11-30-1879 New Orleans, La. (#222)
3-[illegible]-1880 Vincennes, Ind. (#234)
6-4-1880 Council Bluffs, Iowa (#212)
8-9-1880 St. Louis, Mo. (#236)
9-6-1880 Louisville, Ky. (#250)
11-14-1880 New Orleans, La. (#258)
1-28-1881 Delphos, Ohio (#272)
3-10-1881 Oshkosh, Wis. (#274)
3-28-1881 Lena, Ill. (#244)
5-24-1881 Cincinnati, Ohio (#280)
11-5-1881 Frankfort, Ky. (#292)
4-6-1882 Greencastle, Ind. (#302)
7-12-1882 Memphis, Tenn. (#318)
10-13-1882 St. Louis, Mo. (#330)
3-2-1883 Vincennes, Ind. (#234)
8-23-1883 Peoria, Ill. (#364)
9-14-1883 Milwaukee, Wis. (#372)

9-27-1883 Sidney, Ohio (#378)
2-5-1884 Missouri Penetentiary (#393)
2-29-1884 St. Paul, Minn. (#391)
5-4-1884 Milwaukee, Wis. (#396)
11-22-1884 Catlettsburg, Ky. (#417)
1-7-1885 Saint Paris, Ohio (#415)
4-2-1885 Arcanum, Ohio (#425)
5-30-1885 St. Louis, Mo. (#419 and #420)
8-1-1885 Vicksburg, Miss. (#422)
12-8-1885 Pendleton, Oregon (#432)
3-6-1886 Ashtabula, Ohio (#455)
3-10-1886 Waynesville, Ohio (#457)
7-1886 Chattanooga, Tenn. (#461)
9-21-1886 Milwaukee, Wis. (#451)
10-16-1886 Akron, Ohio (#465)
11-18-1886 North Vernon, Ind. (#477)
4-1887 St. Louis, Mo. (#502)
6-1887 Birmingham, Ala. (#520)
1-14-1888 Lake View, Ill. (#527)
2-8-1888 Indianapolis, Ind. (#530)
5-10-1888 Versailles, Ky. (#540)
7-24-1888 North Amherst, Ohio (#541)
11-10-1888 Chattanooga, Tenn. (#558)
11-14-1888 St. Louis, Mo. (#562)
12-3-1888 Canton, Ohio (#549)
12-88 St. Louis, Mo. (#564 and #565)
8-7-1889 Pueblo, Colo. (#583)
8-21-1889 Wilkes-Barre, Pa. (#582)
9-21-1889 Tacoma, Wash. (#577)
12-6-1889 Columbus, Ohio (#574/#596)
12-11-1889 Toledo, Ohio (#593 and #594)
1-7-1890 St. Louis, Mo. (#599 and #600)
8-29-1890 Springfield, Ill. (#622)
12-22-1890 Columbus, Ohio (#610-611)
12-22-1890 Columbus, Ohio (#613-614)
2-23-1891 Dubuque, Ia. (#640)
5-19-1891 Decatur, Ill. (#647)
9-10-1891 Westwood, Ohio (#661)
9-18-1891 Belle Plaine, Ia. (#641)
10-6-1891 St. Louis, Mo. (#651 and #653)
5-11-1892 Chattanooga, Tenn. (#2104)
5-24-1892 Rockford, Ill. (#2105)
7-5-1892 Cincinnati, Ohio (#672)
7-27-1892 Nashville, Tenn. (#2108)
8-31-1892 Milwaukee, Wis. (#2123)
11-19-1892 St. Louis, Mo. (#2126)
8-5-1893 Milwaukee, Wis. (#2177)
10-7-1893 St. Louis, Mo. (#2174)

Chicago, Nov., 1873

Just two years after the city was left in ashes, much of Chicago has been rebuilt, and it is once again a thriving center of commerce. Stiff laws have been passed regarding the fire-resistance of building materials, and the fire department is seeking to put an engine company in every neighborhood so that firemen will no longer have to travel several miles to a fire.

Fire Marshal Robert Williams has not forgotten the superlative performance of our four engines at the Great Fire. He asked me to build him a new engine, and I have been in Chicago with this engine the past few days to demonstrate its capabilities before its acceptance.

The city did accept this engine from me, but city officials have made it clear that they will not be buying any more Ahrens engines. While Fire Marshal Williams feels that no expense should be spared to prevent a repeat of the disaster that nearly eradicated his city two years ago, the politicians are keeping a tight fist on the money.

Why spend 33% more to buy one of our engines, they asked, when they can buy 3 "nearly-as-good" Amoskeag or Silsby engines for the price of two of ours. Apparently, they will need a repeat of the 1871 disaster before they learn that it is worth a few thousand extra to buy a steamer that starts working in under 4 minutes, compared to one that takes over 10 minutes. A fire can cause a lot of unnecessary destruction in those extra six or seven minutes!

Editor's Note: Chicago's 1873 delivery is the first time that the Ahrens records specifically mention that Chris Ahrens delivered a steamer. Chris Ahrens delivered the following Ahrens steamers.

11-1873 Chicago, Ill. (#31)
12-4-1873 Hamilton, Ohio (#32)
1-1-1874 Louisville, Ky. (#33)
7-8-1874 Indianapolis, Ind. (#37)
11-10-1874 Indianapolis, Ind. (#41)
4-5-1875 St. Louis, Mo. (#44)
7-30-1875 St. Louis, Mo. (#148)
10-25-1875 Hillsboro, Ohio (#151)
11-16-1875 Waverly, Ohio (#153)
12-30-1875 Nashville, Tenn. (#158)
1-7-1876 Indianapolis, Ind. (#155)
2-7-1876 St. Louis, Mo. (#156)
5-29-1876 Watertown, Wis. (#161)
3-1-1877 Cleveland, Ohio (#168)
3-9-1877 Louisville, Ky. (#165)
6-9-1877 Uniontown, Pa. (#176)
7-4-1877 Galveston, Texas (#174)
9-1877 Topeka, Kansas (#194)
9-5-1877 Montreal, Quebec (#180/#182)
10-9-1877 Montreal, Quebec (#196)
12-8-1877 Gallipolis, Ohio (#184)
1-8-1878 Paris, Ky. (#188)
5-29-1878 Parkersburg, W. Va. (#182)
6-5-1878 Columbia, Tenn. (#190)
8-20-1879 Louisville, Ky. (#220)
2-7-1880 Chillicothe, Ohio (#232)
4-11-1880 New Orleans, La. (#230)
4-25-1880 New Orleans, La. (#238)
4-25-1880 New Orleans, La. (#240)
7-24-1880 Chicago, Ill. (#248)
9-4-1880 New York, N.Y. (#252)
11-10-1880 Chicago, Ill. (#260 and 262)
12-24-1880 Louisville, Ky. (#270)
4-2-1881 Louisville, Ky. (#276 and 278)
4-19-1881 Rushville, Ind. (#282)
12-22-1882 Louisville, Ky. (#344)
6-26-1883 New York, N.Y. (#358/#360)
6-30-1883 Portsmouth, Ohio (#362)
7-9-1883 New York, N.Y. (#354/#356)
8-30-1883 Louisville, Ky. (#384)
11-1883 Nashville, Tenn. (#380 and #382)
11-27-1883 Harrison, Ohio (#388)
5-29-1884 Louisville, Ky. (#400)
6-1884 Milwaukee, Wis. (#401)
2-4-1885 Oakland, Cal. (#411)
8-13-1885 Minneapolis, Minn. (#437)
11-12-1885 Dayton, Ohio (#445)
12-8-1885 St. Paul, Minn. (#448)
12-18-1885 Bardstown, Ky. (#453)
1-7-1886 Sullivan, Ind. (#435)
2-16-1886 Birmingham, Ala. (#459)

3-10-1886 Waynesville, Ohio (#457)
3-29-1886 St. Louis, Mo. (#467)
5-3-1886 Germantown, Ohio (#475)
5-4-1886 Miamisburg, Ohio (#470)
6-14-1886 San Antonio, Texas (#449)
10-21-1886 Minneapolis, Minn. (#499)
10-21-1886 Minneapolis, Minn. (#500)
2-22-1887 Lebanon, Ohio (#423)
3-26-1887 Deshler, Ohio (#487)
5-3-1887 Milwaukee, Wis. (#505)
8-30-1887 St. Paul, Minn. (#509 and #511)
10-12-1887 Detroit, Mich. (#518)
6-9-1889 Morrow, Ohio (#555)
1-20-1890 Wichita, Kansas (#602)
3-31-1890 Minneapolis, Minn. (#637)
3-31-1890 Minneapolis, Minn. (#638)
4-7-1890 Columbus, Ohio (#607 and #608)
4-17-1890 Quincy, Ill. (#616)
8-16-1890 Detroit, Mich. (#623)
12-17-1890 Wilmington, Ohio (#605)
2-16-1891 Louisville, Ky. (#629)
6-17-1891 Milwaukee, Wis. (#636)
9-10-1891 Westwood, Ohio (#661)
11-2-1891 Madison, Ind. (#660)
4-21-1892 Indianapolis, Ind. (#671)
6-13-1892 San Antonio, Texas (#2106/2107)
8-10-1892 Wheeling, W. Va. (#2110)
11-20-1892 Wheeling, W. Va. (#2130)

Cincinnati, Jan., 1874

I have moved into a new and larger house at 50 Webster St. My new partner, Fred Hawekotte, has also moved; he now lives at 13 Orchard St.

My brother Peter has been transferred from Fire Co. 7 on Webster St., to Fire Co. #5 on Vine between Court and Canal.

Louisville, June 8, 1874

We have broken our own world's record for fastest water from draft. Today, our engine #20 threw water in 3:45, starting from a cold boiler. The old record was 3:57, set by Ahrens #12.

Nashville, Sept. 17, 1874

We have once more beaten our own fastest-water record. Today, our engine No. 38 threw water in 3 minutes 30 seconds, taking 15 seconds off the old record we set in June.

Cincinnati, Nov. 8, 1874

The first public demonstration of our new 1st Class Upright engine (Registered No. 41) was held in Cincinnati yesterday. Until now, all of our Upright engines have been of the 2nd size. This engine set a new world's record for farthest stream, the water hitting the street 309' from the nozzle. The previous record was 297', set by the Latta steamer *Citizen's Gift* in 1854.

St. Louis, April 6, 1875

Our first big order for Ahrens engines was recently placed by the city of St. Louis, Mo. They have ordered seven of our 1st Class engines. The first of these seven engines was tested and accepted here yesterday, and is named *Future City #18.* The second of the seven engines is to follow in about three months, and the remaining five will be delivered this winter.

Cincinnati, April 24, 1875

Today, we put on a public demonstration at Webster and Sycamore, of an experimental new 3rd Class engine. This is the first small engine we have built in our Upright Works. Mr. Latta had built a few 3rd and 4th Class engines during my apprenticeship with him, but I have not built any engines this small since buying the company. Up to this point, I had been building only 1st and 2nd Class engines, mostly for paid fire departments in major municipalities who are likely to place many repeat orders.

The new 3rd Class engine is intended for volunteer fire departments in small towns and villages. This is a new market for C. Ahrens & Co. This type of customer is likely to use one steamer for the next several decades and never place a repeat order. However, there are many more villages in this country than there are cities, so while we might sell 50 big engines apiece to 20 cities, we might also sell one small engine to each of a thousand villages.

While our chief competitors in the big-city market have been Silsby and Amoskeag, we will be facing different competitors in this new market. The principal builders of steam fire engines for small towns are Button (Waterford, N.Y.) and Clapp & Jones (Hudson, N.Y.).

Cincinnati, May 17, 1875

Today, we delivered a new 2nd Class Upright engine to Fire Co. #7, directly across the street from our factory on Webster Street. My brother Peter has been transferred back to Fire Co. 7, as stoker of this engine.

This will be very convenient when prospective customers want to see one of our engines that is actually in service. We need only walk them across the street, to receive testimonials from a fireman who has been very satisfied with the performance of Ahrens engines under actual fire-fighting conditions (we might not tell customers that the fireman is my own brother).

Chattanooga, Tenn., June 14, 1875

We took another 12 seconds off the world's "fastest water" record of 3:30 that we set last September. Our engine No. 146 had water in 3:18.

Cincinnati, Sept., 1875

We have set a new world's record for longest distance throwing a stream of water -- 312' from the nozzle, using our engine No. 150. We set the previous record of 309' last November.

Competition between steam fire engine manufacturers has become as fierce as that between steamers of rival fire companies. Our arch-rival continues to be Silsby of Seneca Falls, N.Y., with Amoskeag (Manchester, N.H.) a close third. Actually, both makes outsell us, but we have repeatedly proven that neither can generate steam faster, nor throw water farther, than our engines. To add to this air of competitiveness, new fire engine manufacturers are constantly springing-up. Most new concerns fall by the wayside, but you never know when one might become a major competitor in a class with Silsby and Amoskeag.

This is why I have decided to be better than our competitors in a new way. It is not enough to simply make a superior engine, for even the best-made engines need repairs sooner or later. We must be there to support fire departments when they need parts and service.

From now on, all of our steam fire engines will carry a 10-year service warranty. If any part of one of our engines breaks during its first ten years of service, we will repair or replace it free. Since we understand that a town cannot be without its fire engines for any extended period, we shall endeavor to repair all engines in the firehouse, the day after we receive a telegram about a problem. If more extensive repairs are required, we will make them at the factory but loan the customer another engine while repairs are made.

We have expanded our factory with a new building at 22 Webster St., and set aside a room in the factory to keep a stock of spare parts for just such emergency repairs. The item which needs replacement most often is the coil in the boiler; in their haste to get water at a fire, engineers often let the boilers overheat until the coils become red-hot and the lead solder-joints melt. So we are especially keeping a supply of boiler-coils, as well as such routine maintenance items as pump valves.

None of our competitors provide this level of service. Our new service policy, together with the superior performance of our engines, should keep us competitive for many years to come.

To implement this new policy, I have assigned Joe Terbeck to make all service calls in the Ohio, Kentucky, and Indiana area. Joe will continue to work in the factory as a pipefitter and a finisher when not out on the road making repairs.

This is quite a responsibility for Joe, as the customer's impression of our service can very well affect whether we receive repeat orders the next time they are in the market for a new steamer. Along with these new responsibilities, Joe has been given a higher salary, and in recognition of his new status, he has moved to a classier neighborhood (his new address is 864 Eastern Ave.)

Editor's Note: Joe Terbeck made the following service calls on Ahrens steamers over the years:

9-7-1875 Madison, Ind. #12
2-19-76 Hamilton, Ohio #2
6-12-76 Owensboro, Ky. #26
3-17-77 Madison, Ind. #5
3-13-78 Galion, Ohio #25
4-16-78 Eaton, Ohio #18
6-17-79 Indianapolis, Ind. #37
6-30-81 Hamilton, Ohio #32
11-17-81 Cambridge City, Ind. #163
5-8-82 Greensburg, Ind. #39
9-16-82 Waverly, Ohio #153
8-25-83 Georgetown, Ky. #170
12-19-84 Marysville, Ohio #180
2-27-85 Huntington, W. Va. #328
7-26-85 Aurora, Ind. #166
8-14-85 Paris, Ky. #188
8-21-85 Eaton, Ohio #18
11-11-85 Newark, Ohio #216
1-28-86 Cynthiana, Ky. #159
5-11-86 Waverly, Ohio #153
8-1-87 Greensburg, Ind. #39
8-3-87 Van Wert, Ohio #167
1-1-88 Peoria, Ill. #364
9-19-88 Marysville, Ohio #180
3-18-89 Rushville, Ind. #282
7-11-89 Catlettsburg, Ky. #417
8-1-90 Madison, Ind. #5
9-3-90 Lawrenceburg, Ind. #340
9-30-90 Indianapolis, Ind. #160
12-3-90 Eaton, Ohio #18
8-17-91 Galion, Ohio #25

Cambridge City, Ind., May 19, 1876

Our engine No. 163 shaved one second off of the fastest-water record for fire engines, throwing a stream just 3:17 after lighting the boiler. We set the previous 3:18 record last June.

Watertown, Wis., June 6, 1876

I spent the entire week in Watertown, convincing the Commissioners of Watertown that our piston engine is faster and more powerful than Silsby's rotary engine. It was not the first time that my "friendly rival" Horace Silsby and I pitted our engines against each other in competition, and will likely not be the last, but I think it was the most exciting competition so far.

The contest began at 1:30 P.M. on the iron bridge across the Rock River that runs through the middle of Watertown. Horace and I both set up our engines on the bridge, and each stretched 250 feet of $2^1/2$" hose from our engines. At 2:18 P.M., Horace lit the fire in his rotary steamer, and I started the firebox on the Upright engine. My engine had water in 4 minutes and 30 seconds, while the Silsby took a minute longer.

The object of this first test was to shoot a stream 100 feet vertically. One of our large engines can do that quite handily, but this small engine failed to achieve the desired result. The Silsby could not deliver a stream to that height, either. After about half an hour, a valve broke in our engine due to having drafted the muck from the river into our pump. I had to make various repairs before the contest could continue.

It took me less than two hours to repair the Upright engine, and by 5:00 I was back on the bridge for a test of the horizontal stream. In this test, we beat the Silsby by throwing a stream more than 20 feet farther than the Silsby did. We both tried the vertical stream again, the Silsby throwing water 160 feet straight up and our engine matching, but unable to exceed, that height.

At 7:00, we tried another contest, this one to see which engine could draft at the farthest distance from the river. Horace Silsby successfully set up draft 1000 feet from the river, but our broken valve gave out again and the Upright could not draft at that distance.

I spent all night putting the Upright back in operating condition for the next two days of competition. On both days, the Silsby and the Ahrens alternated in their victories; our Ahrens engine won about two-thirds of the contests.

Watertown is a community made up largely of German and Irish immigrants. The neighborhood east of the Rock River was settled by Germans, and west of the river by Irish. Naturally, the German population sided with us and the Irish with Horace Silsby. The feelings ran so strongly during the three-day competition that vocal and sometimes-violent arguments broke out between the east-side and west-side residents. Various harsh remarks were made about both engines, many with ethnic slurs behind them. For example, one of Watertown's Irish businessmen called our Ahrens engine a "Dutch sauerkraut machine".

After the contests were over, the Board of Street Commissioners had to decide which engine to purchase. The political ramifications of the decision reminded me of when Cincinnati tried to eliminate its volunteer fire department 24 years ago: all of the politicians were too concerned about their re-election to make a decision that might prove unpopular with any group of voters. If Watertown bought the Silsby, the German east-side would be upset; if they bought the Ahrens, the Irish west-side would be angered.

There were two good arguments in favor of our engine: we had won the majority of the competitions, and the only fire house in town was in the German neighborhood. The arguments in favor of the Silsby were equally-convincing: it had a lower price, and it had not broken down during the contests (our engine had).

The Board of Street Commissioners did not want to decide, because they knew that whatever decision they made would be unpopular. So they appointed a special Fire Committee to decide. The committee did not wish to make this decision either, and turned the matter over to the Committee of Examiners. The Examiners announced that they considered the Silsby to be the better engine, and reported their decision to the Fire Committee. The Fire Committee responded that this was not the right decision, and asked the Board of Street Commissioners to settle the issue.

The Street Commissioners held a long meeting today, and finally reached a decision. They will buy our Ahrens engine and place it in the existing fire house in the German side of town. They will also buy the Silsby, build a fire house for it in the Irish side of town, and organize a second fire company to man it.

Philadelphia, Pa., July 5, 1876

This city put on a spectacular celebration yesterday, to mark the 100th anniversary of the signing of America's Declaration of Independence. The signing was reenacted in the same old congress hall where the original signing took place, followed in the evening by the most dramatic display of fireworks that I have ever seen.

The part of the festivities that interested me the most was the huge industrial exhibition held in Philadelphia, to show how much America has changed in the past century. The newest products and inventions from around the country were on display. I brought one of my Ahrens engines.

One of my purposes in exhibiting here was to try to generate interest in Ahrens engines in the northeastern states. The Northeast has always been a tough market for us, because there are so many steam fire engine manufacturers in the northeastern states. Our western location puts us at an advantage with western customers due to our closer proximity, but by the same token, we are farther from northeastern customers than are our competitors. I hope that my visit will help to make us more competitive in this region.

Another advantage of this industrial exhibition was that it gave me a chance to see the latest offerings of our competitors. Silsby and Amoskeag were represented.

As an inventor and manufacturer, the other industrial exhibits were of interest to me, too. I made a particular point of visiting the two most widely-publicized exhibits.

The first was the giant copper arm of the Statue of Liberty that Bartholdi of Paris is sculpting as a gift from the people of France to the people of the United States. Since this will be the largest statue ever made, new manufacturing methods and new engineering techniques are being developed to create this statue.

The second was Alexander Bell's newest invention, which he calls the "telephone". Bell said that if his invention ever comes into general use, it will be of real benefit to businessmen like myself. Communications with both customers and suppliers will be much faster than the present system of mail and telegrams. Bell has started a small company to manufacture his telephones, and to establish telephone offices similar to the telegraph offices that Western Union operates.

A less-publicized exhibit, but one of genuine interest to me, was Steven Roper's steam-carriage. This is a small two-passenger carriage propelled under its own steam power, much like Mr. Latta's self-propelled fire engine of 1852.

Like Latta, Roper has the problem that a steam-powered road vehicle moves too slowly to be of much practical use. If Roper could find a way to safely propel his carriage at faster speeds, his invention could allow people to travel about a city, or from city-to-city, with much more convenience than horses, stage-coaches, or railroads.

I stopped making self-propelled fire engines for that very reason: horses can bring an engine to the fire faster than a self-propellor can. If Roper or anyone else ever solves the speed problem, perhaps I will resume making self-propelled fire engines.

Cincinnati, Nov. 21, 1876

In our continued effort to broaden the market for Ahrens steam fire engines, Fred Hawekotte and I recently developed a 4th Class engine. We gave the first one its first public trial today. We now make two large sizes of engines (1st and 2nd) for highly-populated residential areas and business districts of major cities, and two small sizes of engines (3rd and 4th) for towns and villages.

The differences in size are easily achieved by simply scaling-down boilers, steam cylinders, and pumps. All parts for engines of the *same* Class are interchangeable (differing only in finish), and many parts are even interchangeable from one Class to another. This interchangeability reduces the spare parts that we must carry (we are one of the few who don't merely sell a fire engine and run, but keep a supply of spare parts and a staff of expert mechanics to service the customer years after the sale has been made).

We are the first, and so far the only, fire engine manufacturer to use interchangeable parts. In constructing two 3rd Class engines last year, one of our assemblers accidentally switched parts of the two engines, and nobody noticed until the mistake was caught in reviewing the paper-work, after both engines had been delivered to their respective customers. Because the parts were switched unintentionally, that was a very good test of just how interchangeable our parts are.

April 10, 1877
I have been granted my second patent, number 189,414. This patent is for the steam-powered "Upright" fire engine pump that I developed six years ago. My first patent was granted six years ago, for an improved coil boiler.

Chapter 5

Cincinnati, June 30, 1877
C. Ahrens & Co. has been reorganized as Ahrens Manufacturing Co., and incorporated under the laws of Ohio. I am the president of this new corporation, and Fred Hawekotte is the vice-president. We will continue to operate out of the same factory at 22, 24, and 26 Webster St.

Because the "Upright" engine has become quite popular, the new corporation will manufacture this style exclusively. The old three-wheel "Click" engine will be dropped from the product line, as it has been two years since any fire department ordered a Click engine from us. We will continue to service both Click and Upright engines, and Joe Terbeck will continue to repair both styles of engines wherever he is needed.

It is ironic that the old C. Ahrens & Co. tried so hard and so unsuccessfully to compete with Amoskeag and Silsby in the northeast, yet the first engine made by Ahrens Manufacturing Co. was delivered early this month to the northeastern community of Uniontown, Pa. (near Pittsburgh).

Both Fred Hawekotte and I moved recently. Fred had been living in the office-building of our factory (22 Webster St.) since we added this building to the factory last year. His new address is 53 Webster St. (corner of Mansfield). My own new home is at 21 Orchard St., where I live with my wife, four children, and my brother.

Speaking of my brother, Peter has recently been transferred from the number 7 firehouse across the street from our factory. He is now the engineer for Fire Co. #6 at Pearl & Martin. They have an 1870 Ahrens Registered No. 7. I have lost a valuable sales ally: I can no longer walk a prospective customer across the street and have my brother proclaim the merits of our engine.

Galveston, Texas, July 4, 1877
I demonstrated the new engine we built for Galveston, at the Galveston firemen's picnic today. The occasion, of course, was Independence Day, a celebration of America's 101st birthday. My old rival Horace Silsby arrived from Seneca Falls, N.Y. with one of his engines to compete with us. Firemen's picnics are a great sales opportunity for us, as they are filled with competitive spirit (such as tugs-of-war between various fire department teams), and competitions between us and other makes of steam fire engines are an appropriate part of the festivities. I plan to attend more of these picnics in future.

Montreal, Quebec, Sept. 6, 1877
I arrived here with two of our small engines yesterday (a 3rd Class and a 4th Class). Montreal is host to a huge international industrial exposition, and these two engines will be displayed and demonstrated at the exposition. A third engine, a 2nd Class now under construction, will be sent up to the exposition next month. These industrial expositions give us a chance to display our engines to a much larger audience than visits to individual communities do. At these exhibits, our engines are seen not only by firemen, but also by the politicians who appropriate the funds for purchase of new fire engines, and also by the voters who ultimately decide what level of fire protection their homes and families deserve.

By exhibiting in Canada, I also hope to establish contacts and sales-leads among firemen in that country. If demand warrants it, perhaps we will establish a branch factory somewhere in Canada, or license an existing Canadian company to manufacture Ahrens engines.

Cincinnati, Feb. 24, 1878
My fifth child was born today, a daughter whom we have named Christine after myself.

Cincinnati, April 27, 1878
Today, I held the first public test of a new style of engine that Fred Hawekotte and I developed. The test was held at the corner of Fifth and Walnut. Up to now, all of our Upright engines have had a one-cylinder piston pump mounted perpendicular to the frame. This new style of engine places two Upright pumps back-to-back, and mounted parallel to the frame.

The frame of this new two-cylinder engine is also different. Up to now, all of our engines have had straight-frames. This new engine has a Crane Neck frame. That is, the front of the frame curves upward in an inverted-horseshoe shape, so that the front wheels can steer under the frame. This allows the engine to maneuver around tighter street-corners than the straight-frame engine. This is not a new idea, as Mr. Latta built a few Crane Neck engines when I worked for him.

These two innovations will greatly expand our product-line. Every fire department is faced with slightly different firefighting problems than any other department. Now a fire department will be able to purchase whatever combination of pump (single or double), frame (straight or crane-neck) and Class (1st, 2nd, 3rd, or 4th) best suits the particular fire company's firefighting needs.

Cincinnati, May 30, 1878
I have recently hired Dick Raymond to work on the assembly-floor, building our steamers, and to assist in delivering new engines to fire departments. Yesterday, he went to Parkersburg, W. Va., to deliver their new engine.

Editor's Note: Dick Raymond made the following deliveries over the years:

5-29-1878 Parkersburg, W. Va. (#182)
3-26-1879 Omaha, Nebraska (#210)
7-7-1879 Newark, Ohio (#216)
10-24-1879 Milwaukee, Wis. (#224)
12-11-1879 Montgomery, Ala. (#226)
3-5-1880 Vincennes, Ind. (#234)
6-4-1880 Council Bluffs, Iowa (#212)
7-24-1880 Chicago, Ill. (#248)
9-6-1880 Louisville, Ky. (#250)
11-13-1880 Wausau, Wis. (#266)
12-24-1880 Louisville, Ky. (#270)
1-28-1881 Delphos, Ohio (#272)
3-10-1881 Oshkosh, Wis. (#274)
3-28-1881 Lena, Ill. (#244)
4-2-1881 Louisville, Ky. (#276 and 278)
10-16-1886 Akron, Ohio (#465)

Cincinnati, July 4, 1878

Today's Independence Day parade in Cincinnati was a special occasion; this year the city is celebrating its 75th anniversary. An elegant open coach led the parade to carry the city's most prominent citizens, including the mayor. I had the great honor of being invited to ride in this leading carriage, because the parade committee felt that the many innovations I developed have prevented this city's destruction by fire.

Cincinnati, March 26, 1879

I recently hired Hiram H. Nieman of 126 Harrison Ave. as Secretary-Treasurer of Ahrens Mfg. Co., Inc. He is a partner with Ignatius Hengge (of 718 Central Ave.) in H.H. Nieman & Co. jewelers, located at 556 Main St. Nieman is in Omaha right now, delivering a new engine for us.

Editor's Note: Hiram Nieman made the following deliveries over the years:

3-26-1879 Omaha, Nebraska (#210)
5-1879 New Orleans, La. (#214)
9-25-1880 Chicago, Ill. (#254)
11-10-1880 Chicago, Ill. (#260 and 262)
10-13-1882 St. Louis, Mo. (#330)
6-30-1883 Portsmouth, Ohio (#362)
3-17-1885 Chattanooga, Tenn. (#418)
11-12-1885 Dayton, Ohio (#445)
3-10-1886 Waynesville, Ohio (#457)
5-4-1886 Miamisburg, Ohio (#470)
10-12-1887 Detroit, Mich. (#518)
8-1888 Detroit, Mich. (#538)
9-18-1889 Dayton, Ohio (#589)
12-6-1889 Columbus, Ohio (#574/#596)
9-10-1891 Westwood, Ohio (#661)

Cincinnati, Dec. 15, 1879

My sixth child, and fourth daughter, was born today. Her name is Clara.

Chicago, July 24, 1880

Seven years ago, when I delivered Chicago's first Ahrens engine, city officials declared that they would never again buy an Ahrens steamer, because the Ahrens price is 33% higher than a Silsby. Now there is a new city administration that sees the wisdom of paying a little more for an engine that not only takes less time to start throwing water on a fire, but also can throw that stream of water a longer distance.

So Dick Raymond and I are in Chicago today delivering a new 2nd Class steamer that will be placed in service at Engine 7. In addition, we already have orders on hand for four 3rd Class engines that we will deliver to Chicago later this year. Chicago has plans to standardize on Ahrens engines, just as Cincinnati, Louisville, and St. Louis have already done.

New York, Sept. 4, 1880

Today I delivered the first Ahrens engine to go in service in New York City. Until now, New York has been buying Amoskeag engines almost exclusively. This is a real triumph for me, as I recall how violent was the opposition to our steam fire engines when Mr. Latta and I were here with the *Miles Greenwood* engine 25 years ago. And now that New York has bought an Ahrens engine, perhaps other communities in the area (who have traditionally bought Amoskeag and Silsby engines) will follow New York's example and buy an Ahrens. In the meantime, our sales continue to be very strong throughout the western and southern states, where the manufacturers from New York and New England are not as competitive.

Cincinnati, Sept., 1881

Our two newest employees at Ahrens Manufacturing Co. are both named George. George Washington Krapp is working for Hiram Nieman in the Treasurer's office. The other George is my eldest son, George Fred Ahrens (age 18), who is an apprentice machinist at the factory.

West Belleville, Ill., Oct. 8, 1881

It was just ten years ago today that the Great Chicago Fire began, and over the next three days destroyed most of the city of Chicago. I went to Illinois with one of my engines then, and I am in Illinois with another Ahrens engine now. This time, my visit is under happier circumstances: I have just delivered the first of our 5th Class engines to the town of West Belleville.

Cincinnati, Oct., 1882

My brother Peter has been transferred back to Fire Co. #7, so I once more have my favorite "satisfied customer" across from the factory.

I recently moved to 77 Euclid Ave., in the Corryville district on the north side of Cincinnati. One our Ahrens engines (#186) is quartered at Fire Co. No. 19, not far from my new home.

My son Fred is now assisting with the delivery of new steamers, as well as continuing his duties as a machinist at the factory.

Editor's Note: George Fred Ahrens delivered or repaired the following Ahrens engines.

9-27-1883 Sidney, Ohio (#378)
10-3-1883 Delphos, Ohio (#376)
11-1883 Nashville, Tenn. (#380 and #382)
11-27-1883 Harrison, Ohio (#388)
3-4-1884 Louisville, Ky. (#397)
5-29-1884 Louisville, Ky. (#400)
11-22-1884 Catlettsburg, Ky. (#417)
1-1885 Racine, Wis. (#412)
1-7-1885 Saint Paris, Ohio (#415)
3-17-1885 Chattanooga, Tenn. (#418)

4-2-1885 Arcanum, Ohio (#425)
8-21-1885 Eaton, Ohio (repairs to #18)
11-12-1885 Dayton, Ohio (#445)
12-18-1885 Bardstown, Ky. (#453)
12-31-1885 Seymour, Ind. (#433)
1-7-1886 Sullivan, Ind. (#435)
2-16-1886 Birmingham, Ala. (#459)
5-3-1886 Germantown, Ohio (#475)
5-4-1886 Miamisburg, Ohio (#470)
10-16-1886 Akron, Ohio (#465)
10-21-1886 Minneapolis, Minn. (#499)
10-21-1886 Minneapolis, Minn. (#500)
11-18-1886 North Vernon, Ind. (#477)
2-9-1887 Gallipolis, Ohio (repairs to #184)
2-22-1887 Lebanon, Ohio (#423)
3-1-1887 Portland, Ind. (#479)
4-14-1887 Los Angeles, Cal. (#493)
5-10-1887 Phoenix, Ariz. (#506)
8-25-1887 Rushville, Ind. (repaired #282)
10-12-1887 Detroit, Mich. (#518)
12-5-1887 Ripley, Ohio (#481)
2-8-1888 Indianapolis, Ind. (#530)
5-10-1888 Versailles, Ky. (#540)
6-26-1888 Milwaukee, Wis. (#546/#547)
7-24-1888 North Amherst, Ohio (#541)
8-1888 Detroit, Mich. (#538)
9-20-1888 San Diego, Cal. (#561)
11-10-1888 Chattanooga, Tenn. (#558)
12-3-1888 Canton, Ohio (#549)
6-30-1891 Louisville, Ky. (#646)

Cincinnati, Jan. 24, 1883

Now that a telephone exchange has been established in Cincinnati, I recently had a telephone installed at the Ahrens factory. Fire departments that urgently need repairs on one of their Ahrens engines can now contact us instantly (assuming, of course, that their town has a telephone exchange). We can then express repair parts, or send Joe Terbeck to make repairs, or even ship a temporary replacement engine. Customers who notify us by telephone will find that their fire engines are out-of-service for a much shorter time than when they contact us by mail or telegram.

Also, if we do not have a part in stock, we can telephone one of our suppliers to send us the needed part by railway express. If we have a choice between a supplier who has a telephone and one who does not, we will order from the one who *has* telephone service when speed is essential.

Another recent development is that my youngest son, John, has started working for me. Ever since I hired my eldest son Fred as a machinist, John has been asking why he can't also help out at my factory. Quite frankly, the reason is the difference in their ages: Fred was 18 when he started working for me, and at that time John was only 9. John is 10½ now, which is still too young to safely work as a machinist. But I am allowing John to assist John McMahon in delivering two new engines to Lawrenceburg, Ind.

Editor's Note: John P. Ahrens delivered the following steamers for his father's company.

1-24-1883 Lawrenceburg, Ind. (#340)
1-24-1883 Lawrenceburg, Ind. (#348)
3-14-1883 Sheboygan, Wis. (#350)
10-3-1883 Delphos, Ohio (#376)
12-4-1883 Ottawa, Kansas (#386)
5-17-1889 Mount Sterling, Ky. (#580)
9-18-1889 Dayton, Ohio (#589)
12-6-1889 Columbus, Ohio (#574/#596)
4-7-1890 Columbus, Ohio (#607 and #608)
4-17-1890 Quincy, Ill. (#616)
8-1890 Denver, Colo. (#628)
8-28-1890 Duluth, Minn. (#626)
10-28-1890 Grand Rapids, Mich. (#632)
12-22-1890 Columbus, Ohio (#610/#611)
12-22-1890 Columbus, Ohio (#613/#614)
4-27-1891 Dallas, Texas (#643 and #644)
5-19-1891 Decatur, Ill. (#647)
7-23-1891 Superior, Wis. (#648)
2-29-1892 Indianapolis, Ind. (#669)
3-3-1892 Indianapolis, Ind. (#670)
3-16-1892 Anacortes, Wash. (#2103)
4-11-1892 Evansville, Ind. (#2101/2102)
12-13-1892 Piqua, Ohio (#2141)
8-2-1893 Clarksville, Tenn. (#2146)

Cincinnati, Oct. 23, 1883

I was granted my third patent today. Patent number 287,211 is for an improved version of the steam generator that I patented 12 years ago.

Cincinnati, Jan. 22, 1884

My fourth patent is number 292,396 for a pump valve.

I have moved out of Corryville. My new home is at 1236 Vine, which is half of a duplex house.

My son Fred has moved up from machinist to mechanical engineer. In that position, he will assist Joe Terbeck in making service-calls around the Ohio-Indiana-Kentucky area.

During the recent Ohio River flood, our engine No. 171 of Covington, Ky., set a new world's record for endurance by pumping 191½ hours (7 days, 23½ hours) without a single shut-down for any reason. Our engine No. 346, also of Covington, nearly tied that record at 189½ hours.

Milwaukee, Wis., May 4, 1884

Today our engine No. 396 threw a stream 325'. This breaks the old world's record for longest distance of 312', set by our engine #150 in 1875.

Oakland, Cal., March, 1885

My brother Peter and I have just completed the delivery and acceptance-test of the first Ahrens engine on the Pacific coast. Named "Alameda", it will serve as Engine 5 in Oakland.

Cincinnati, Aug. 11, 1885

Patent number 324,207 has been granted to me for the platform-springs running gear that I developed for Upright engines. With these springs, a steam fire engine's boiler and pump stay level (horizontal) even when operating on an uneven road-surface, because the springs absorb all of the tilt. This eliminates the performance-reduction commonly experienced when steamers are operated on inclines.

Cincinnati, Sept. 13, 1885

Mary and I are the proud parents of our seventh child, born today. Her name is Florence.

I have purchased the other half of our duplex house at 1236 Vine. Our new address is 1234

Vine. This double house is really needed now that 10 members of the Ahrens family share one roof: my wife and I, our five daughters and two sons, and my 44-year-old bachelor brother Peter.

Saint Paul, Minn., Aug. 2, 1886
I have just delivered a new 1st Class engine to this city. It includes two new innovations.

The first is that it is equipped with spare wheels. The rough terrain over which the firemen must respond can break even the best-constructed of wooden wheels. Carrying an extra front and an extra rear wheel means repairs can be made quickly and the engine promptly returned to service. Spare wheels are an idea that I feel will prove popular for the time and trouble they save for drivers of horse-drawn vehicles.

The second innovation is that the Saint Paul engine carries its own heater, rather than pipes connecting to a stationary heater permanently mounted on the fire-house floor. This new type of heater is mounted on the side of the boiler, and is powered by gasoline. A small tank of gasoline is also mounted on the boiler.

Instead of the traditional method of keeping the boiler hot in the fire-house at all times (a needless waste of fuel), the engineer activates the gasoline heater only when an alarm is received. By the time the engine arrives at a fire, the combination of the heated boiler-coils from the gasoline heater, and the fire that the stoker built in the fire-box enroute to the fire, will have produced enough steam to start the pumps.

I chose gasoline as the fuel for this heater, rather than such traditional fuels as whale-oil or coal, because gasoline ignites more rapidly to produce heat sooner, and is a less expensive fuel than coal. Last year, all of the scientific journals reported that a fellow German, Gottlieb Daimler, used gasoline to propel his first self-powered vehicle, and met with much better success than steam-propelled carriages ever have (so much success, in fact, that he has already established a company to manufacture his gasoline carriages to meet the demand). I think that Daimler may have unlocked the secret to self-propelled vehicles that eluded Mr. Latta 25 years ago, Steven Roper 10 years ago, and all the others who have experimented with steam-propulsion. I have certainly found gasoline to be inexpensive and efficient for my purposes (throwing water on a fire as quickly as possible), and I don't doubt that it will become an increasingly popular fuel.

Cincinnati, Aug. 18, 1886
My older brother Hermann died yesterday. He was born in Brettbeck Feb. 22, 1829, and is survived by his widow Rebecca (Eggers) Ahrens and eight children, the eldest of whom is his son John Herman. Hermann helped me emigrate 33 years ago; without his help, I would not be the successful manufacturer that I am today. Hermann was a success, too, owning the Brighton Grocery and Feed Store at Western and Bank streets here in Cincinnati. We both came a long way, geographically and economically, from the farm in Bredbeck where we grew up.

I recently became a member of St. John's church, in Mount Auburn, after many years in the 3rd German Protestant church of Cincinnati.

Cincinnati, March 26, 1887
We delivered the first of our 6th Class Upright engines to Deshler, Ohio, today. The 6th Class engine opens up two new markets for us. The first is small villages for whom even our little 5th Class engines are too big and too expensive, and who might otherwise buy a chemical engine. The second is the industrial fire brigade, which has become quite popular as large factories become an increasingly common sight; factories with their own fire brigades reduce the burden on the community's firefighting resources.

Cincinnati, April 11, 1887
Our engine No. 506 set a new drafting record today, pumping at a 24'6" lift. We used a Nathan & Dreyfuss Ejector (siphon) to accomplish this.

Cincinnati, Sept., 1887
My eldest daughter, 22-year-old Lettie, recently married George W. Krapp. Mr. Krapp has been the assistant treasurer of Ahrens Mfg. Co. for the past 6 years, and has been courting Lettie for over a year.

Cincinnati, Nov. 30, 1887
We have a new sales and service representative on the Pacific coast. He is Vincent Chloupek, the engineer of the Ahrens steamer at Oakland's Engine 5. Today we shipped him a 4th Class engine that he will deliver to Fresno City, Cal.

In addition to Vincent Chloupek, our other sales and service representatives are A.E. Foote in Milwaukee, Wis.; Merrick & Ferguson in Louisville, Ky.; and Perry & Dumont in Nashville, Tenn. We do not have a representative in the northeast, and that is one of many reasons why Ahrens sales in the Middle Atlantic states (N.Y., N.J., Pa., Del., Md., and Va.) and New England states (Maine, N.H., Vt., Mass., R.I., and Conn.) are practically non-existant.

Cincinnati, June 13, 1888
Vincent Chloupek continues to represent us in northern California, and we have added a new representative in southern California. He is Major W.C. Furry of Los Angeles, and today we shipped him an engine that he will deliver to the city of Los Angeles.

Our secretary-treasurer, Hiram H. Nieman, has moved from 126 Harrison Ave. to Glenway Ave., in the expensive Clifton district. His Nieman jewelry business has been doing quite well, and the move reflects his improved financial status.

Cincinnati, Dec. 11, 1889
The latest addition to the Ahrens Mfg. Co. staff is Ed Steinway. He delivered two of our engines to Toledo today. [*Steinway was later Fire Marshal of the Cincinnati Fire Dept.--ed.*]

Cincinnati, Aug. 28, 1890
I have promoted my son Fred from "mechanical engineer" to foreman of the assembly-floor. He has moved out of our home at 1234 Vine and established his own home at 168 Jefferson Ave. My son John still lives with me at 1234 Vine, and

works for me as a machinist.

We now have a sales and service representative in Minnesota, John Black of Minneapolis. Today, John delivered one of our engines to Duluth, Minn. We are now represented in Louisville, Nashville, Milwaukee, Minneapolis, San Francisco, and Los Angeles, as well as the home-office sales staff here in Cincinnati.

Because of the excellent work of our sales representatives, the following cities have now standardized on Ahrens engines: Cincinnati, Columbus, Toledo, Indianapolis, Louisville, St. Louis, Chicago, Milwaukee, Minneapolis, St. Paul, Nashville, Memphis, Chattanooga, Dallas, New Orleans, Los Angeles, and Seattle. This is in addition to hundreds of villages, towns, and small cities that use at least one of our engines, or an engine of another make but rebuilt by us.

Cincinnati, Oct. 20, 1890

I have hired George Zink of 654 Race St. as a steamfitter. He is working with Joe Terbeck in our repair-service program. Today we sent him on his first service call, to put new steam pipes in the boiler of the 1875 Ahrens at Oxford, Ohio.

Editor's Note: George Zink was not only a pipe-fitter for Ahrens Mfg. Co., but continued in that role for the Ahrens Fire Engine Co. (1905-1910) and even into motor apparatus at Ahrens-Fox. Zink installed booster-tank piping on Ahrens-Fox pumpers in the Teens and Twenties. Following is a list of Ahrens steamers that he repaired.

10-12-1890 Oxford, Ohio (#45)
11-1-1890 Pekin, Ill. (#405)
3-4-1892 Aurora, Ind. (#394)
6-20-1892 Portsmouth, Ohio (#362)
8-2-1893 Greensburg, Ind. (#39)
12-26-1893 Lawrenceburg, Ind. (#348)
5-16-1894 Aurora, Ind. (#166)
11-16-1894 Lebanon, Ohio (#423)
10-17-1895 Madison, Ind. (#5)
4-24-1897 Portsmouth, Ohio (#362)
4-4-1898 Aurora, Ind. (#394)
6-29-1898 Madison, Ind. (#660)
7-1-1899 Midway, Ky. (#170)

Belle Plaine, Iowa, Sept. 18, 1891

Our engine No. 641 set a new world's record for fastest water at draft (starting from a cold boiler) at 3 minutes 5 seconds. The previous record of 3:17 was set by our engine No. 163 at Cambridge City, Ind., in 1876.

Fred Hawekotte has moved to 251 Senator Place, Clifton.

Cincinnati, Nov. 2, 1891

My son Fred has been promoted from foreman to plant superintendent at my factory. He also has a new home at 2 Nixon, west of Jefferson.

Replacing my son as shop foreman is 30-year-old Charles Hust Fox. Today I sent him to Madison, Ind., to deliver a new engine.

Charles H. Fox was born in Cincinnati Dec. 16, 1860. His mother was Catherine Hust, and his father "unknown Mr. Fox". This created quite a scandal at the time; some gossipers wondered if it was more than coincidence that Cincinnati's most famous judge was also named Charles H. Fox.

Be that as it may, Fox was raised by his grandfather, Captain Jacob Hust of Cincinnati's Mohawk Fire Co. No. 12. He never knew his father at all, and met his mother only briefly when he was seven and she was temporarily living with her father and her son. Captain Hust was a partner in Stahl & Hust livery stable at 78 Hamilton Rd., and his house (where Fox grew up) was next door at 80 Hamilton. Jacob Hust's brother, Henry, was a partner with Jacob Fuldner in a second livery stable, located at 16-18 Locust.

But it was his grandfather's association with the Cincinnati Fire Department, much more than the family's livery stables, that would have the most profound influence on Charles Fox.

Fox began working in 1873, at the age of 12, but spent all of his free time visiting the local fire house.

His first taste of the destructive power of fire occurred at the $180,000 Wilson Saddlery fire of Jan. 10, 1879. There, he was assisting the Cincinnati firemen when he noticed a fire-weakened wall about to fall on him. He jumped out of the way, and escaped death by inches. The fireman who had stood next to Fox was not so lucky, because he had been burdened with a fully-charged hose across his lap and, unable to move, was buried in a pile of bricks.

The following year (1880), Fox joined Fire Co. No. 7, across the street from my factory on Webster St. My brother Peter was engineer at that company, and that is when I first met Mr. Fox.

In 1882, Fox became a stoker at Fire Co. 7. He then moved out of his grandfather's house into a home of his own at 14 Orchard. We were now neighbors, as I lived at 21 Orchard at the time. Within a year, Fox moved again, this time to 487 Sycamore, just around the corner from the No. 7 fire house where he worked.

It was about that time that a local resident visiting the fire house inspired Fox's thirst for engineering knowledge. For the next two years, Fox spent every free minute up in the fire house hay loft studying every book he could find about mechanical engineering.

All of that studying paid off in 1885, when Fox was promoted to engineer-of-steamer and transferred to Fire Co. 13. He then moved to 16 Dayton St.

When Fire Co. 13's quarters needed remodeling in 1887, Fox submitted plans for the work, and his proposal was accepted. Soon, he became the fire department's architect, designing all of the new fire stations in the city.

In 1889, Fox became an Assistant Chief of the Cincinnati Fire Dept. and master mechanic of the fire department repair-shop.

Early this year, Fox was promoted to assistant fire marshal. His office was at Fire Co. No. 4 on Sycamore between 7th and 8th. Soon after that, I hired him to work for me.

Cincinnati, Dec. 5, 1891

The largest engine we have ever built was completed today. It is a Double Extra 1st Class for Chicago, our engine Registered Number 666.

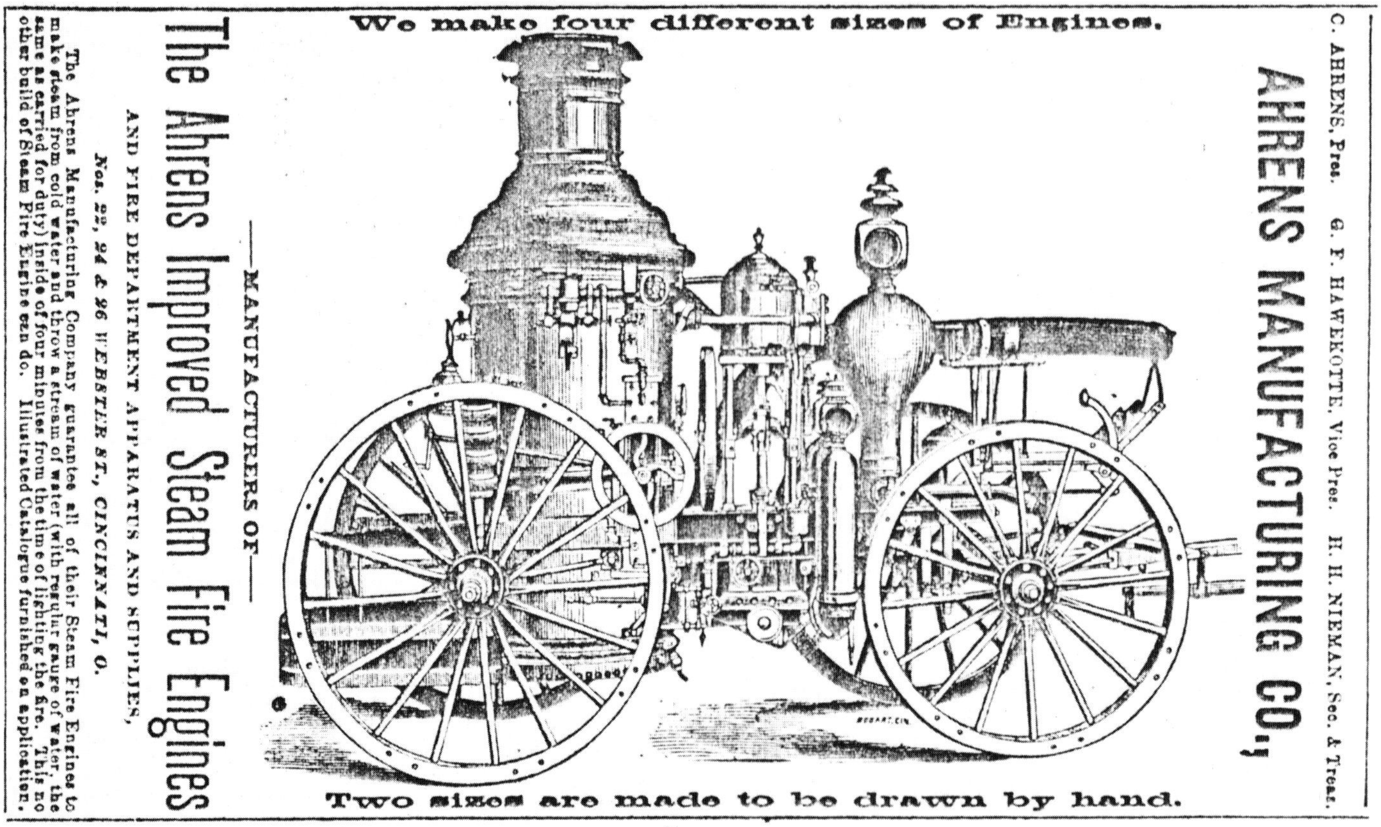

Ahrens Manufacturing Co. advertisements from Cincinnati city directories. Above is the 1878 issue featuring a Single Straight Frame engine -- note that Hiram H. Nieman was already listed as secretary-treasurer. Below is 1881 ad featuring a Double Crane Neck engine and boasting of one of the many contests Ahrens engines won. (Courtesy Public Library of Cincinnati & Hamilton County, Ohio)

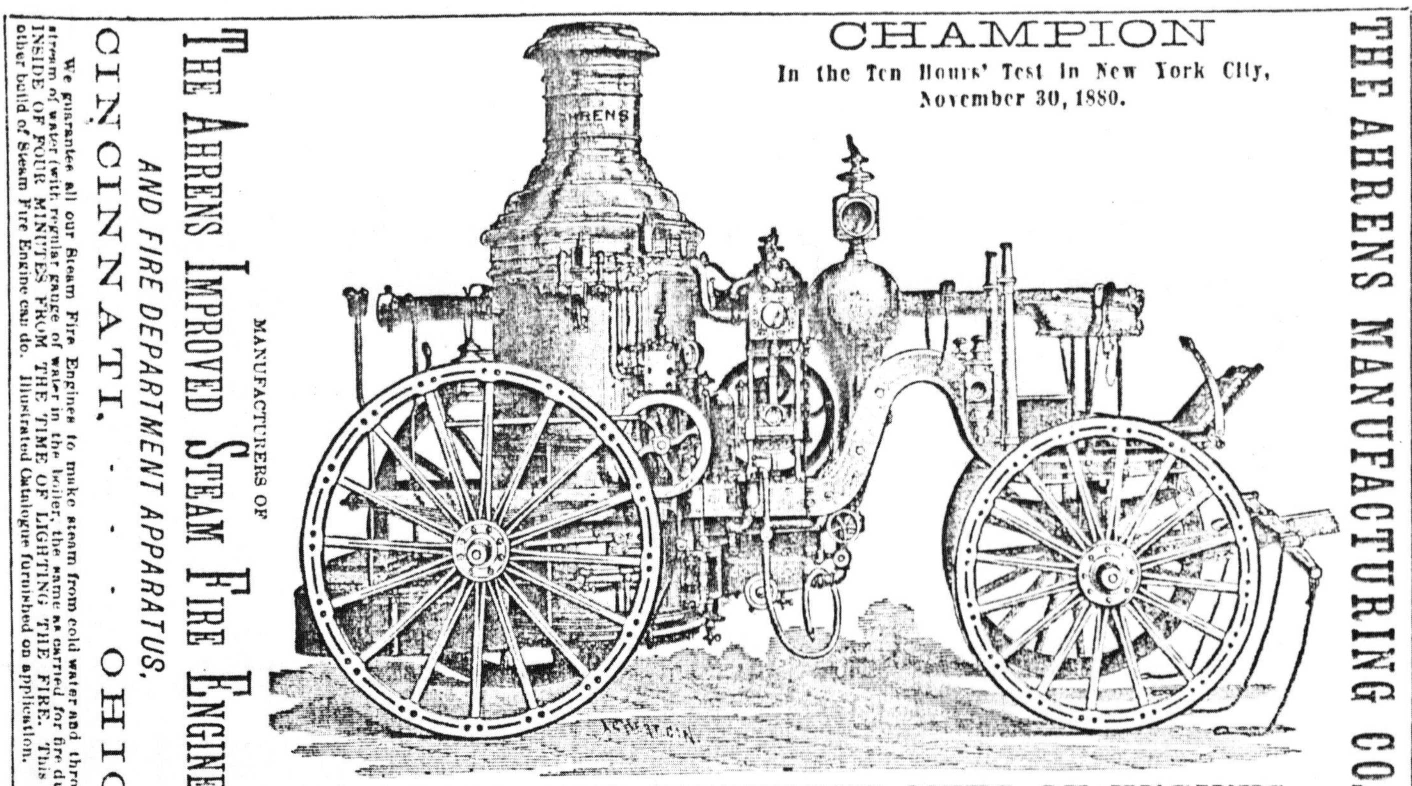

Chapter 6

Cincinnati, Dec. 12, 1891

Today, Ahrens Mfg. Co. was merged with three of our competitors: Silsby Manufacturing Co. of Seneca Falls, N.Y.; Clapp & Jones Mfg. Co. of Hudson, N.Y.; and the Button Fire Engine Co. of Waterford, N.Y. The merged company is called American Fire Engine Co., and has a working capital of $600,000. It is incorporated in both New York and Ohio.

The Clapp & Jones and Button plants are to be closed, and Silsby will take on the manufacture of those makes. Our Ahrens plant will continue to operate autonomously as the Cincinnati branch plant, and will be making only Upright engines, just as we have for the past 16 years. Fred Hawekotte and I will continue to run the Cincinnati plant as we have always done, so really the change in our business will be virtually imperceptible except for the increased sales volume that the merger is expected to bring.

Our former secretary-treasurer, Hiram H. Nieman, has been elected president of American Fire Engine Co. DuBois Collier of Clapp & Jones is vice president, Charles T. Silsby treasurer, and William S. Silsby secretary. My own title is Mechanical Engineer. The company managers are my partner G. Fred Hawekotte, my long-standing rival Horace Silsby, DuBois Collier of Clapp & Jones, and George E. Holroyd of Button.

Cincinnati, July 6, 1892

Yesterday, Fred Hawekotte delivered the last steamer to bear the old Ahrens Mfg. Co. nameplate and Registered Number. It is #672, delivered to Cincinnati Fire Dept. From now on, all engines built on Webster St. will say "Ahrens, Built by American Fire Engine Co., Cincinnati, O." instead of "Ahrens Manufacturing Co."

Engines built in the Silsby plant will bear the Silsby, Button, or Clapp & Jones nameplate, followed by the words "Built by American Fire Engine Co., Seneca Falls, N.Y."

Registered numbers of engines from both plants will be consecutive, starting at 2101. This number was arrived at by adding the highest Registered Numbers used on each of the four makes (our numbers went up to 672, Silsby's to just over 1000, etc.). The total came to almost 2100, so 2101 was chosen as the starting Registered Number for American Fire Engine Co. Numbers are allocated in blocks of ten: we will build numbers 2101-2110, Seneca Falls will make engines 2111-2120, etc. We have already shipped seven engines under this new numbering system.

Cincinnati, Sept., 1892

We recently delivered our first "Extra 1st Class" Ahrens engine (#2123) to Milwaukee, Wis. We now offer eight sizes of engines: Double Extra 1st, Extra 1st, 1st, 2nd, 3rd, 4th, 5th, and 6th. In addition, we offer two frame styles and two styles of pump. This makes for a virtually endless variety of engines available, to satisfy the peculiar needs of each customer.

Cincinnati, Nov. 3, 1892

Charles H. Fox married my daughter Lillie yesterday.

Seneca Falls Feb., 1893

I was elected president of American Fire Engine Co. Charles H. Fox is now foreman of the Cincinnati plant; his new home is on Glendora Ave. north of Nixon, in Cincinnati.

Cincinnati, April 18, 1893

I have been granted U.S. Patent number 495,868 for the Ahrens Tubular steam boiler.

Cincinnati, Nov. 6, 1893

Charles H. Fox has developed a steamer that carries its own supply of hose. This eliminates the need for a separate hose wagon or hose carriage. This new type of engine is based on a crane-neck Ahrens Upright, with a large wooden hose-box mounted between the frame rails in front of the pump.

Yesterday we shipped the first of these pump-and-hose combinations (#2147) to the Columbian Exposition at Jackson Park, Chicago, Ill.; three more just like it will follow. We previously delivered three Upright engines to the Exposition. A World's Fair Fire Brigade has been organized as a battalion of the Chicago Fire Dept. to man these engines. The Exposition commemorates the 400th anniversary of Columbus landing in America.

All future Columbian engines will be built in Cincinnati, because it requires less tooling for us than it would for Seneca Falls. Our main product remains the Upright engine, as it has for 22 years.

Cincinnati, May 1, 1894

I have been granted U.S. Patent number 519,013 for the Ahrens Vertical Flue steam boiler.

We have developed a new style of steamer called the "American". This is basically an Upright engine, except that the pump has been redesigned. The most noticeable difference is that the flywheel has been moved from the side to the center of the pump, which requires fewer moving parts, increasing its efficiency. The first of these was built in the Seneca Falls plant for Erie, Pa.; we built the second one here in Cincinnati for St. Joseph, Mo. We will be building both "Upright" and "American" engines here, while Seneca Falls will build American, Silsby, Button, and Clapp & Jones engines.

Charles H. Fox's new home is at 1272 Vine.

Cincinnati, May 20, 1894

At 59, I am grandfather of three! The latest is Elvira Fox, from my daughter Lillie and Charles H. Fox. The other two are Clarence Krapp and Christopher Krapp, from my daughter Lettie and her husband George W. Krapp.

Our first two export engines were recently shipped to Manila, in the Philipine Islands.

Seneca Falls, Feb., 1895

At the annual meeting of the American Fire Engine Co., I was re-elected President of the company. My long-time business partner, G. Fred Hawekotte, has been appointed Cincinnati branch manager. He is in charge of manufacturing, sales, and service at the Cincinnati factory. Herman H. Weiel is vice-president, Charles T. Silsby treasurer, and William S. Silsby secretary.

I have moved further up Vine Street hill, from 1234 Vine to 3010 Vine.

Cincinnati, May 12, 1896

My son-in-law (Charles H. Fox) and I have been jointly granted U.S. Patent number 559,881 for an automatic relief valve we developed.

Cincinnati, Sept. 29, 1896

Charles H. Fox has been granted U.S. Patent number 568,438 for a new type of steam boiler.

Cincinnati, Oct. 20, 1896

Charles H. Fox has been granted U.S. Patent number 569,966 for a new type of steam boiler.

Cincinnati, Feb., 1897

My son-in-law, Charles H. Fox, has been made superintendent of the Cincinnati plant, while my son Fred is superintendent of the Seneca Falls plant. I am one of the managers of American Fire Engine Co.

The city of Cincinnati has renumbered the buildings on Webster St. Our factory that was 22, 24, and 26 Webster is now 216, 218, and 220 Webster. The No. 7 firehouse at 15 Webster is now 209-211 Webster.

Cincinnati, May 3, 1898

Charles H. Fox was granted Patents 603,482 and 603,502 for his Vertical Water Tube boiler.

Cincinnati, Sept. 19, 1898

My son-in-law, Charles H. Fox, has licensed his Fox Vertical Water Tube boiler (Patents 603,482 and 603,502) to American Fire Engine Co.

American Fire Engine Co. is to use the Fox boiler exclusively on all new steamers built, unless a customer requests some other type of boiler. Royalties are to be paid by American to Mr. Fox on Jan. 1, April 1, July 1, and Oct. 1 of each year, for as long as the agreement is in effect.

On all new steamers built by American Fire Engine Co. from now until April 3, 1904, Fox is to be paid $50 per boiler. On that date, the royalty drops to $45 per boiler. On Oct. 18, 1909, the royalty again decreases, to $40 per boiler, until the patents expire on May 3, 1915.

Fox boilers that are fitted to existing steamers require American to pay a royalty of 5% of the selling price of the boiler. American expects to sell these boilers for about $850 each, which makes Fox's royalty $42.50 per boiler.

American Fire Engine Co. plans to use this boiler as the basis for a new style of engine called the "Metropolitan". This engine will be the same as the "American" except that it will use the Fox vertical water-tube boiler in place of the Ahrens coil and tubular types. Metropolitans for customers along the eastern seaboard (from Maine to Florida) will be built in Seneca Falls; we will build Metropolitans for all customers in the west and southwest, and for export.

Seneca Falls, Feb. 14, 1899

At the annual meeting of American Fire Engine Co., I was elected to the board of directors. Charles H. Fox is the new manager of the Cincinnati plant. He is also vice president of Schlemmer Co., a plumbing and heating manufacturer at 809 Charlotte, of which my son John is the Manager.

Cincinnati, Sept. 26, 1899

Charles H. Fox has been granted Patent #633,644 for a heater to keep a fire engine boiler warm in the firehouse. We have used heaters on our engines for years, but Fox's patent represents a much-improved version of this heater.

Seneca Falls, Aug. 6, 1900

Charles E. Locke announced today that controlling interest in the American Fire Engine Co. has been acquired by a group of five Wall Street financiers (John H. Flagler, Fred W. Shibley, August Hecksher, Alfred C. Bedford, and George H. Robinson). None of the five have any experience in the fire apparatus industry, so only time will tell if we can remain competitive under this new ownership. I remember what a disaster it was when my first business partner (30 years ago) had neither experience, nor interest, in the needs of fire departments.

The same five financiers who own American Fire Engine Co., have also put together a new conglomerate called International Fire Engine Co. This new company was created by merger of the following fire engine manufacturers: LaFrance Fire Engine Co. (Elmira, N.Y.), Thomas Manning Jr. & Co. (Cleveland, Ohio), Fire Extinguisher Manufacturing Co. (Chicago, Ill., makers of Babcock, Champion, and Hale apparatus), Charles T. Holloway & Co. (Baltimore, Md.), Gleason & Bailey Manufacturing Co. Ltd. (Seneca Falls, N.Y.), and Macomber Chemical Fire Extinguisher Co. (Worcester, Mass.).

To avoid the trust-busters, American and International will operate as two separate companies, even though both companies belong to the same five financiers.

My son John has moved out of our house at 3010 Vine, into a place of his own at 422 East 3rd. He is not home much, as he is a travelling salesman for American Fire Engine Co.

Charles H. Fox has moved to a flat at the northeast corner of Vine and Rochelle.

Seneca Falls, Feb. 13, 1901

At the annual meeting of American Fire Engine Co. today, I was elected Second Vice-President. Charles T. Silsby is President and Treasurer, Charles E. Locke (who represents the interests of the owners) is First Vice President, and William S. Silsby is Secretary.

Cincinnati, April 10, 1901

My third grandson was born today. His name is John Krug Ahrens. His first name is from his father (my son), John P. Ahrens (born Aug. 11, 1872). His middle name is from his mother's maiden name, Louise Krug (born March 30, 1874).

Cincinnati, Nov. 14, 1901

My fifth grandchild was born today, George W. Krapp, Jr. His mother is my daughter Lettie, and father is George W. Krapp, Sr. After 14 years of marriage, this is their third child.

Cincinnati, Feb. 4, 1902
Another steam-generator patent has been awarded to Charles H. Fox. It is patent #692,436. This patent is not subject to the 1898 licensing agreement with American Fire Engine Co.

Seneca Falls, Feb. 11, 1902
At the annual meeting of American Fire Engine Co. today, I was re-elected Second Vice-President of the company. In fact, all of last year's officers were re-elected. Also, George W. Krapp was made superintendent at Cincinnati.

My son John has moved to 130 Malvern Place. Charles H. Fox has moved into our home at 3010 Vine with his wife (my daughter) Lillie.

Cincinnati, May 20, 1902
Charles H. Fox's newest patent is #700,320 for a "steam generator for locomotives".

Another new style of engine has been developed. It is somewhat similar to the little one-cylinder 6th Class Uprights we used to make before the 1891 merger, except that the pump is completely enclosed in a dust-proof casing. Some of the working parts are so small that even the slightest amount of dirt in the oil will seriously impair their operation.

This new style of engine is called the Cosmopolitan, and is intended for small villages, industrial fire brigades, and the export market. We built the first Cosmopolitan here, and recently delivered it to Xenia, Ill. All Cosmopolitan engines are to be built here in Cincinnati.

Cincinnati, July 28, 1902
My son Fred has been appointed superintendent of American Fire Engine Company's main factory in Seneca Falls, N.Y. My son-in-law, George W. Krapp, is superintendent at the Ahrens branch factory here in Cincinnati. In addition, my son John has become sales manager for American Fire Engine Co.

The headquarters office of American Fire Engine Co. has been moved from Seneca Falls to New York City, where my son-in-law Charles H. Fox is General Superintendent of the company.

The city of Cincinnati has again renumbered the addresses on Webster St. Each lot has been divided in half (that is, two numbers for each old number). So without expanding, our factory is now 216, 218, 220, 222, and 224 Webster.

Cincinnati, the first city in the world to adopt steam fire engines, and the first to organize a paid fire department, recently introduced another first -- the world's first fire engine propelled by a gasoline motor.

This novel fire engine belongs to the Underwriters Salvage Corps, not to the Cincinnati Fire Dept. The Captain of the Salvage Corps, J.J. Conway, donated his new Winton automobile for the project. The roadster body was removed, and a fire engine (squad-car) body mounted.

The new body was designed by Jack Conway and my son-in-law, Charles H. Fox, and was built for the salvage corps by Thomas B. Hanauer Bicycle Co. of Cincinnati. Inside this iron body are two bench seats parallel to the frame. Outside the body are wicker "picnic baskets" for carrying salvage covers and assorted firefighting tools. Because this body is longer than the standard Winton body, the driver's seat and steering wheel had to be moved from behind to above the hood.

Captain Conway named this new fire engine *Margeurite* after his four-year-old daughter. The newspapers published a photo of little Margeurite Conway christening the fire engine by smashing a champagne bottle against its radiator.

Cincinnati, Oct. 25, 1902
Hiram Nieman invented a protector for the flues of a steam fire engine's boiler. It was granted patent number 485,112 today.

Cincinnati, April 7, 1903
Patent No. 724,564 is for Charles H. Fox's newest steam generator.

Seneca Falls, April 23, 1903
At the annual meeting of the American Fire Engine Co. today, the Silsby brothers announced their resignations. Their remarks summed-up how many of us feel regarding the incredible mismanagement since the company was acquired by people more intent on making a profit than on serving the needs of the customers. None of the five owners has any real understanding of the fire apparatus industry.

The resignations of the Silsbys only make matters worse for those of us who remain. For the first time, none of the company's officers are experienced fire-engine men. George R. Bidwell is the new President, Charles E. Locke is Treasurer, and William Dinsmore is Secretary; all three have a Wall Street background.

Seneca Falls, Aug. 17, 1903
At a special meeting of the American Fire Engine Co. today, I was re-elected Second Vice-President of the company.

Seattle, Wash., Aug. 25, 1903
Today one of my old 1889 Upright engines, Registered #597, set a world's record for fastest water. It threw a stream exactly 3 minutes after lighting the boiler. The old record of 3:05 was set by our engine #641 at Belle Plaine, Iowa, in 1891. The rules for this record require an engine to start from a cold boiler and drained pump, and operate from draft rather than a hydrant.

Seneca Falls, Sept. 10, 1903
At a special meeting of the American Fire Engine Co. today, it was announced that James R. Clarke has been named President, and George L. Olney is Treasurer. My son John is the new Sales Manager. Our sister company, the International Fire Engine Co., has been in deep financial trouble for quite some time, and recently filed for bankruptcy. So the mission of these new officers is to prevent the same thing happening to American Fire Engine Co.

Seneca Falls, Oct. 1, 1903
International Fire Engine Co. has been placed in receivership. A reorganization plan was drawn today, calling for International Fire Engine Co.

and American Fire Engine Co. to merge. The strong finances of American would be used to prop-up the weak International Fire Engine Co. The company will be called American-LaFrance Fire Engine Co., since its main components would be American Fire Engine Co. of Seneca Falls and Cincinnati, and the LaFrance Fire Engine Co., Elmira, N.Y. (part of International).

The reorganization plan would raise money by mortgaging all of the company's factories (including my factory on Webster St.) to Manufacturers Hanover Trust Co. for $1 million. In addition, the stockholders are to deposit $7.5 million of stock at Manufacturers Hanover as security on an additional $1 million cash-advance. That is, $8.3 million of assets in the bankrupt company will be exchanged for $2 million in working-capital.

This plan must be approved by a majority of stockholders before it takes effect.

Cincinnati, Oct. 10, 1903

Charles H. Fox has cancelled his 1898 licensing agreement with American Fire Engine Co. for the Fox Water Tube boiler. His reason was that he has not received his quarterly royalties in over a year. He is also worried that American Fire Engine Co.'s merger with the bankrupt International Fire Engine Co. will make it difficult to collect his royalties in future.

The new American-LaFrance company will have to use another type of boiler on their steamers. Fox will watch American-LaFrance closely, before he decides whether to assign his agreement to the new company.

Cincinnati, Oct. 23, 1903

Charles H. Fox has resigned his post as general superintendent at the Elmira, N.Y., plant of International Fire Engine Co. He was simply fed-up with fighting to collect royalties that were due him from a nearly-bankrupt company, and even reimbursement for travelling expenses is a real battle. He has not been reimbursed for $650 incurred on Aug. 1, 1902, nor for any business expenses since then. With the American-LaFrance reorganization underway, the future of the company is uncertain, so Fox has retained legal counsel to represent his interests in his patents.

Cincinnati, Oct. 31, 1903

Charles H. Fox was nominated to the board of directors of International Fire Engine Co., but he declined. His resignation of last week still stands.

Cincinnati, Nov. 20, 1903

International Fire Engine Co. has hired Charles H. Fox as a Consulting Engineer at $2500 per year. This leaves him free to pursue other business interests, and to spend time with his ailing wife (my daughter Lillie), who is in the tuberculosis hospital at Saranac Lake, N.Y.

I had a glimpse recently of a very interesting paper debating what rights, if any, American-LaFrance might have to the Fox boiler patents that were licensed to American Fire Engine Co. The original 1898 licensing agreement stated that the agreement was not assignable to the company's successors. Yet the laws of N.Y. State say that in a merger (such as the impending merger of

Portrait of Ahrens family members who worked for Ahrens Manufacturing Company. Standing: George W. Krapp, son-in-law of Chris Ahrens (left); John P. Ahrens, son of Chris (right). Seated, L to R: G. Fred Ahrens, eldest son of Chris; Chris Ahrens; Peter Ahrens, brother of Chris; and Charles H. Fox, son-in-law of Chris Ahrens.
(Courtesy Cincinnati Historical Society)

American Fire Engine Co. into the International Fire Engine Co.), all rights of the merged company become the property of the company into which the merger was made. The question was, whether Fox could be forced to assign his patent, or whether persuasion was the better way.

A related issue was whether the exclusive-use clause forced American-LaFrance to use Fox boilers on all steamers, or only those built in the former plants of American Fire Engine Co. The concern was that the Elmira plant would no longer be permitted to manufacture the popular LaFrance boiler.

Elmira, N.Y., Jan. 2, 1904

Today, the stockholders of both American Fire Engine Co. and International Fire Engine Co. met here to approve the planned reorganization of International, and American's merger with International to form American-LaFrance Fire Engine Co. The reorganization/merger plan was drafted on Oct. 1 last year.

Cincinnati, Jan. 30, 1904

Charles H. Fox has reinstated his 1898 licensing agreement with American Fire Engine Co., whereby they are to use Fox boilers exclusively and pay him royalties on each boiler built.

Elmira, N.Y., May 2, 1904

The American Fire Engine Co., Cincinnati and Seneca Falls, and the International Fire Engine Co. of New York, were both legally dissolved as corporations today. The new American-LaFrance Fire Engine Co. began operations here in Elmira today, owning all of the assets of the two dissolved companies. James R. Clarke is president of the new company, George L. Olney (formerly treasurer of International Fire Engine Co.) vice-president, and William G. McIntyre Treasurer.

Saranac Lake, N.Y., Sept. 17, 1904

My daughter Lillie died of tuberculosis today. She was just 34. Her husband (Charles H. Fox) and I are making funeral arrangements for her; she is to buried in the family plot at Spring Grove cemetery in Cincinnati.

Cincinnati, Oct. 25, 1904

My sons-in-law (Charles H. Fox and George W. Krapp), sons (John and Fred), and I have begun to think about following the example of the Silsby brothers a year-and-a-half ago, and pulling out of the American-LaFrance "fire engine trust". Our plan is to remove Ahrens Manufacturing Co. from the conglomerate, and reorganize it under the name "Ahrens Fire Engine Co." to avoid any possible legal disputes over the rights to the old name "Ahrens Manufacturing Co."

Since Fox is the only one of us who is not a direct employee (he is under contract as a consultant), he is the only one free to pursue this idea and lay the groundwork for our eventual departure from the company. Today, he secretly had one of the draftsmen make some blueprints for him, one of which (Ahrens Fire Engine Co. Blueprint #4) is for the crank-shaft, journal boxes, and fly wheel of a Double Extra 1st engine we propose to produce at Ahrens Fire Engine Co. Fox is taking a risk with these blueprints, as we do not yet wish to tip our hand to the American-LaFrance management about what we are doing. Luckily, nobody noticed the four blueprints labeled "Ahrens Fire Engine Co." on one of the American-LaFrance drafting tables!

Cincinnati, Nov. 29, 1904

James R. Clarke, president of American-LaFrance Fire Engine Co., has discovered our plans to leave American-LaFrance and manufacture steam fire engines in competition with them. Their lawyers are now working to secure the patent rights to the Fox boiler so that we cannot use it on our engines. What Mr. Clarke doesn't know is that this is fine with us, since we are developing a speedier boiler to use on our engines.

Cincinnati, Dec. 9, 1904

Having been dissatisfied for some time now with the way that American-LaFrance Fire Engine Co. conducts its business, we disassociated our Ahrens Mfg. Co. from American-LaFrance today. Officially, I have resigned along with my sons John and Fred, my sons-in-law George W. Krapp and Charles H. Fox, and all of the employees of the Webster St. factory. We plan to rename the company Ahrens Fire Engine Co., and build steam fire engines in competition with American-LaFrance.

We weren't really ready to resign yet, as we do not yet have our own manufacturing facility. But because Mr. Clarke discovered our plans, we felt that it was best to resign now, rather than facing the awkward situation of having our employer know that we plan to compete with them.

I will let my sons and sons-in-law run the day-to-day activities of the company, as I am getting too old for this (I just turned 70 this week), but I will be available to them as a consultant whenever they need the benefit of my 50 years experience in the fire engine business.

Since our old factory on Webster St. belongs to American-LaFrance (or more accurately, to the bank which holds American-LaFrance's mortgages), we are building a new factory at Alfred and Cook streets. Until this is completed, we will operate in rented quarters at the rear of 1118 Elm St. Meanwhile, the Webster St. factory stands empty, because its employees and management have resigned from American-LaFrance to work for our new Ahrens Fire Engine Co.

Cincinnati, Dec. 22, 1904

Ahrens Fire Engine Co. blueprint #5 is for the connecting rods, yokes, and pistons of the Double Extra 1st engine that we began designing (in blueprints 1 through 4) before we resigned from American-LaFrance.

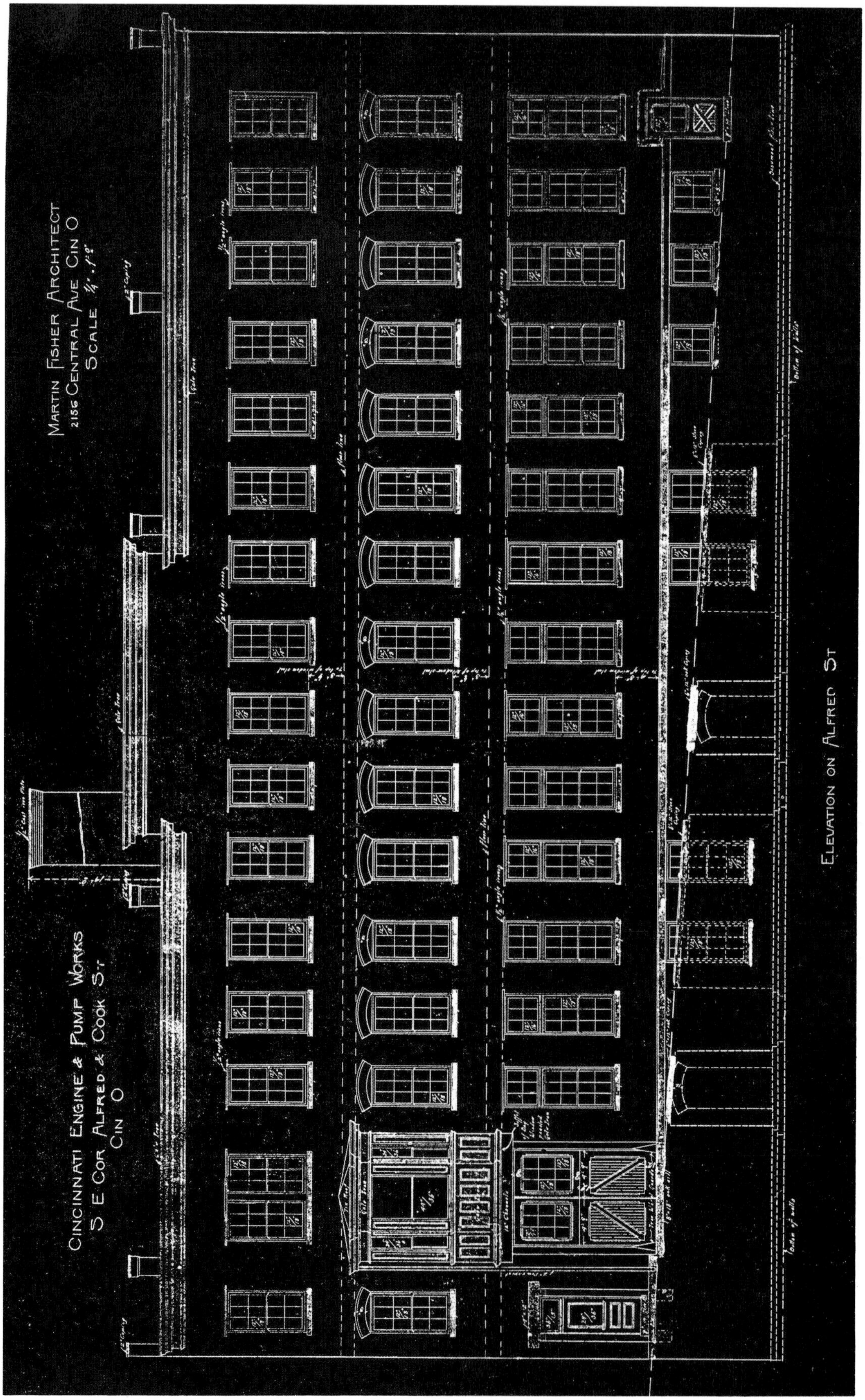

The factory which Ahrens-Fox Co. now occupies, at Alfred & Cook Streets, was built for them in 1905 when the company was called Cincinnati Engine & Pump Works. These original architect's drawings are still in Ahrens-Fox's extensive blueprint file today. Above, the front of the building, on Alfred Street. Sub-assemblies were made on the upper floors, while final assembly occupied the ground floor; completed steamers and motor apparatus rolled out the door under the bay window at left. Next page, the Cook Street side, where such specialties as the pattern room, machine shop, blacksmith shop, and paint shop were located. This building has changed little in 80 years.

(Courtesy Curt Nepper, Ahrens-Fox Co.)

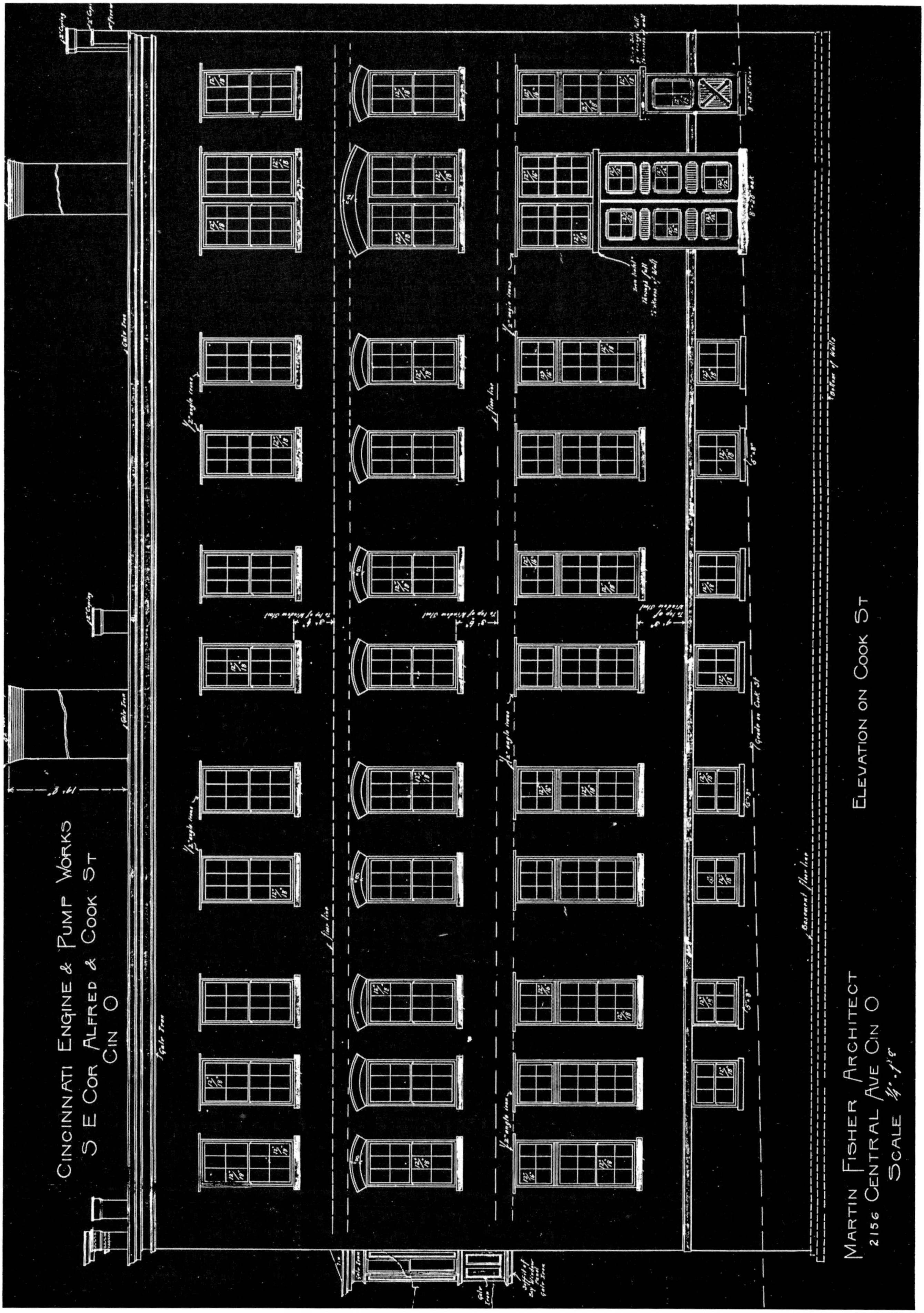
Cincinnati Engine & Pump Works
S E Cor Alfred & Cook St
Cin O
Elevation on Cook St
Martin Fisher Architect
2156 Central Ave Cin O
Scale 1/4 = 1'0"

Chapter 7

Cincinnati, Jan. 19, 1905

Charles H. Fox received a check from American-LaFrance last week, for royalties owed to him on his Fox vertical water tube boiler patents. The check, dated Dec. 31, 1904 was for $1282.50, representing payment for approximately 30 boilers built within the past 3 months.

It was only after he cashed it that I discovered an American-LaFrance memo of Dec. 29, 1904, about this check. It seems that American-LaFrance's lawyers very carefully and cleverly worded this check, so that cashing the check would deprive Mr. Fox of certain legal rights.

The check said "For royalties on Fox boilers made, sold, and delivered from the American Fire Engine Co. plants of the American-LaFrance Fire Engine Co. (the possessor by statutory merger of the property, franchises, and business of the American Fire Engine Co.) for 3 months ending on 31 Dec., 1904."

According to the memo, the wording "boilers made, sold, and delivered" indicates that by cashing the check, Fox surrenders the right to royalties on boilers that were made but **NOT** delivered.

"The American Fire Engine Co. plants of the American-LaFrance Fire Engine Co." is also a careful choice of words. In cashing the check, Fox agreed to assign the patent license from American Fire Engine Co. to American-LaFrance, thereby forever preventing him from objecting to the assignment of the 1898 licensing agreement.

Those few words also signify Fox's approval for plants that did not belong to American Fire Engine Co. to manufacture other types of boilers, as long as the American Fire Engine Co. plants continue to make only Fox boilers (Mr. Fox thereby loses considerable royalties that he might have had were *all* of American-LaFrance's plants required to manufacture Fox boilers exclusively).

The Dec. 29 memo makes it quite clear that these words were carefully chosen, so that cashing the check would signify consent to all of this.

The memo does not mention it, but it occurs to me that this choice of words could even be construed to mean that Fox is permitting American-LaFrance's plants that did **NOT** formerly belong to American Fire Engine Co., to manufacture Fox boilers without paying royalties. There is no evidence that American-LaFrance intends to do this, and if they did, their lawyers would be hard-pressed to prove in court that Fox acquiesed in letting *anyone* make his boilers without paying him royalties.

The words "for 3 months ending Dec. 31, 1904" have an equally-devious purpose. They signify that Mr. Fox is accepting American-LaFrance's word for it that $1282.50 represents all the royalties that are due to him for boilers delivered in October, November, and December.

Finally, they added the words "for royalties in suspense" on the check. Cashing the check waives Fox's right to the back-royalties that were never paid to him, for boilers built during International Fire Engine Co.'s receivership and reorganization.

Cincinnati, Feb. 3, 1905

Charles H. Fox has completed a series of drawings for fire-doors for Ahrens Fire Engine Co.'s steamers. These will be in three sizes: 6 1/2" x 10 1/2" for boilers under 34"; 7" x 11 1/2" for boilers 34" to 40"; and 7 1/2" x 12 1/2" for boilers over 40".

We have decided to use the same standard boiler diameters that I developed over 30 years ago, as those have long since become the standard of the industry. That is, 4th Size engines will use 30" boilers, 3rd Size 32", 2nd 34" to 36", 1st 36" to 38", Extra 1st 40", and Double Extra 1st 42".

Since suction-hose sizes have also become standard throughout the fire engine industry, we will be using 3 1/2" suctions on 4th Size engines, 4" on 3rd Size, 4 1/2" on 2nd, 5" on 1st, 5 1/2" on Extra 1st, and 6" on Double Extra 1st.

Deciding on standard dimensions for our engines is an important first step toward tooling-up for manufacturing.

Cincinnati, Feb. 10, 1905

American-LaFrance's lawyers continue to hamper our efforts to launch Ahrens Fire Engine Co. Having solidly secured Mr. Fox's acquiesence in giving American-LaFrance exclusive rights to the Fox boiler, they have now taken us to court to challenge our right to call our company Ahrens Fire Engine Co. They claim that in acquiring American Fire Engine Co. they acquired the exclusive right to build and sell fire engines with the "Ahrens" name.

While this is being debated in court, we have renamed our company Cincinnati Engine and Pump Works, and set up shop in rented quarters at the rear of 1118 Elm St. American-LaFrance still stubbornly refuses to sell us the old Webster St. plant, even though it is unused. So instead, we are building a modern $70,000 factory at the corner of Alfred & Cook streets. Martin Fisher, the architect, has promised that the building will be ready by May 1.

My eldest son Fred (3205 Glendora Ave.) is president of Cincinnati Engine & Pump Works, and my son John is sales manager (the same position he held with American Fire Engine Co. until last year). One of my sons-in-law (Charles H. Fox) is vice-president, and another son-in-law (George W. Krapp) is secretary and treasurer.

40-year-old William Henry Shafer is the chief engineer of Cincinnati Engine & Pump Works (he was born in Allentown, Pa. Oct. 7, 1864). He is hard at work developing new boilers, pumps, etc., so that our new engines will not infringe any of American-LaFrance's patent rights. In fact, today he applied for a patent on a new helical water-tube boiler that he developed. We believe that Mr. Shafer's work will result in an engine which generates steam faster, discharges more water, and throws streams farther than the Metropolitan.

Cincinnati, Feb. 18, 1905

Today, William H. Shafer applied for a patent on his new inclined water-tube boiler. It is a more-efficient boiler than the helical water-tube type he recently developed; we plan to use this boiler on our engines when production begins.

Cincinnati, March 6, 1905
William H. Shafer has developed, and applied for a patent on, a radial water-tube boiler.

Cincinnati, April 5, 1905
Our lawyers dropped in at the 1118 Elm factory to discuss American-LaFrance's case against us. They nearly hit the ceiling because a draftsman was making an axle drawing labeled "Ahrens Fire Engine Co." (our blueprint #162). They advised us that, until the court decides whether we can call our company Ahrens, everything we do (from blueprints to correspondence) must use the name "Cincinnati Engine & Pump Works".

Cincinnati, April 20, 1905
William H. Shafer has developed a new type of pump barrel for a steam fire engine pump, and today applied for a patent on it under the name "Cincinnati Engine & Pump Works".

Cincinnati, April 29, 1905
Charles H. Fox drew a blueprint for cylinder drain-cocks. What is unique about this blueprint is that it is the first and only drawing to bear the name "Cincinnati Engine & Pump Works", rather than the "Ahrens Fire Engine Co." name we were using prior to our court battle with American-LaFrance. This blueprint is not numbered.

Fox also filed for another patent today, this one is for a new type of throttle valve.

Cincinnati, June 22, 1905
We recently won our court case against American-LaFrance. We have the right to use the name Ahrens Fire Engine Co., and today we started using that name on our blueprints again (drawing #300, for a suction coupling). The only changes in management, from the Cincinnati Engine & Pump Works, are that my brother Peter is now Chief Engineer, and William H. Shafer now heads the drafting staff.

Cincinnati, July 3, 1905
At long last, production has started at the new Ahrens Fire Engine Co. factory, corner of Alfred and Cook. We are calling our new type of engine the "Continental", and the first one (a 3rd Size) is now under construction for East St. Louis, Ill. William H. Shafer made the boiler drawing for it today, it is a 32" (diameter) x 64" (high) version of the inclined water-tube boiler that he invented.

Cincinnati, Aug. 10, 1905
Today, William H. Shafer applied for a patent on his new steam fire engine pump.

We have now developed all of the major components to build a steamer that is superior to American-LaFrance's Metropolitan, or any other steamer now on the market.

In addition to the 3rd Size engine we have been building for East St. Louis, Ill., we recently received an order from Indianapolis, for another 3rd Size. Work has begun on that second engine.

Cincinnati, Aug. 16, 1905
The first advertisement for Ahrens Fire Engine Co. was published in today's issue of *Fire & Water Engineering* magazine. They also published a very nice article about our new company.

Many predicted that it would be impossible to launch a new fire engine company in competition with American-LaFrance's "fire engine trust", but we ignored the predictors of doom and have at last met with success. Actually, it is not really a new company at all, as we have almost the same corporate name (and already the same reputation for quality), the same management, and the same employees, as the Ahrens company that has been building the finest fire engines for 36 years.

Cincinnati, Sept. 12, 1905
William H. Shafer was granted his first three patents today: #799,196 for his helical water-tube boiler; 799,197 for his inclined water-tube boiler; and 799,198 for his radial water-tube boiler.

Cincinnati, Oct. 11, 1905
Cincinnati has ordered a 2nd Class Continental engine from us, and today the first blueprint for it was completed. This will be our third Continental engine. We recently delivered our first Continental to East St. Louis, Ill., and will deliver our second to Indianapolis next month.

Cincinnati, Nov. 7, 1905
Patent No. 803,889 has been issued to Charles H. Fox for his new throttle valve.

Cincinnati, Nov. 28, 1905
Patent No. 805,586 has been issued to William H. Shafer for his new pump barrel.

Cincinnati, March 25, 1906
Another grand-daughter was born today. Her name is Louise Krug Abbihl, after her aunt Louise Krug Ahrens (wife of my son John P. Ahrens). Her parents are my daughter Clara (born Dec. 15, 1879) and her husband Robert Abbihl (born July 16, 1878).

My son Fred has moved to 359 Thrall Ave. My son John has been promoted from sales manager to secretary of Ahrens Fire Engine Co.

Cincinnati, April 19, 1906
William H. Shafer has developed a water-tower fire engine, and today applied for a patent on it. A water-tower is basically a giant pipe (called the mast) that can be raised to about the fifth floor of a building, and that several steamers can pump into at once to deliver a single, powerful stream at that height. Shafer's water-tower differs from all others in that the nozzle at the top of its pipe has two pneumatic cylinders on it for tilting and swiveling in virtually any direction.

Cincinnati, April 28, 1906
William H. Shafer has invented an aerial ladder that is raised by a combination of springs and cables. He applied for a patent today. Since we are in the business of making steam fire engines, Ahrens Fire Engine Co. will not be manufacturing Mr. Shafer's aerial ladder, nor his water-tower.

Cincinnati, May 29, 1906
William H. Shafer has been granted Patent No. 821,674 for the Continental fire engine pump.

Cincinnati, June 1, 1906
William H. Shafer invented an inclined radial water tube boiler, and applied for a patent today.
At a recent test in Toledo, one of our 1st size Continental engines (Reg. No. 107) set a new world's record for discharge from a 1st Size steamer by pumping 1090 gallons per minute. The National Board of Fire Underwriters only require 800 GPM from a 1st Size engine.
We also recently beat the record for a 3rd Size engine, when one of our 600 GPM Continentals (Reg. No. 106) delivered just over 700 gallons per minute at Galion, Ohio.

Cincinnati, Aug. 22, 1906
William H. Shafer has invented a gasoline-propelled fire engine, or as he calls it, a "traction fire engine". It uses an in-line six-cylinder gasoline motor, with an in-line six-cylinder piston pump mounted directly above it, and the pistons of the motor in continuous engagement with the pistons of the pump. A special gear-shift is provided to switch the gasoline motor out of road (or driving) gear and into pump gear when the fire engine is pumping water. Shafer applied for a patent on it today.

Dallas, Texas, Oct., 1906
A new 2nd size Continental steamer, Reg. No. 115, was tested at Elm & Houston streets during the International Assn. of Fire Engineers convention here. Although rated at 700 gallons per minute, our engine discharged 930 GPM. This is a new record for discharge from a 2nd size engine.
We have been building Continentals for less than a year now, and already our engines hold the world's discharge records for 1st, 2nd, and 3rd size steam fire engines.

San Diego, Cal., Dec. 11, 1906
Two months ago, one of our Continental engines at Dallas, Texas, set a new world's pumping record for 2nd size fire engines, at 930 gallons per minute. Today, another of our Continentals (Reg. No. 118) broke our own record, delivering 975 GPM, on test at the San Diego city hall.

Cincinnati, June 18, 1907
Charles H. Fox has agreed to give up his royalty on the Fox boiler of a Metropolitan steamer that American-LaFrance is building for New Rochelle, N.Y. He is continuing to receive royalties on all other Fox boilers that American-LaFrance makes, even while serving as vice-president of American-LaFrance's biggest competitor.

Cincinnati, Nov. 12, 1907
William H. Shafer has been granted Patent No. 870,646 for the water-tower that he invented.
Radnor Fire Co., Wayne, Pa., recently placed the world's first gasoline-propelled pumping engine in service. It was built by Waterous of St. Paul, Minn. Until now, all motor fire engines have been chemical engines, hose cars, or squads, because nobody but our own William Shafer had discovered a practical way to harness the power of an internal-combustion motor to driving a fire pump. Actually, Shafer's motorized pumper (if anyone ever builds it) would be superior: it uses only one motor, while the Waterous uses two (one for propulsion, and one for the pump).

Cincinnati, Nov. 19, 1907
William H. Shafer has been granted Patent No. 871,147 for a chemical fire extinguisher.

Cincinnati, Dec. 24, 1907
William H. Shafer has been granted Patent No. 874,546 for the aerial ladder that he invented.

Cincinnati, Jan. 14, 1908
William H. Shafer has been granted Patent No. 876,404 for his (motorized) traction fire engine.

Cincinnati, Jan. 28, 1908
William H. Shafer has been granted Patent No. 877,601 for his inclined radial water tube boiler.

Cincinnati, March 3, 1908
Charles H. Fox has been granted Patent No. 880,513 for "an advertisement...having...marks to indicate the positions of perforations for filing the advertisement in a loose-leaf binder".

Cincinnati, April 14, 1908
William H. Shafer has been granted Patent No. 884,769 for a force-pump. American-LaFrance has sold its long-unused Cincinnati plant.

Cincinnati, Dec. 18, 1908
My brother Peter died at 3:00 this morning due to effects of an operation on his ear last night.

Cincinnati, June 15, 1909
C.J. Cross of American-LaFrance has expressed concern that the steam take-off which W.S. Nott Co. of Minnesota use on their steam fire engines, is a direct violation of the Fox patents. American-LaFrance threatens to discontinue royalties unless Mr. Fox takes legal action to protect American-LaFrance's interests in his patents.
We recently had the Underwriters Salvage Corps' 1902 Winton in our factory for rebuilding. Named "Margeurite", it was the world's first gasoline-propelled fire engine The wicker "picnic basket" equipment boxes on both sides often fell off when turning sharp corners. So we built steel tool-boxes instead, and riveted them to the body.
We have a new Chief Engineer at Ahrens Fire Engine Co., in place of my late brother Peter. He is Arnold Neuenschwander, former master mechanic of the fire department in Louisville, Ky. As master mechanic, he was quite familiar with Ahrens engines (the only make Louisville uses), so he will be a valuable addition to our company.
My son John has moved from 130 Malvern Place to 2316 Nelson Ave.

Cincinnati, March 29, 1910
Charles H. Fox has been granted Patent No. 953,166 for a steam fire engine boiler.

Cincinnati, June, 1910
A new Continental steamer, with my son John as engineer, broke the world's record for discharge from a 1st size (800 gallons per minute) steamer. This engine delivered 1160 GPM, breaking the record of 1090 GPM set by another Conti-

nental at Toledo four years ago.

Cincinnati, June 30, 1910
American-LaFrance recently entered the motor fire engine field with delivery of a chemical car to Lenox, Mass.; it is their Registered No. 1.

Because fire engines propelled by gasoline motors are becoming increasingly popular, we recently hired two brothers who are experienced automotive engineers. Ben and Emil Graf helped us to develop an automobile pumping-engine of our own, and today Charles H. Fox applied for a patent. We will soon begin construction of the porototype. Of course, we will also continue to manufacture "Continental" steam fire engines.

Cincinnati, Aug. 9, 1910
The Ahrens Fire Engine Co. was reorganized as the Ahrens-Fox Fire Engine Co. Officers are Charles H. Fox, pres.; John P. Ahrens, vice-pres.; and George W. Krapp, secretary-treasurer.

My son Fred is no longer associated with the family fire engine business, as he has purchased the Union Iron Works of 3323-3331 Colerain Ave., renamed it Ahrens Iron Works Co., and established himself as President and Treasurer. J.F. Walton is vice president, and L. Dubrow, Jr., is secretary.

Charles H. Fox moved out of our home at 3010 Vine, to 3547 Shaw Ave.

Cincinnati, Aug. 23, 1911
The first Ahrens-Fox automobile pumper (Reg. #500) was tested for the first time, in front of the Fire Co. 43 firehouse at the corner of Spring Grove Ave. and Alfred St. The test was to see if it could throw water over the 100-foot flagpole on the fire station. It threw a $1\frac{1}{2}$" stream twice that height, and two $\frac{7}{8}$" streams over the flagpole.

I have not been active in the family fire engine business for about 7 years, although I have been (and continue to be) available as a consultant. This was not enough -- I was getting bored and restless. So I have established the Ahrens Realty Co. at 111 East 4th, with myself as President, long-time friend and partner G. Fred Hawekotte as vice president, and son-in-law George W. Krapp as secretary/treasurer.

Cincinnati, Sept. 7, 1911
Speed tests of the first Ahrens-Fox automobile pumper were conducted over a 14.2-mile course today. The test took 50 minutes, averaging 15 miles per hour on hills and 34 MPH on levels. It was next tested on the Cincinnati Automobile Club's mile-long Stanley Ave. incline, carrying 14 men and weights equivalent to a full hose load. The hill-climb was made in $2\frac{1}{2}$ minutes, averaging 24 MPH.

This new fire engine uses a six-cylinder Hirschell-Spillman T-head gasoline motor of $5\frac{3}{4}$" bore x $6\frac{1}{2}$" stroke, 1015 cubic inch, 80 horsepower. This motor is coupled to a Brown & Sharpe transmission. Pump is a front-mounted two-cylinder Ahrens-Fox piston type, based on the Continental steam fire engine pump, with a $6\frac{1}{2}$" bore x 4" stroke. It is on a 147-inch Ahrens-Fox chain-drive chassis, and is designated Model "A".

Cincinnati, Sept. 14, 1911
My sister, Mary Ahrens Embhoff, died yesterday. She was born in Brettbeck Aug. 12, 1839, and emigrated to Cincinnati in 1858.

Milwaukee, Wis., Sept. 22, 1911
Ahrens-Fox displayed two new fire engines at the four-day International Association of Fire Engineers convention, which ended today. The first was a Double Extra 1st size Continental steamer (Registered #187) built for Beaumont, Texas. It delivered over 1300 gallons per minute.

The other engine, the Model A pumper prototype that we hoped would pump 700 gallons per minute, failed to meet its Underwriters rating of 450 GPM at 120 pounds-per-square-inch pressure, delivering 244 GPM at 185 psi and 340 GPM at 93 psi. When this engine returns to Cincinnati, work will begin on developing a new pump that will give it the desired 700 GPM capacity.

Denver, Colo., Oct., 1912
Ahrens-Fox developed a new version of its Continental steam fire engine, with a two-wheel electric tractor. The prototype was recently demonstrated at the International Association of Fire Engineers convention here, and has since been sold to the Denver Fire Department. The tractor was made for Ahrens-Fox by the Couple Gear Freight Wheel Co. of Grand Rapids, Mich. Ahrens-Fox now offers horse-drawn, gasoline-propelled, and electric-powered fire engines.

American-LaFrance, which sold its Cincinnati plant four years ago, is back in town. They recently opened a sales office at 230 West 4th Ave., Cincinnati.

Cincinnati, 1913
The Ahrens Company's long-standing rival, Amoskeag of Manchester, N.H., recently delivered its last steam fire engine; they have been building steam fire engines since 1859. This leaves only our Ahrens-Fox company and American-LaFrance in the steam fire engine industry, all other fire engine manufacturers having switched to motor fire apparatus production exclusively.

Cincinnati, Sept. 26, 1913
I made my will yesterday, dividing my effects and real-estate among my wife, six surviving children, and grand-daughter Elvira Fox.

Cincinnati, 1914
My son John has moved to 258 Ludlow Ave.

Cincinnati, 1916
My son John has moved to 3450 Cornell Place.

Cincinnati, Sept., 1916
The last Continental steam fire engine, Reg. No. 205, was recently delivered to American Fire Co. #1, Lansford, Pa. Called a "Continental Special", it consists of an Ahrens-Fox radial water tube boiler and two-cylinder steam-powered piston pump, mounted on a four-wheel chain-drive White truck chassis with a 6-cylinder T-head motor.

The "Continental" name was dropped from the motor fire apparatus line last year. From now on, no new Ahrens-Fox apparatus will use the

"Continental" brand name.

After 64 years, Cincinnati's steam fire engine industry has quietly come to an end. I have been a part of that industry since its inception, so it is sad to me to see its demise. But the gasoline motor has replaced the steam boiler, and the fire engine company which I founded in 1868 -- now in the capable hands of my children -- is already a leader in the new automobile fire apparatus technology. Only one manufacturer (American-LaFrance) is still making steamers.

Cincinnati, June 15, 1917

At 9:00 yesterday morning, I was out in the side yard of my home at 3010 Vine Street, watering my garden with a hose. My wife Mary was in the house, reading a book. I heard a voice shout "Hey, Ahrens," and turned to see who was calling me. I then heard a loud explosion and felt a sharp pain in my chest. My knees buckled, and I collapsed on the lawn. When I came to, I was lying in a hospital bed, with bandages wrapped tightly about my chest.

The doctors told me that I had been shot in the chest with a pistol, and the bullet had punctured my esophagus. That explained why I was having such trouble breathing! The doctors said I was lucky, because if the bullet had lodged only an inch away, it might have entered my heart and killed me instantly. They asked me who was so impatient to see an 82-year-old man die that they couldn't wait for it to happen naturally? Since the last thing I had seen was my assailant's face, I was able to answer that.

When I was running the Ahrens Manufacturing Co. in the 1880s and 1890s, it was generally known around Cincinnati that I was quite successful financially. So, many other German-American businessmen came to me for loans to help start businesses of their own. In many cases, if their idea had merit, I invested in their business, usually with a handsome return for myself.

One such business I had helped finance back then was Henry Bremer's cigar factory. A few days ago Bremer, who is now 65, visited me asking for a second loan. Since I retired 12 years ago and turned the company over to my children, I am not as well-off as I was when I advanced him the first loan all those years ago. I tried to explain that to him, but he kept arguing.

I was not at all convinced that, even if I had the money, Bremer's cigar business was a good place to invest it. When I turned-down his request for money last week, he stormed out of my house in a rage, threatening "revenge". Bremer made good that threat, as it was his face I saw when the bullet hit my chest.

Police arrested Bremer last night. He kept ranting about how he was going bankrupt, but his company's books showed that his business was doing extremely well. Bremer's employees told police that he had recently been suffering from all sorts of delusions, including those about the financial health of his company.

Bremer told police that at one time one of my daughters was interested in marrying one of his relatives; this is apparently another of his delusions. He also referred to me as his "best friend", when actually we were never more than business acquaintances.

Because of these and other delusions Bremer expressed, a group of alienists who examined Bremer have committed him to a mental hospital.

Meanwhile, I am beginning my long recuperation that will probably take several weeks. The gunshot, although painful, will not be fatal.

New York, N.Y., June 25, 1919

My name is Fred W. Shepperd. I am the publisher of *Fire & Water Engineering*, the trade journal of the fire apparatus industry. Seven months ago, I asked my friend Chris Ahrens of Cincinnati to turn his years of diaries about the steam fire engine industry, into a book. He has been working diligently on that book ever since.

Yesterday morning at 5:45 a.m., 84-year-old Chris Ahrens died of nephritis, leaving his book unfinished. I feel it is now my duty to turn his unfinished manuscript and voluminous notes into a book, and publish it for him.

I am publishing Chris Ahrens' book posthumously, as a tribute to a man who was a close friend to me personally. But more than that, I am publishing it as a "thank you" from the firefighters of the world whose task was greatly simplified by the numerous innovations which he invented. Chris Ahrens will be sorely missed, and it is my hope that this book will give him the immortality he so richly deserves.

List of Cincinnati-Built Engines

1. Sources

A major source of steamer information was correspondence with fire departments, fire buffs, and libraries in cities that once used Cincinnati-built steamers. The historical section at the Cincinnati Public Library was also a prime source of general research material, including old Cincinnati city directories. William T. King's *History of the American Steam Fire Engine* was also a source, although relied upon sparingly due to the many inaccuracies it seems to contain.

Latta material (1852-1867) is based on lists of Latta steamers compiled by Cincinnati native John White of the Smithsonian, and obtained from Harold S. Walker and John M. Peckham. Information was also found in contemporary issues of *Scientific American*, and in an 1860 Latta advertising brochure at New York Public Library. Special thanks to John Peckham for his write-up on Abel Shawk's use of the Buchanan patent, excerpts from which appear (with permission) both here and in the main text of this book.

Records of Ahrens steamers built 1868-1892 are extracted (in condensed form) from the detailed hand-written Ahrens sales and service records called *Book #1: Destination of Ahrens Engines*, the original of which is now at Firemen's Home, Hudson, N.Y. A complete copy of that book was obtained through courtesy of John Peckham and the late Armand Bolton. Additional

Ahrens material was obtained from original Ahrens advertising brochures at the Cincinnati Historical Society, and from various Ahrens steamer lists supplied by Harold S. Walker.

Records of American steamers (1892-1903) come from lists supplied by John Peckham, H.S. Walker, and Jack Robrecht. Much of it originates from the extensive archives of the American-LaFrance company.

Records of Continental steamers (1905-1916) come from lists supplied by Harold S. Walker, and from information extracted from the 172 original Continental blueprints still at Ahrens-Fox Co. Other sources were contemporary issues of *Fire & Water Engineering* magazine at New York Public Library, and Continental advertising brochures in the author's collection.

Photographs used in these lists are as credited. Special thanks to Wayne Sorensen, Dr. Lester Williams, Harold S. Walker, and Bruce Neal. Also thanks to the experts at Camera Mart, Sunnyvale, CA, who took the effort to make the best possible quality photo enlargements.

2. Abbreviations Used in This List

The number hanging in the left margin is the Registered Number, that is, the manufacturer's serial number.

Dates are listed American style (month first), not European style (day of the month first). That is, 3-5-1883 is March 5, 1883, not May 3, 1883.

The symbol "#" before a number means "number" (for example, #425), while the same symbol after a number means the steamer's weight in pounds (for example, 7300#).

The first paragraph of each listing describes the specifications of that steamer. The successive paragraphs detail the machine's history. The remainder of this explanation of abbreviations applies to the specification paragraphs.

The first number used is the steamer's *Class*, or size, as follows:

DX1 (Double Extra 1st)	=	1000 GPM
X1 (Extra 1st)	=	900 GPM
1 (1st Class)	=	800 GPM
2 (2nd Class)	=	700 GPM
3 (3rd Class)	=	600 GPM
4 (4th Class)	=	500 GPM
5 (5th Class)	=	400 GPM
6 (6th Class)	=	300 GPM

GPM means Gallons Per Minute.

The *Class* is followed by a 3-letter *Pump & Frame Description*:

DCN = Double (2-cylinder) pump, Crane Neck
DSF = Double (2-cylinder) pump, Straight Frame
SCN = Single (1-cylinder) pump, Crane Neck
SSF = Single (1-cylinder) pump, Straight Frame
SRT = Single (1-cylinder) pump, Round Tank

The *Pump & Frame Description* is followed by the *Means of Propulsion*:

/HD = Horse Drawn
/MD = Manually Drawn
/MD-HD = Equipped for horse *and* manual draft
/SP = Self Propelled (its own steam power)
/TD = Tractor Drawn (gasoline *or* electric)

The *Means of Propulsion* is followed by the *Type of Wheels*:

/A = spoked wooden wheels by Archibald Wheel Co., Lawrence, Mass.
/H = Home-made spoked wooden wheels (made by Ahrens Co.'s own wheelwright).
/I = Iron spoked wheels of unknown make
/S = Sarven's Patent spoked wooden wheels, by New Haven Wheel Co., New Haven, CT
/W = Wooden spoked wheels, unknown make.

The term "air pump" refers to an auxiliary feature that seems to have been used only on the *Upright* and *Columbian* models. Attached to the right side of the boiler, the steam-powered air pump forced a constant supply of compressed air into the air chamber of the pump. As the compressed air in the air chamber expanded, it forced the water out through the discharge gate and into the hose between strokes of the pump, reducing the pulsation associated with a piston type pump.

Contemporary reports indicate that the air chamber and air pump combination dramatically reduced pulsation in the hose compared to a piston pump equipped only with an air chamber, so it is not clear why this feature was not offered on later Metropolitan and Continental models.

Two types of air pumps were offered: a piston air pump (with flywheel) that had to be started manually, and an automatic air pump (no flywheel) that operated automatically whenever steam from the boiler was driving the fire pump.

Most, but not all, steamers were equipped with brakes. These brakes consisted of two blocks which, when activated, pressed against the outer rim of both rear wheels, using the friction against the wheel to stop. On Click steamers and a few Uprights, the brake was operated by a hand lever controlled by the driver of the horses, much like an early automotive emergency brake. On Columbian, Metropolitan, Continental, and most Uprights, the brake was operated when the driver stepped on a steel bar atop the foot board. On a few Uprights, the brakes were controlled by the engineer on the rear platform (coal bunker) using a hand lever; in these cases the brake pressed against the rear edge of the rear wheels (the usual location was in front of the rear wheels).

Poles for pulling a steamer came in four basic types. The front end of a *drop pole* dropped to the ground when the horses were unhitched. A *stiff pole* remained in a straight horizontal position when the horses were unhitched. A *manual pole* (indicated by /MD at the start of the first paragraph in each listing) was similar to the stiff pole, but replaced the trees and neck yoke of a horse hitch with a reel of rope for men to grab onto when pulling by hand. A *combination pole* (indicated by /MD-HD at the start of the first paragraph in each listing) combined the features of a stiff pole for horses with the rope-reel setup of a manual pole.

3. Model Names

The series of abbreviations for Class, description, pump/frame type, and propulsion (in the first paragraph of each listing) is always followed by the Model name:

American = Revised Upright, distinguished from other models by flywheel in center of pump, gradual curve in Crane Neck frame, and patent plate attached to frame just below pump.

Click = 3-wheel Latta with horizontal pump and cross-hatched (i.e. grate-like) frame.

Columbian = An Upright with addition of a hosebox between pump and driver's seat.

Cosmopolitan = Very small steamer built for export, industrial fire brigades, villages, etc. Resembled 6th Class Upright, except working parts of pump enclosed in dust-proof casing.

Continental = Similar to Upright, except smokestack atop boiler was short and wide-brimmed. Flywheel had very wide rim, and the steam chests extended to completely encase the steam cylinders. Ornament atop air chamber was a small globe, not like ornament on the Upright.

Metropolitan = Similar to American, but used Fox tube boiler, instead of Ahrens coil boiler.

Upright = 4-wheel Ahrens with vertical pump. Characterized by tall smokestack atop boiler. Crane Neck version had sharp curve in frame, resembling an inverted horseshoe. Air chamber ornament resembled an hourglass except lower half was larger than upper half. Flywheel and steam chest were transverse to frame on single-pump Upright, parallel to frame on double-pump. Some double-pump Uprights had two flywheels.

4. Dimensions

When a dimension is shown as ? it means that existing records do not provide that dimension.

Boiler dimensions are always listed as diameter x height (for example, 38" x 67").

Pump and cylinder dimensions are always listed as bore x stroke (for example, 5" x 7"); the word *cylinder* refers to the steam cylinder, while *pump* refers to the cylinders of the pump.

Wheel dimensions are always listed as number-of-spokes x diameter-in-inches (for example, 16 x 56 is 16 spokes x 56" diameter), front wheels listed first. *Track* means the width from outside-edge to outside-edge of wheels (for example, $56^1/2$").

Suction hose dimensions are always listed as diameter x length (for example, 4" x 20'). *Pre-connect* means that one end of the suction hose was carried always attached to the pump, and the hose was wrapped across the front of the steamer. If the word *pre-connect* is not used, it means that the suction hose was carried on brackets on the side of the steamer, and had to be manually attached to the pump each time it was used.

Results of timed pumping competitions are always given in minutes and seconds (for example, 5:23 is 5 minutes and 23 seconds). Stream distances are always measured horizontally, unless the text specifically says "vertical stream" or "stream height" (for example, *threw one 1" stream 160'* means a single stream of water thrown 160' horizontally from the end of a 1" nozzle).

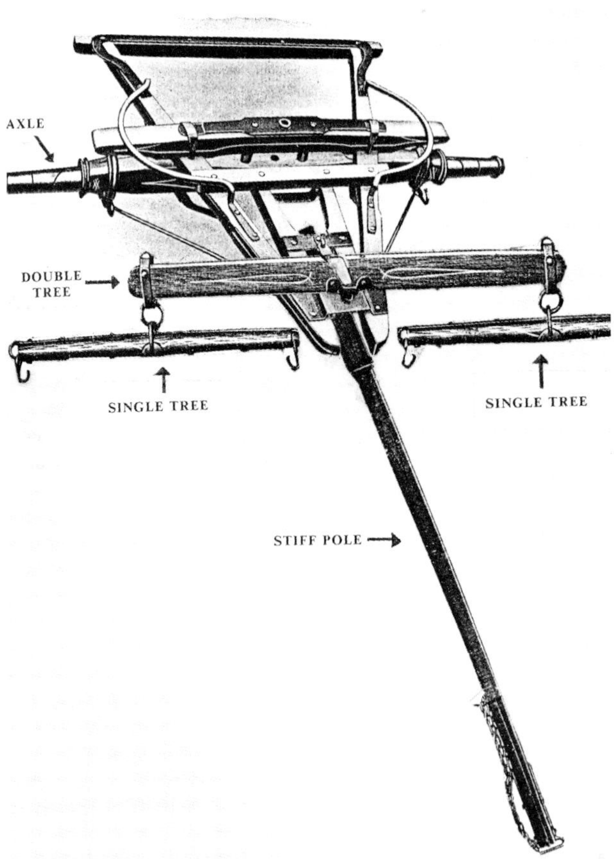

Stiff pole rigging for two horses. Although the illustration is for a farm-wagon, Ahrens used a similar system on many of their steam fire engines. The horse pole remained horizontal when the horses were unhitched, and could be removed by releasing one pin. Horse-harness was buckled to the hooks on the single-trees.

(Ed Hass collection)

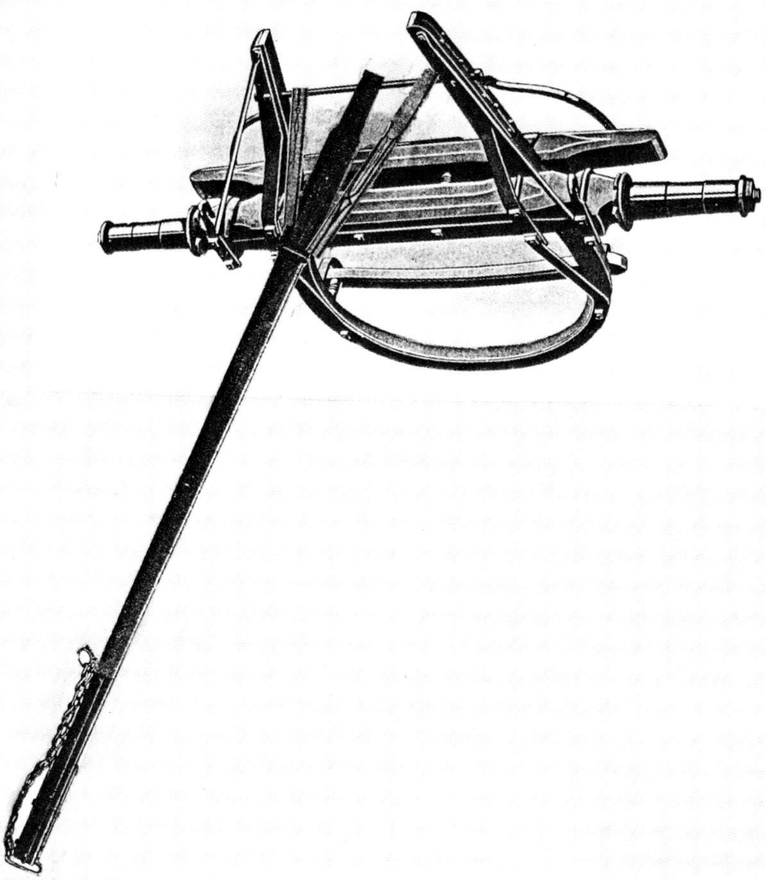

Another popular horse-hitch was the drop-pole, illustrated here. The front end of this pole dropped to the ground when the horses were unhitched. Drop poles were sometimes fitted with a folding brace underneath, to keep them horizontal while the horses were being hitched.

(Ed Hass collection)

CONTINENTAL THREE HORSE HITCH

The Continental Three Horse Hitch is our own design. Either pole can be raised to a full perpendicular; the locks are automatic and have a positive action. The parts are exceptionally well fitted, and great strength, minimum weight and simplicity are the points which commend it. Intended primarily as a high class pole for our "Continental" Steamers, we also adapt the Hitch to any other apparatus requiring three horses.

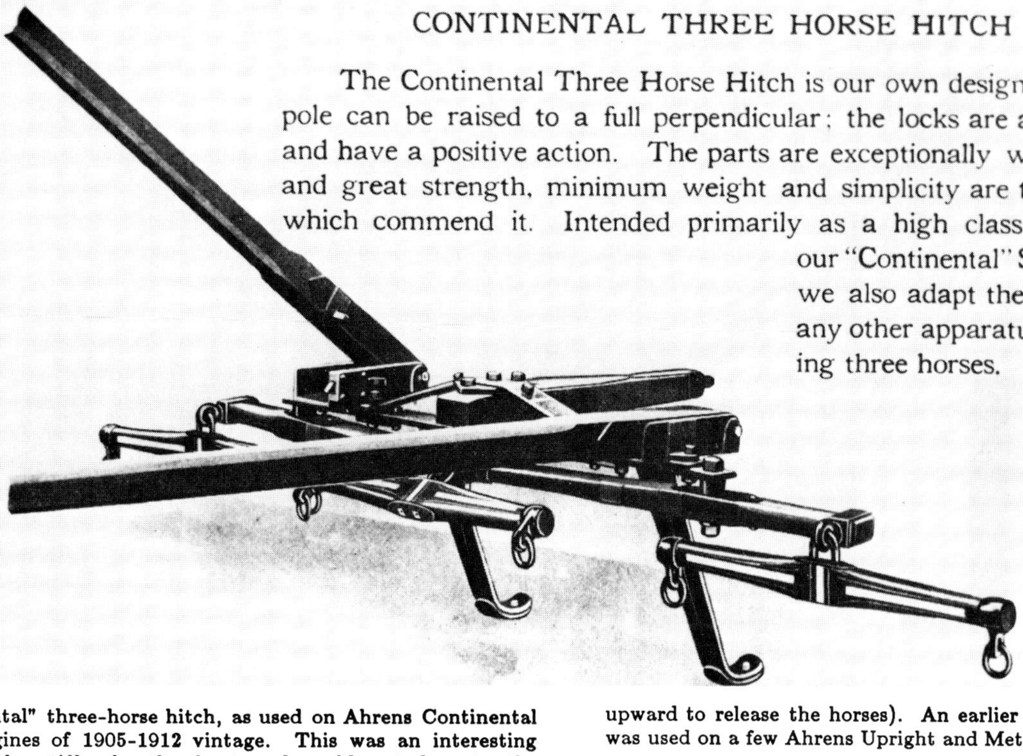

The "Continental" three-horse hitch, as used on Ahrens Continental steam fire engines of 1905-1912 vintage. This was an interesting variation on the stiff-pole: the front end could not drop to the ground, yet the pole was not rigidly horizontal (it could swing upward to release the horses). An earlier style of three-horse hitch was used on a few Ahrens Upright and Metropolitan steamers.

(Ed Hass collection)

Latta & Shawk Steamers

1852

William King's *History of the American Steam Fire Engine* states that Latta and Shawk were partners in the steam fire engine business until 1855. Some contemporary reports (including one by Abel Shawk himself) refer to "Latta, Shawk & Co." although there is no record that such a corporate entity ever existed. In fact, it seems that the two were never really partners at all. Latta already had a patnership with his brothers making locomotives and stationary engines, and Shawk appears to have been more a customer than a partner of Latta's. That is, Latta built the steam mechanisms that Shawk needed to power his steam fire engines.

Latta & Shawk apparently worked together on only two steamers: one prototype that was dismantled, and one steamer actually placed in service. The fact that Shawk, not Latta, was the principal designer and builder of these two steamers probably explains why neither was built at Latta's "Buckeye Works" factory.

Neither of the two Latta & Shawk steamers seems to have had a serial number. In order to differentiate these two steamers from each other, and to approximate the chronological order in which they were built, the author has arbitrarily assigned serial numbers to them for this list.

The idea for the steam fire engine originated with Abel Shawk, but Latta designed and built the steam machinery for it. Subsequently, Latta took an interest in manufacturing steamers (at his Buckeye Works), made a number of improvements in Shawk's original design, and patented many features of steam fire engine design such as the boiler, chassis, wheels, etc.

1 2 SSF/SP/W Click. Square "Buchanan Coil" boiler (iron) built by Abel Shawk. One 4" x 24" (used) D.L. Farnham hand engine pump. One 8" x 24" cylinder by Alexander Latta. Frame from old fire department ladder wagon. Assembled in 1851 by Alexander Latta, Abel Shawk, Miles Greenwood, and Richard G. Bray at Eagle Iron Works, 396 Walnut (on Miami Canal).

3-2-1852: Tested for City Council and 3000 citizens of Cincinnati, Ohio. Steam in 4:10. Through 350' of hose, threw one $^{3}/_{4}$" stream 130' in 5:00.

Experimental prototype. Dismantled and not sold. Boiler sold to City of Cincinnati to power the machinery of the city shops.

2 1 SSF/SP/W Click. Square coil boiler. One 6" x 24" pump. Two 5" x 10' pre-connect suctions. One 10" x 24" cylinder. Weight 22000#. Built in shop of John H. McGowan.

12-15-1852: Construction completed.

12-22-1852: Tested at Front St. & Broadway, Cincinnati, Ohio. Threw one $1^{1}/_{2}$" stream 225' in 7:00.

1-1-1853: Tested in Cincinnati. Steam in 5:00, moving under own power in 10:00, reached cistern at Broadway & 2nd in 15:00, threw two 1" streams in 19:00. Beat best throw of Cincinnati's most powerful hand pumper (a Hunneman). Threw 6 streams 175' each. From six siamesed lines, threw one $1^{1}/_{2}$" stream 224'. Threw three streams over 4-story Broadway Hotel.

2-1-1853: In service with Cincinnati's first all-paid fire company. Captain was R.G. Bray, former Chief of Cincinnati's volunteer fire companies, who donated half of the steamer's $10,000 purchase price out of his own pocket. Name *Uncle Joe Ross* (named for councilman who persuaded Cincinnati city council to appropriate the other $5000).

2- -1853: Fought first fire, in warehouse at 3rd & Main, Cincinnati.

3- -1853: Drained flooded cellar of Cincinnati fire house at the rate of 819 GPM.

4-1-1853: All volunteer fire companies brought into the paid department, making Cincinnati the first paid professional fire department. City councilman Miles Greenwood, a former volunteer fireman, became Cincinnati's first professional fire chief.

Test on 9th Street. Steam in 4:00. Threw one $1^{3}/_{4}$" stream 238'.

5- -1853: Responded 8 blocks to fire on Sycamore street, supplied six 600' lines in 12:30 (4 lines to hand pumpers, 2 to fire).

11- -1853: Rebuilt, reduced 1100# weight.

12-16-1854: Responded three blocks and threw one $1^{1}/_{4}$" stream in 13:30 at Cooper's Shop fire, 8th & Sycamore, Cincinnati.

12-6-1855: Demonstrated at Cincinnati for officials of Chicago, Ill. Boiler exploded at 180# steam pressure, killing engineer & two spectators. Afterward, the *Joe Ross* was rebuilt and returned to service.

4-1-1859: Annual report of Chief Enoch G. Megrue says that the *Joe Ross* was still in reserve, but recommends canabilizing it for spare parts for Cincinnati's other Lattas.

Above: Maiden portrait of the second Latta & Shawk steamer built, and the first one put into actual firefighting service. Named the "Uncle Joe Ross", it was built in 1852 and used in Cincinnati. The "Ross" was a 1st Class self-propelled Click (i.e., three-wheel) engine.

Below: The "Uncle Joe Ross" explodes at Cincinnati, during a demonstration for Chicago fire officials, December 6, 1855. Rebuilt, it returned to service, but was eventually canabilized for spare parts.
(Ed Hass collection)

Abel Shawk

"Young America Works"
11 East 7th St., Cincinnati
1855-1856

The 1853 Cincinnati City Directory shows Abel Shawk as a machinist at a shop on the east side of Ohio Ave. Apparently, he was no longer involved with the Latta brothers in the steam fire engine business at that time. Reputedly, Shawk had temporarily left the fire engine business following a disagreement with A.B. Latta about the design of steam fire engines. Shawk received a patent for his steam fire engine boiler on Sept. 20, 1853. Abel Shawk's brother, George W. Shawk, was listed in the 1853 directory as a maker of mathematical instruments at 29 East 4th St.

In 1855, Shawk founded *Young America Works*, steam fire engine builders, at 11 East 7th St. (between Main and Sycamore). Abel hired his brother George as a machinist for this company. The Shawk brothers' steamer was a completely new design, featuring four wheels (as opposed to Latta's three-wheeler), a horizontal piston pump mounted lengthwise (Latta steamers had transverse-mounted horizontal pumps), and a round-tank frame (Latta used a cross-hatched straight frame). While *Young America Works* apparently built only three steamers, some elements of Shawk's new design later appeared in Reanie & Neafie and round-tank Amoskeag steamers.

Although the last Shawk steamer was built in 1856, Shawk was still listed as a steam fire engine builder in the 1859 Cincinnati city directory, with his factory still at 11 East 7th.

In 1856, Shawk's home address was listed as "head of Vine Street." This changed to "south side of 9th between West Row and John" in 1858, 269 West 9th in 1859, 275 West 9th in 1860, and 296 George in 1863. It is not clear what became of Shawk after 1863, but John Peckham's research reveals that Shawk may have died in "a lunatic asylum near St. Louis" in 1873.

Like the Latta & Shawk steamers, those built at Young America Works do not seem to have had serial numbers. The author arbitrarily numbered them to differentiate one from another and to indicate approximate chronological order.

1 4 SRT/HD/W. Square boiler. Rated 500 GPM. Four discharge gates. One ? x 10' suction, squirrel-tailed around boiler. One steam cylinder.

5-23-1855: Tested at County Prison, Philadelphia, Pa. Name *Young America.* Steam in 3:00, water in 11:36. Threw one 1" stream 172', one $1^1/4$" stream 175'. Pumped 397 GPM. City officials praised it as "equal to at least 7 of our first-class [hand] engines".

Although Abel Shawk designed and helped build the world's first successful steam fire engine in 1852, he did not open his own steam fire engine factory until 1855. This 4th Class steamer, named "Young America", was the first product of that factory. Built in 1855, it was placed in service in Philadelphia in 1856.
(Ed Hass collection)

6-3-1855: Tested on Dock Street at Delaware River, Philadelphia. Through 100' of hose, threw a $1^{1}/_{2}$" stream 170'. Through 325' of hose, threw a $1^{1}/_{2}$" stream 175'. Through 680' of hose, threw a 1" stream 74'. Through 925' of hose, threw a $^{3}/_{4}$" stream 66'.

8-16-1855: Sold for $9500 to the citizens of Philadelphia; donated to city of Philadelphia. A paid fire company was to be formed to man it, but given to volunteers instead.

8-23-1855: The Fire Department Committee of the Philadelphia City Council appropriated $5000 to put the steamer *Young America* in service.

11-22-1855: Succumbing to political pressure from the volunteer fire companies, who did not wish to be replaced by steamers and paid men, the city council cut the August appropriation to only $2500. That smaller sum included construction costs for the *Young America* firehouse. $250 was spent on mechanical improvements to the Shawk steamer, $1300 for hose, $350 for a hose tender, and $1500 for the firehouse at Front & Noble streets. Total cost: $3400.

1- -1856: In service. Philadelphia's first steamer, used by Philadelphia Hose Co. #1 (volunteer, not paid). Junked 1857.

2 3 SRT/HD/W. Square boiler.

10- -1855: In service at St. Louis, Mo. Name *Old Union #2.*

10-10-1870: Replaced by Ahrens #8.

3 3 SRT/HD/W. Square boiler.

2- -1856: In service at St. Louis, Mo. Name *Missouri #5.*

1-31-1876: Replaced by Ahrens #156.

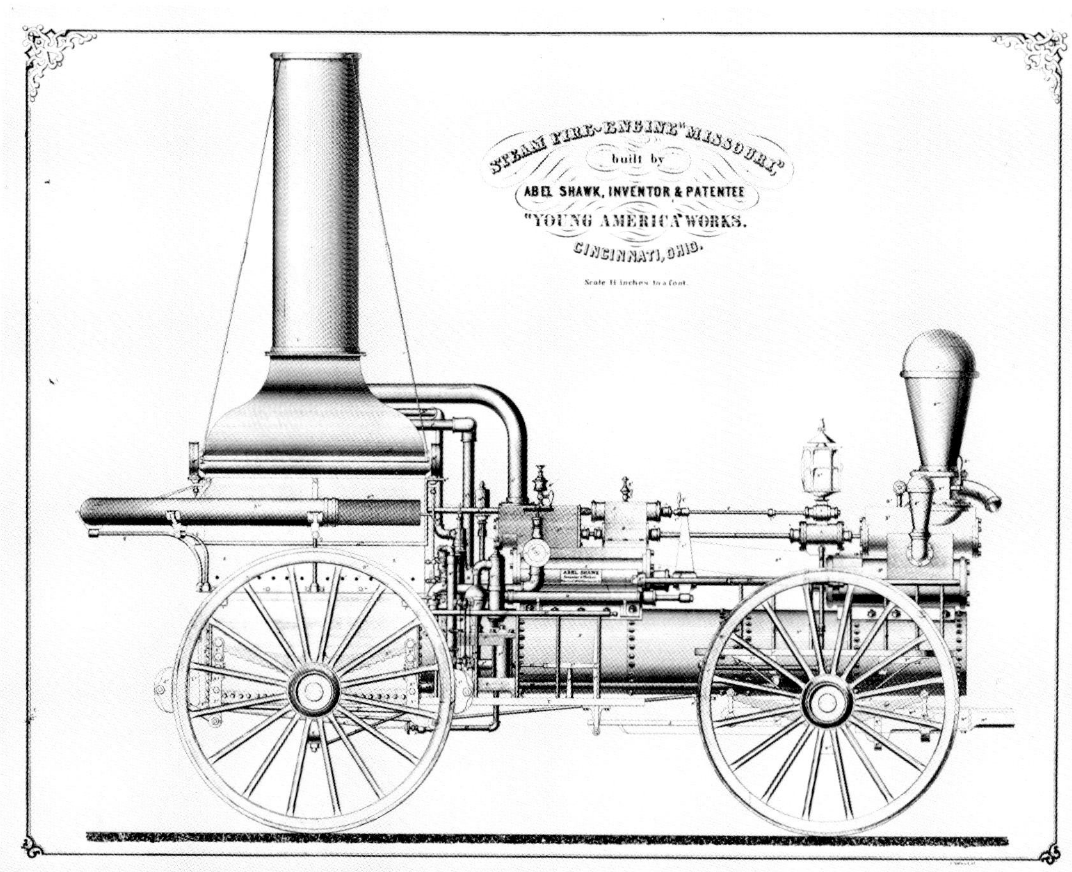

The last of only three steam fire engines built in Abel Shawk's Young America Works, "Missouri #5" was placed in service at St. Louis, Mo., in 1856. This is a reproduction of Shawk's original engineering drawing for that 3rd Class engine.

(Ed Hass collection)

A.B. & E. Latta
"Buckeye Works"
179 Race Street, Cincinnati
1853-1862

Latta steamers do not seem to have had serial numbers. In order to differentiate these steamers from one other, and to approximate the chronological order in which they were built, the author has arbitrarily numbered them for this list (1 and 2 were the two which Latta and Shawk built in places other than Buckeye Works).

3 1 SSF/SP/W Click. Square coil boiler. One $6^{5}/8$" x 24" pump. Two 5" x 10' pre-connect suctions. Two 11" x 24" cyls. Weight 11820#.

Spring 1854: Cincinnati's citizens donated $13400 to purchase a new steamer.

7-27-1854: On test in Cincinnati, raised steam in 3:10, threw one $1^{1}/2$" stream 297'. In service at Fire Co. #3, 117 West 6th St. (between Vine and Race), Cincinnati. Name *Citizen's Gift*. Engineer of this steamer during its entire career was Finley Latta, brother of the builder.

3-16-1874: Replaced by Ahrens #27. Chris Ahrens preserved patent plate, steam gauge, and decorative shield & antlers commemorating a contest victory. These parts donated to Ohio Mechanics Institute.

1981: All remaining parts of *Citizen's Gift* donated to Cincinnati Fire Museum, where they are now displayed.

4 ? SSF/HD/I Click. Square coil boiler.

1854: In service, Harrisburg, Pa. Name *Hope #2*. This steamer is on John Peckham's list (from John White), but accuracy is in doubt, especially 1854 date (possibly 1864 Button?).

5 1 SSF/HD/I Click. Square coil boiler. One $6^{5}/8$" x 24" pump. Two 5" x 10' pre-connect suctions. One 11" x 24" steam cylinder. Weight 12000#.

12-16-1854: Built as demonstrator at Washington, Baltimore, Philadelphia, New York, and Boston. Name *Miles Greenwood* (named for Cincinnati fire chief).

2-8-1855: Demo at City Hall Park, New York, N.Y. Competed against *Mankiller*, a double-deck side-stroke "hayrack" Henry Waterman hand pumper of Empire Eng. 42, New York. Waterman threw water faster than Latta's $8^{1}/2$ minutes, but the volunteers were exhausted in minutes, while the Latta kept operating for hours. On the first test, the Waterman threw a stream 189' versus 182' for the Latta; second test, Waterman 200' vs. Latta 185'. Both machines threw water over the Statue of Justice atop City Hall, about 150' above street level.

3- -1855: Demo at Boston, MA. They bought it for $12000. Instead of creating a paid fire company to man it, as Cincinnati did, Boston turned it over to the volunteer firemen, who were opposed to the steamer and let it be known that they would never allow the steamer to fight a fire in Boston.

4-29-1855: A fire in Boston caused over $1 million damage, partially because the firemen stubbornly refused to use the steamer.

11- -1856: On a routine drill of the Boston fire department, top of the steamer's pump blew off, allegedly due to sabotage by firemen. The $12000 steamer was proclaimed *"utterly incapable of further action"* and junked before it ever fought even one fire!

6 1 SSF/HD/W Click. Square coil boiler. One $6^{5}/8$" x 24" pump. Two 5" x 10' pre-connect suctions. Two $11^{1}/2$" x 24" steam cylinders. Weight 18000#.

3- -1855: Sold to New Orleans, La. Name *Young America*. Price $12186.

1858: Annual report of Fire Chief John F. Gruber of New Orleans revealed how strong was the feeling of the volunteer firemen against steam fire engines. Steamers threatened to end the rowdy good times and political clout the volunteers then enjoyed. Gruber claimed that the Latta's 18,000-pound weight destroyed pavements (yet he said nothing about the more-destructive effects of far-heavier commercial wagons that used the streets much more frequently). He also stated that the Latta had a habit of sinking when parked in swampy soil, as if the steamer was the only vehicle on earth that would sink should anyone be foolish enough to drag it into a swamp. Gruber even claimed that he could not use the Latta at a pier fire, because no wharf could support its weight -- those were the same wharves that supported the entire cargo of ocean-going steamships before loading.

Another doubtful claim was $6000 a year in repair bills on the Latta, when cities like Cincinnati and St. Louis reported maintenance costs under $200 a year (of course, it may well have cost $6000 a year if Chief Gruber constantly let the Latta sink in swamps and fall into the Gulf). Gruber's claim that a hand pumper was faster and more powerful than a Latta also contradicted the experience of other cities, where Latta steamers routinely had water several minutes before the hand tubs, delivered as much water as 7 to 10 hand engines, and kept operating for hours after the operators of hand pumpers dropped from exhaustion.

1859: Based on Chief Gruber's report, the

Above: Alexander Latta at the controls of his 1854 1st Class self-propelled Click steamer "Citizen's Gift". It was the third steam fire engine Latta built, but the first built in his own Buckeye Works. This steamer served Fire Co. #3 in Cincinnati until scrapped in 1874; only a handful of parts remain today, and they are displayed at the Cincinnati Fire Museum.

Below: Top view of Latta's 1854 Click steamer "Citizen's Gift". A single pivoting front wheel was mounted under the wheel-housing at left, between the steam cylinders. Latta built a working patent model of this engine, which was discovered in the Patent Office 80 years later and is now displayed at the Smithsonian.

(Ed Hass collection)

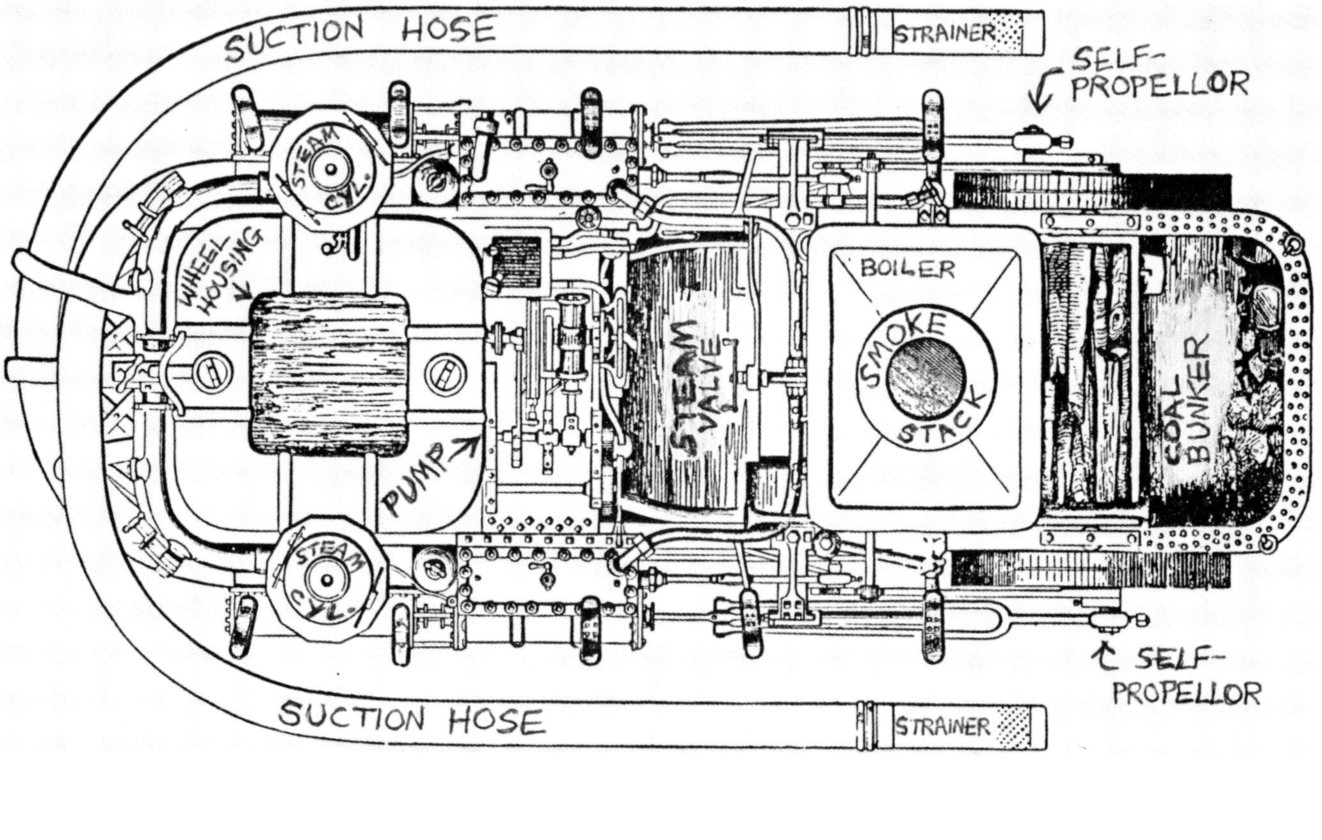

Young America was scrapped. It was not, however, replaced by a hand pumper as Chief Gruber wished, but by the steamer *John Adams*, built in New Orleans. Chief Gruber had won the battle but lost the war.

7 1 SSF/HD/I Click. Square coil boiler. One $6\frac{5}{8}$" x 24" pump. One 4" x 20' pre-connect suction. Two steam cylinders with signal lamps atop. Weight 12000#.

11- -1854: In service at Fire Co. #4, Sycamore betwen 7th and 8th, Cincinnati, Ohio. Name *A.B. Latta.*

1863: Fire Company renamed *Mississippi #4.*

1867: Replaced by Lane & Bodley #21.

8 2 SSF/HD/I Click. Square coil boiler. One 6" x 14" pump. 4" x 20' pre-connect suction. Two steam cylinders, lamps atop. Weight 9000#. The first lightweight Click built.

1855: In service at Marion Fire Co. #8, Cutter between Laurel and Betts, Cincinnati, Ohio. Name *Marion.* Price $5500.

1-1-1870: Replaced by Ahrens #4.

9 2 SSF/HD/I Click. Square coil boiler. One 6" x 14" pump. One 4" x 20' pre-connect suction. Two steam cylinders with signal lamps atop. Weight 9000#.

1855: In service at Deluge Fire Co. #10, 3rd and Lawrence, Cincinnati, Ohio. Name *Ferguson Clements.* Price $5500.

1-1-1869: Replaced by Ahrens #1.

10 2 SSF/HD/I Click. Square coil boiler. One 6" x 14" pump. One 4" x 20' pre-connect suction. Two steam cylinders with signal lamps atop. Weight 9000#.

1855: In service at Mohawk Fire Co. #12, Vine St. at Hamilton Rd., Cincinnati, Ohio. Name *Miles Greenwood.* Price $5500. Captain was Jacob Hust, whose grandson (Chas. Hust Fox) later became a son-in-law of Chris Ahrens and president of Ahrens-Fox.

12-30-1871: Replaced by Ahrens #17.

11 2 SSF/HD/I Click. Square coil boiler. One 6" x 14" pump. One 4" x 20' pre-connect suction. One steam cylinder. Weight 10000#.

1856: In service at Wasington Fire Co. #1, Vine between Columbia and Front, Cincinnati, Ohio. Name *Washington.* Price $5500.

12 3 SSF/HD/I Click. Square coil boiler. Weight 7500#.

1856: In service Fire Co. #2, Evansville, Ind.

1875: Was still in service, per letter from Evansville to Ahrens Mfg. Co.

13 4 SSF/HD/I Click. Square coil boiler. Weight 5000#.

1856: In service at Columbus, Ohio. Threw a stream 200'.

14 3 SSF/HD/I Click. Square coil boiler. Weight 7500#.

1856: In service at St. Louis, Mo. Name *George Kyler #1.*

6-25-1875: Replaced by Ahrens #147.

15 3 SSF/HD/I Click. Square coil boiler. Weight 7500#.

1857: In service at St. Louis, Mo. Name *David Moore #3.*

11-8-1875: Replaced by Ahrens #152.

16 3 SSF/HD/I Click. Square coil boiler. Weight 7500#.

1857: In service at St. Louis, Mo. Name *John M. Winn #4.*

8-20-1875: Replaced by Ahrens #149.

17 1 SSF/HD/I Click. Square coil boiler. One $6\frac{5}{8}$" x 24" pump. Two 5" x 10' pre-connect suctions. Two $11\frac{1}{2}$" x 24" steam cylinders. Weight 12000#. Name *Eclipse #1.*

3-1-1857: In service, Eng. 1, Louisville, Ky. Paid fire department organized to man it and city's hand pumpers, making Louisville only the second city to have a fully-paid fire depart-ment (Cincinnati was first). Later renamed *P.B. Atwood #1.*

3-9-1877: Replaced by Ahrens #165.

18 3 SSF/HD/I Click. Square coil boiler. Weight 7500#.

9-14-1857: In service at St. Louis, Mo. Name *Underwriter #6.* Same date, St. Louis became the third city with a paid fire department. The paid department, under Fire Chief H. Clay Sexton, began with 6 steamers in service (two Shawks and four Lattas).

7-30-1875: Replaced by Ahrens #148.

19 3 SSF/HD/I Click. Square coil boiler. Weight 7500#.

1858: In service at St. Louis, Mo. Name *Deluge #7.*

10-1-1875: Replaced by Ahrens #150.

20 2 SSF/HD/I Click. Square coil boiler. One 6" x 14" pump. One 4" x 20' pre-connect suction. One steam cylinder. Weight 10000#.

The "Miles Greenwood", an **1854 1st Class Latta (Number 5), was** demonstrated at Washington, Baltimore, Philadelphia, and New York, before it was sold to Boston, Mass. **Boston firefighters were so** opposed to the steamer that they never let it **fight a fire, and it was** junked in 1856. Note that this **Latta was not self-propelled.**

(Courtesy John Peckham)

One of Cincinnati's **lightweight horse-drawn Latta steamers. It is** not known when this one was built nor which Fire Company it **served. Note that this engine had no self-propellor.**

(Ed Hass collection)

1858: In service, Western Fire Co. #14, Fifth between Smith and Mound, Cincinnati, Ohio.

5-17-1875: Replaced by Ahrens #10.

21 2 SSF/HD/I Click. Square coil boiler. 6" x 14" pump. One 4" x 20' pre-connect suction. Weight 10000#. Name *Joseph A. Gilliss #2.*

1858: In service at Engine 2, Louisville, Ky.

4-2-1881: Replaced by Ahrens #276.

22 3 SSF/HD/I Click. Square coil boiler. 7500#.

1858: In service at Engine 3, Louisville, Ky.

12-24-1880: Replaced by Ahrens #270.

23 2 SSF/HD/I Click. Square coil boiler. One 6" x 14" pump. One 4" x 20' pre-connect suction. One steam cylinder. Weight 10000#.

1859: In service at Engine 4, Louisville, Ky.

4-2-1881: Replaced by Ahrens #278.

24 2 SSF/HD/I Click. Square coil boiler. One 6" x 14" pump. One 4" x 20' pre-connect suction. One steam cylinder. Weight 10000#.

1859: In service at Engine 5, Louisville, Ky.

9-6-1880: Replaced by Ahrens #250.

25 4 SSF/HD/I Click. Square coil boiler. Weight 5000#.

1860: In service at Toledo, Ohio.

26 4 SSF/HD/I Click. Square coil boiler. Weight 5000#.

1860: In service, Engine 2, Indianapolis, Ind.

11-12-1874: Threw one stream 18' farther than the best throw of a new Silsby steamer.

3-8-1876: Replaced by Ahrens #160.

27 2 SSF/HD/I Click. Square coil boiler. One 6" x 14" pump. One 4" x 20' pre-connect suction. One steam cylinder. Weight 10000#.

1861: In service at West End Fire Co. #2, Cincinnati, Ohio.

3-29-1877: Replaced by Ahrens #34.

28 2 SSF/HD/I Click. Square coil boiler. One 6" x 14" pump. One 4" x 20' pre-connect suction. One steam cylinder. Weight 10000#.

1861: In service, Jefferson Fire Co. #5, Vine between Canal and Court, Cincinnati, Ohio.

8- -1871: Replaced by Ahrens #11.

29 2 SSF/HD/I Click. Square coil boiler. One 6" x 14" pump. One 4" x 20' pre-connect suction. One steam cylinder. Weight 10000#.

1861: In service at Eastern Fire Co. #6, Pearl and Martin, Cincinnati, Ohio.

9-1-1870: Replaced by Ahrens #7.

30 2 SSF/HD/I Click. Square coil boiler. One 6" x 14" pump. One 4" x 20' pre-connect suction. One steam cylinder. Weight 10000#.

1861: In service at Northern Fire Co. #7, Webster between Main and Sycamore, Cincinnati, Ohio. Name *Phil Sheridan.*

1873: Engineer on this steamer was J. Peter Ahrens, younger brother of Chris Ahrens.

7-1-1874: Replaced by Ahrens #10. Chris Ahrens saved the smokestack from the top of the boiler, including the *Phil Sheridan #7* nameplate, and donated it to the Ohio Mechanics Institute in Cincinnati.

1981: Smokestack and nameplate donated to Cincinnati Fire Museum, where they are now on display.

31 2 SCN/HD/I Click. Square coil boiler. One steam cylinder. First Latta on four wheels (technically not a Click model, since "Click" meant 3-wheel), also first with Crane Neck frame, otherwise similar to other Lattas.

1861: In service at Memphis, Tenn.

Lane & Bodley

John and Water Streets, Cincinnati
1863-1868

In 1863, the Latta brothers retired from the steam fire engine business to form the Aerated Bread Company, operating in the same factory (179 Race St.) where they had previously built steam fire engines. They then sold their old fire engine company, including all patent rights, to Lane & Bodley, a machine shop at John & Water streets in Cincinnati. Lane & Bodley hired Latta's former apprentice, Chris Ahrens, as their Superintendent of Fire Engine Construction.

Lane & Bodley continued to make Latta steamers until Colonel Philander P. Lane died in 1868. At that time, Lane's partner, Joseph T. Bodley, sold the fire engine division (including patent rights) to machinist Chris Ahrens (residence 17 Woodward St.) and saloon-keeper Henry Lachtrop (residence 468 Main St.).

Like the steamers built by Latta & Shawk,

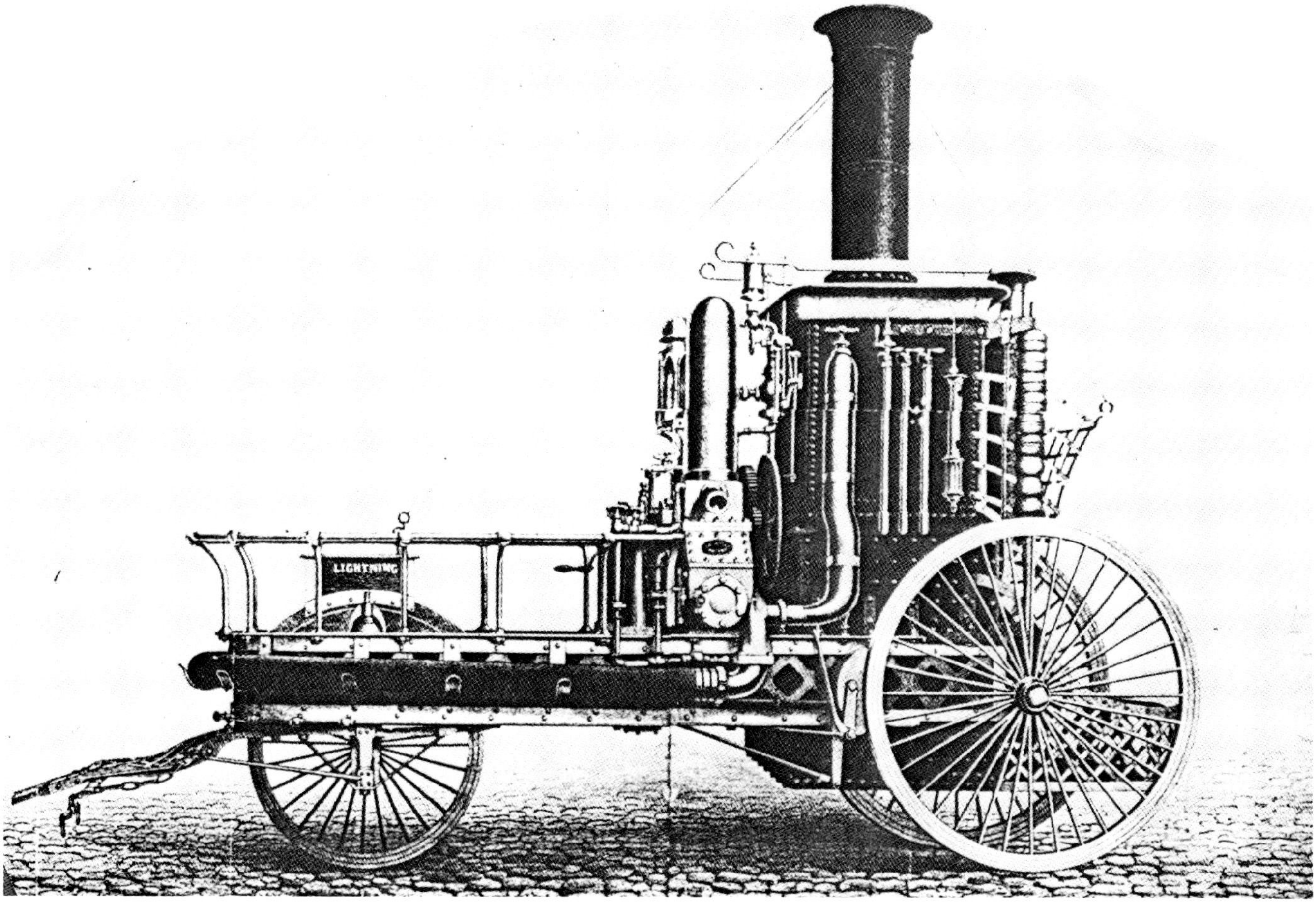

It is not known what city used this c-1860 Latta horse-drawn steamer, named "Lightning". This looks like one of Latta's smaller engines, either 3rd or 4th Class.

(Ed Hass collection)

The last steamer which the Latta brothers built before selling the business to Lane & Bodley, was also their most unusual. It featured a Crane Neck frame and four wheels, rather than their customary straight-frame, three-wheel design. This 2nd Class steamer #31 went to Memphis, Tenn., in 1861. Civil War blockades made its delivery from Cincinnati extremely difficult.

(Wayne Sorenson collection)

Young America Works, and Buckeye Works, those built by Lane & Bodley do not appear to have had serial numbers. The author has numbered them here to differentiate one from another, and to establish approximate chronology. Note that on some of the entries copied from John White's Lane & Bodley list, the author has indicated his suspicions that White may have been listing Ahrens steamers and not Lane & Bodleys at all.

1 ? SSF/HD/I Click. Square coil boiler.

1863: In service at Engine 1, Dayton, Ohio. Not certain that this was built.

2 ? SSF/HD/I Click. Square coil boiler.

1863: In service at Engine 2, Dayton, Ohio. Not certain that this was built.

3 ? SSF/HD/I Click. Square coil boiler.

1863: In service at Engine 3, Dayton, Ohio.

10-9-1871: Taken to Great Chicago Fire with Ahrens steamers #1, #3, and #14 from Cincinnati. Operated there for at least 24 hours, possibly as much as 38 hours.

11-12-1885: Replaced by Ahrens #445.

4 ? SSF/HD/I Click. Square coil boiler.

1863: In service, Engine 1, Nashville, Tenn.

2-26-1895: Replaced by Ahrens #190.

5 ? SSF/HD/I Click. Square coil boiler.

1863: In service at Cleveland, Ohio.

6 ? SSF/HD/I Click. Square coil boiler.

1863: In service at Battle Creek, Mich., per the John White list (obtained from Harold Walker and John Peckham). White may have been in error, as Battle Creek had an 1881 Ahrens #290. Ahrens steamers were often called "Latta", even in Ahrens advertising, so this was a possible source of confusion.

7 ? SSF/HD/I Click. Square coil boiler.

1864: In service at Hamilton, Ohio, again per the John White list. There is no record that Hamilton had a Latta, possibly White mistook Hamilton's Ahrens #2 for a Latta.

8 2 SSF/SP/I Click. Square coil boiler. One 6" x 14" pump. One 4" x 20' pre-connect suction. Two steam cylinders.

1864: In service at Cincinnati, Ohio. Name *John F. Torrence*. Price $7000.

9 3 SCN/SP/I Click. Square coil boiler. 7500#.

1865: In service at Engine 6, Louisville, Ky.

8-20-1879: Replaced by Ahrens #220.

10 ? SSF/HD/I Click. Square coil boiler.

1865: In service at Grand Rapids, Mich. Not certain that this was built.

11 2 SCN/SP/I Click. Square coil boiler. One 6" x 14" pump. One 4" x 20' pre-connect suction. Two steam cylinders. Second of only three Crane Neck, 4-wheel Click steamers built (the other two were for Memphis).

1865: In service at Washington Fire Co. #8, St. Louis, Mo. Name *Hampton Woodruff*.

10- -1871: Replaced by Ahrens #15.

12 ? SSF/HD/I Click. Square coil boiler.

1866: In service at Akron, Ohio. Not certain that this was built.

13 ? SSF/HD/I Click. Square coil boiler.

1866: In service at Leavenworth, Kansas. This may be 1875 Ahrens #157.

14 2 SCN/HD/I Click. Square coil boiler. 6" x 14" pump. 4" x 20' pre-connect suction. One steam cylinder. Last of three 4-wheel Crane Neck Click steamers built. Weight 10000#.

1866: In service at Memphis, Tenn.

15 ? SSF/HD/I Click. Square coil boiler.

1866: In service at Galveston, Texas. This could be wrong, may be 1877 Ahrens #174.

16 ? SSF/HD/I Click. Square coil boiler.

1866: In service at Quincy, Ill. This could be wrong, may be 1874 Ahrens #35.

17 ? SSF/HD/I Click. Square coil boiler.

1866: In service at Omaha, Nebraska. This could be wrong, may be 1879 Ahrens #210.

18 ? SSF/HD/I Click. Square coil boiler.

1866: To Racine, Wis. This is doubtful; 1885 Ahrens #412 reputedly first Racine steamer.

19 ? SSF/HD/I Click. Square coil boiler.

1866: In service at Oshkosh, Wis. This could be wrong, may be 1881 Ahrens #274.

20 ? SSF/HD/I Click. Square coil boiler.

1866: In service at St. Paul, Minn. This could be wrong, may be 1884 Ahrens #391.

21 3 SSF/HD/I Click. Square coil boiler. Weight 7500#.

1867: In service at Mississippi Fire Co. #4, Cincinnati, Ohio. Name *Theodore Chambers*.

5-1-1878: Replaced by Ahrens #198.

22 ? SSF/HD/I Click. Square coil boiler.

1867: In service at Newark, Ohio. This could be wrong, may be 1871 Ahrens #14 that was loaned to Newark, and/or Newark's own 1879 Ahrens #216.

23 ? SSF/HD/I Click. Square coil boiler.

1867: In service at Engine 9, St. Louis, Mo. This is probably 1872 Ahrens #19.

24 ? SSF/HD/I Click. Square coil boiler.

1867: In service at Engine 10, St. Louis, Mo. This is probably 1872 Ahrens #23.

C. Ahrens & Co.

24 Webster St., Cincinnati, 1868-1870
Chris Ahrens and Henry Lachtrop, Proprietors

1 2 SSF/HD/I Click. Square 42" x ?? coil boiler. One $5^1/_2$" x 14" pump, signal lamp atop pump. One 4" x 20' pre-connect suction. One 10" x 14" cylinder. Driver's seat.

1-1-1869: In service at Deluge Fire Co. #10, Cincinnati, Ohio. Name *A.B. Latta*.

10-9-1871: Pumped 38 hours at Great Chicago Fire, with Ahrens #3 and #14.

4-24-1878: Boiler rebuilt, pump overhauled.

5-24-1881: Replaced by Ahrens #228. Probably served Relief Fire Co. #2 from 5-26-1881 to 3-1-1882.

2 2 SSF/HD/I Click. Square 42" x ?? coil boiler. One $5^1/_2$" x 14" pump. One 4" x 20' pre-connect suction. One 10" x 14" cylinder. No seat.

4-23-1869: Delivered to Fire Co. #3, Hamilton, Ohio.

1-9-1875: Returned to factory for new coil and repairs. Afterward, raised steam in 4:30.

2-19-1876: Joe Terbeck (Ahrens pipefitter) installed new coil in boiler at Hamilton.

9-7-1881: New boiler coils.

1899: Sold for $100 to Mr. Wolff, scrap iron dealer, Hamilton, Ohio.

3 1 SSF/HD/I Click. Square ?? x ?? coil boiler. One $6^1/_2$" x 14" pump. One $4^1/_2$" x 20' pre-connect suction. One $11^1/_2$" x 14" cylinder.

7-1-1869: In service at Brighton Fire Co. #13, Cincinnati, Ohio.

10-9-1871: Pumped 38 hours at Great Chicago Fire, with Ahrens #1 and #14.

11-12-1877: Replaced by Ahrens #10. To Washington Fire Co. #1, Vine St. between Front and Second.

5-1-1883: Replaced by Ahrens #322.

5-4-1887: Sold to Fleischman Distillery, Cincinnati, Ohio.

4 1 SSF/HD/I Click. Square ?? x ?? coil boiler. One $6^1/_2$" x 14" pump. One $4^1/_2$" x 20' pre-connect suction. One $11^1/_2$" x 14" cylinder.

11- -1869: Steam up for 36 days non-stop at Exposition Building, 14th St. at Canal, Cincinnati, Ohio.

1-1-1870: In service at Marion Fire Co. #8, Cutter between Laurel and Betts, Cincinnati.

4-1-1882: Replaced by Ahrens #314.

5 2 SSF/HD/W Upright. Round $35^1/_2$" x 67" coil boiler. 6" x 8" pump, one inlet. 4" x 20' suction. 10" x 8" cylinder. Wheels 16 x ?? front, 16 x ?? rear. Weight 6110#.

3- -1870: First test at Orphan Asylum, Canal at 14th St. Through 500' of hose, threw one stream 271'.

3- -1870: Demonstrated at Louisville, Ky.

4-1-1870: Demonstrated at Indianapolis, Ind. Fair Play Fire Co. #1 of Madison, Ind., was so impressed that they bought it on the spot.

4-9-1870: In service at Fair Play Fire Co. #1, Madison, Ind. Threw one stream 283'10".

11-19-1875: Pump rebuilt at Madison for pre-connect suction.

3-17-1877: Joe Terbeck installed new boiler coils at Madison.

8-1-1890: Joe Terbeck installed new boiler coils at Madison.

10-17-1895: George Zink (Ahrens pipefitter) installed new boiler coils at Madison.

8-3-1904: New boiler coils installed at Madison by American-LaFrance.

1920: Traded-in on new Ahrens-Fox Model J-M-4 #958.

BRANCH OFFICE AND WORKS AT CINCINNATI.

The factory buildings of Ahrens Manufacturing Co. at 22-26 Webster St. (now East 14th), Cincinnati. Top, as they looked in the 1890s, viewed from Cincinnati's No. 7 firehouse. Bottom, as they look today. Note that the three-story office building at left no longer stands, and the site of the No. 7 firehouse is now the parking lot in foreground. Building in center is still recognizable, despite its concrete facade and bricked-over windows, while building at right appears unchanged in 90 years.

(Ed Hass collection)

Ahrens #1, an 1869 2nd Class Click steamer, is shown in front of Deluge Fire Co. #10's firehouse at 3rd & Lawrence, Cincinnati, Ohio. This engine was named "A.B. Latta". The decorative smokestack distinguishes an Ahrens from its drab ancestors that Latta and Lane & Bodley built. This was one of three Ahrens engines used at the Great Chicago Fire of 1871.

(Courtesy Steve Hagy)

C. AHRENS & CO., 24 and 26 WEBSTER ST., CINCINNATI, O. 1st and 2d CLASS LATTA BOILERS.

Fire Co. #3 of Hamilton, Ohio, used this 1869 2nd Class Ahrens Click steamer, Registered Number 2. It was scrapped when a new steamer replaced it in 1899.

(Courtesy Steve Hagy)

C. Ahrens & Co.
24 and 26 Webster St., Cincinnati, 1870-1871
Chris Ahrens and Henry Lachtrop, Proprietors

6 2 SSF/HD/W Upright. Round 35 1/2" x 67" coil boiler. One 6" x 7 3/4" pump, one inlet. One 4" x 20' pre-connect suction. One 10" x 7 3/4" cylinder. Weight 5500#.

7- -1870: Demonstrated at Saint Louis, Mo. Steam in 3:00. Threw one 1 1/8" stream 246'6", two streams 203' each.

8-1-1870: In service at Walnut Hills Fire Co. #16, Cincinnati, Ohio. Steam in 3:00.

1-25-1878: New boiler coils.

9-5-1884: Replaced by Ahrens #410. To Spare.

11-1-1885: To Fire Co. #12 in place of Ahrens #17.

1-10-1899: Scrapped at Fire Co. #13. Ahrens bought 1121# scrap brass @ 10c/lb.

7 2 SSF/HD/W Upright. Round 35 1/2" x 67" coil boiler (iron). 6" x 7 3/4" pump, one inlet. 4" x 20' pre-connect suction. 10" x 7 3/4" cyl.

9-1-1870: In service at Queen City Fire Co. #6, Martin and Pearl streets, Cincinnati, Ohio. Name *Queen City*. Engineer on this steamer from 1877 to 1882 was J. Peter Ahrens, brother of the builder.

1- -1874: New Ahrens coil boiler.

6-27-1879: New pump.

5-9-1888: New boiler coils.

12-5-1899: Scrapped at Fire Co. #1. Ahrens bought 733# of scrap brass @15c/lb. and donated *Queen City #6* nameplate to Ohio Mechanics Institute, Cincinnati.

1981: *Queen City #6* nameplate from this Ahrens donated to Cincinnati Fire Museum, where it is now displayed.

8 2 SSF/HD/W Upright. 35 1/2"x67" round coil boiler (iron). 6" x 7 3/4" pump, one inlet. 4" x 20' pre-connect suction. 10" x 7 3/4" cyl. Wheels 16 x ??, 16 x ?? rear. 57 1/2" track.

10-10-1870: In service at Engine 2, St. Louis, Mo. Replaced 1855 Shawk *Old Union #2*.

C. Ahrens & Co.
24 and 26 Webster St., Cincinnati, 1870-1871
Chris Ahrens and Frank Kamman, Proprietors

9 2 SSF/HD/W Upright. Round 35 1/2" x 67" coil boiler (iron). 6" x 8" pump, one inlet. 4" x 20' pre-connect suction. 10" x 8" cylinder.

1-21-1871: Delivered by Frank Kamman to Alton, Ill. Water in 4:03. Threw one 1" stream 280', two 1" vertical streams 110'. Name *James T. Drummond*. Price $5000.

1-23-1871: In service at Alton Fire Co. #1, Alton, Ill.

12-16-1876: Sold to Belleville, Ill., for $2000.

4- -08: Rebuilt for Belleville Fire Dept. by Harrison Machine Works, Belleville, Ill. New 34" x 64" Radial Inclined Water Tube boiler supplied by Ahrens Fire Engine Co.

10 2 SSF/HD/W Upright. Round 35 1/2" x 67" coil boiler (iron). 6" x 8" pump, one inlet. 4" x 20' pre-connect suction. 10" x 8" cylinder. Wheels 16 x ?? front, 16 x ?? rear.

1-1-1871: In service at Storrs Fire Co. #17, Cincinnati, Ohio.

10-20-1871: To Jefferson Fire Co. #5, and Ahrens #11 of Fire Co. #5 to Fire Co. #17.

5-30-1874: Replaced by Ahrens #36. To Northern Fire Co. #7.

5-17-1875: Replaced by Ahrens #43. To Western Fire Co. #14.

11-12-1877: To Brighton Fire Co. #13. Replaced Ahrens #3.

1-28-1878: New boiler coils.

9-1-1883: New boiler coils.

6-2-1887: Replaced by Ahrens #503.

2- -1897: Dismantled by Cincinnati Fire Dept. Ahrens bought scrap copper and brass.

11 2 SSF/HD/W Upright. Round 35 1/2" x 67" coil boiler (iron). 6" x 8" pump, one inlet. 4" x 20' pre-connect suction. 10" x 8" cylinder. Wheels 16 x ?? front, 16 x ?? rear.

4-1-1871: Tested in Louisville, Ky., against the steamer *Cotton Blossom*. The Ahrens won, but the *Cotton Blossom* purchased because it was built in Louisville. After only 3 fire runs, *Cotton Blossom #7* was replaced in 1872 by Ahrens #20 and scrapped.

8- -1871: In service at Jefferson Fire Co. #5, Cincinnati, Ohio.

10-20-1871: To Storrs Fire Co. #17, and Ahrens #10 of Fire Co. #17 to Fire Co. #5.

7-8-1876: New boiler coils.

7- -1894: Replaced by Ahrens #322. Loaned to Midway, Ky., while their Ahrens (#170) was repaired.

12 2 SSF/HD/W Upright. Round $35^{1}/_{2}$" x 67" coil boiler (iron). 6" x 8" pump, one inlet. One 4" x 20' pre-connect suction. 10" x 8" cylinder. Wheels 16 x ?? front, 16 x ?? rear.

5-8-1871: Tested at Cincinnati, Ohio. Water in 3:57. Threw one stream 276'.

5-17-1871: In service at Washington Fire Co. #2, Madison, Ind.

7-9-1875: Air pump and two more suction inlets installed, entire steamer repainted.

9-7-1875: Joe Terbeck installed new boiler coils at Madison.

9-12-1888: Joe Terbeck installed new boiler coils at Madison.

8-1-1899: Replaced by Metropolitan #2658 and traded-in. Sold to L&M Kronenberger scrapyard, Madison, Ind.

13 2 SSF/HD/W Upright. Round $35^{1}/_{2}$" x 67" coil boiler (iron). One 6" x 8" pump. One 4" x 20' pre-connect suction. One 10" x 8" cylinder. Wheels 16 x ?? front, 16 x ?? rear.

6-15-1871: In service at Reliable Fire Co. #18, Cincinnati, Ohio.

3-29-1877: Replaced by Ahrens #172. To Union Fire Co. #9 in place of Ahrens #34.

2-25-1882: New boiler coils.

4-17-1890: New Ahrens coil boiler.

14 2 SSF/HD/I Upright. Round $35^{1}/_{2}$" x 67" coil boiler (iron & silver). One 6" x 8" pump, signal lamp atop air chamber. One 4" x 20' pre-connect suction. One 10" x 8" cylinder.

7-1-1871: Completed and given its first test.

9-13-1871: Went on display at the Cincinnati Exposition Building.

10-9-1871: Pumped 38 hours at the Great Chicago Fire, with Ahrens #1 and #3.

10-20-1871: Demonstrated at Kansas City, Mo. Water in 4:18. Through 600' of hose, threw one $1^{1}/_{4}$" stream 216'6" horizontally, 100' vertically.

Ahrens #14, a 2nd Class Upright, was built in 1871 as a demonstrator. It was one of three Ahrens steamers used at the famous Great Chicago Fire that Mrs. O'Leary's cow started. Before it returned to Cincinnati from the fire, this Ahrens was demonstrated at Kansas City, Mo. It was finally sold to Cincinnati's Corryville Fire Co. #19, and dubbed the "Lincoln".

(Harold S. Walker collection)

11- -1871: In service at Corryville Fire Co. #19, Cincinnati, Ohio. Name *Lincoln.*

4-18-1876: New boiler coils.

11-5-1877: Replaced by Ahrens #186. To Extra Engine "C".

3- -1879: New Ahrens coil boiler. Loaned to Newark, Ohio, while Ahrens #216 was built. When #216 delivered 7-7-1879, #14 returned to Cincinnati Fire Dept.

C. Ahrens & Co.

24 and 26 Webster St., Cincinnati
Chris Ahrens, Proprietor, 1872-1873

15 2 SSF/HD/W Upright. Round 35 1/2" x 67" coil boiler (iron). One 6" x 8" pump. One 4" x 20' pre-connect suction. One 10" x 8" cylinder. Wheels 16 x ?? front, 16 x ?? rear. 57 1/2" track.

10- -1871: In service, Eng. 8, St. Louis, Mo. Replaced 1865 Lane & Bodley #11.

1895: Replaced by Metropolitan #2337.

16 2 SSF/HD/I Upright. Round 35 1/2" x 67" coil boiler (brass & silver). One 6" x 8" pump. One 4" x 20' pre-connect suction. One 10" x 8" cylinder. 57 1/2" track.

10- -1871: In service, Eng. 11, St. Louis, Mo.

6-10-1884: Replaced by Ahrens #398.

17 2 SSF/HD/I Upright. 35 1/2" x 67" round coil boiler (brass). One 6" x 8" pump. One 4" x 20' pre-connect suction. One 10" x 8" cyl.

12-30-1871: Delivered to Mohawk Fire Co. #12, Cincinnati, Ohio. Name *Miles Greenwood.* Replaced 1855 Latta #10.

5-20-1875: Demonstrated at 5th & Main streets, Cincinnati. Water in 4:15, vs. 3:50 for Ahrens #43 and 4:40 for Ahrens #36.

1-29-1880: New boiler coils.

11-1-1885: Replaced by Ahrens #6. To Cumminsville Fire Co. #20.

1-10-1899: To Co. #12. Replaced Ahrens #6.

3-1-1902: Replaced by Ahrens #280.

6-18-1902: Dismantled at Fire Co. #1.

1982: Working model of this Ahrens steamer donated to Cincinnati Fire Museum, where it is now displayed.

18 2 SSF/HD/I Upright. Round 35 1/2" x 67" coil boiler (iron). One 6" x 8" pump, one inlet. One 4" x 20' pre-connect suction. One 10" x 8" cylinder. Weight 6200#.

2-21-1872: Delivered to Eaton, Ohio. Name *Eaton #1.* Also one Ahrens hose reel, 500' of hose. Price (steamer, reel, & hose) $6000.

2-21-1872: Tested at Eaton. Water in 4:00. Threw one 1 1/8" vertical stream 125' through 1220' of hose; two 7/8" streams 105' up each.

8-28-1873: Had water on a fire in 6:00, vs. 53:00 for Silsby of Richmond, Ind. Pumped through 1000' of hose to the Silsby, faster than the Silsby could pump it on the fire.

2- -1875: Block of stores caught fire at Eaton, potential $100,000 damage. Ahrens had fire under control in 4:00 and extinguished in 10:00.

4-16-1878: Joe Terbeck installed new boiler coils at Eaton.

4-11-1884: Rebuilt at factory. Ahrens #272 loaned to Eaton while repairs were made.

8-21-1885: Joe Terbeck and G. Fred Ahrens installed new boiler coils at Eaton.

2-27-1888: Pumped 60 hours non-stop at fire.

12-3-1890: Joe Terbeck installed new boiler coils at Eaton.

19 2 SSF/HD/I Upright. Round 35 1/2" x 67" coil boiler (brass). One 6" x 8" pump, three inlets. One 4" x 20' pre-connect suction. One 10" x 8" cylinder. 57 1/2" track.

4-27-1872: Tested & accepted by Engine 9, St. Louis, Mo. Name *Mound.*

12-1-1875: Replaced by Ahrens #154. To Engine 12.

10-13-1880: Replaced by Ahrens #330.

20 2 SSF/HD/I Upright. Round 35 1/2" x 67" coil boiler (brass). One 6" x 8" pump, three inlets. One 4" x 20' pre-connect suction. One 10" x 8" cylinder. 59" track.

6-14-1872: Tested & accepted by Louisville, Ky. Name *George W. Levi #7.* Replaced their *Cotton Blossom* steamer (see Ahrens #11).

6-8-1874: Tested at 6th & Lexington, Louisville. Water in 3:45 through 50' of hose with 1 1/4" nozzle.

9-30-1875: New boiler coils and rear springs.

6-30-1891: Replaced by Ahrens #646.

21 2 SSF/HD/I Upright. $35^{1}/_{2}$" x 67" round coil boiler (brass). 6" x 8" pump. 4" x 20' pre-connect suction. 10" x 8" cyl. $57^{1}/_{2}$" track.

7-30-1872: Tested at Lima, Ohio, against a Silsby rotary steamer, both through 300' of hose with $1^{1}/_{8}$" nozzle. Ahrens threw one stream 214' in 4:14, vs. Silsby 187' in 9:22.

8-3-1872: Tested at Chicago, Ill., against Silsby rotary, Amoskeag, and Button. Each pumped through 250' of hose with $1^{1}/_{4}$" nozzle, but only the Ahrens started with a cold boiler, other three boilers were already hot at the start of the test. Nevertheless, the Ahrens got water in fastest time of the four. Ahrens threw one stream 271'6", vs. 222' for the Amoskeag. The Silsby and Button were so badly beaten that it was not considered worthwhile to measure their streams.

8-7-1872: Tested & accepted by Saint Louis, Mo. Name *New Franklin #13*. Price $5000. Steam in 2:58. Threw a $1^{1}/_{4}$" stream 280' through 50' of hose; two streams 250' each.

9-22-1882: New boiler, front & rear springs, and front axle installed by Fostoria Foundry & Machine Works, Fostoria, Ohio.

1904: Replaced by Metropolitan #2880.

22 2 SSF/HD/I Upright. Round $35^{1}/_{2}$" x 67" coil boiler (brass). One 6" x 8" pump, three inlets. One 4" x 20' pre-connect suction. One 10" x 8" cylinder. 59" track.

9-13-1872: Tested & accepted by Engine 8, Louisville, Ky. Name *James Leach.*

6-8-1874: Tested at 13th & Broadway, Louisville. Water in 3:48 through 50' of hose with $1^{1}/_{4}$" nozzle.

1-15-1876: New boiler coils.

9-22-1884: New rear springs.

8-17-1896: New rear springs.

2-13-1888: Replaced by Ahrens #467.

23 2 SSF/HD/I Upright. Round $35^{1}/_{2}$" x 67" coil boiler (brass). One 6" x 8" pump, three inlets. One 4" x 20' pre-connect suction. One 10" x 8" cylinder. $57^{1}/_{2}$" track.

10-26-1872: Tested & accepted by Engine 10, St. Louis, Mo.

5-4-1876: Replaced by Ahrens #162. To Engine 14.

1900: Replaced by Metropolitan #2747.

24 2 SSF/HD/I Upright. Round $35^{1}/_{2}$" x 67" coil boiler (brass). One 6" x 8" pump, three inlets. One 4" x 20' pre-connect suction. One 10" x 8" cylinder. 59" track.

12-2-1872: Tested & accepted by Engine 9, Louisville, Ky. Name *J.A. Koach.*

1-15-1876: New boiler coils.

2-23-1888: Replaced by Ahrens #535.

1907: Replaced by new Continental steamer. To Engine 11.

1915: To Spare at Engine 2.

25 2 SSF/HD/S Upright. Round $35^{1}/_{2}$" x 67" coil boiler (iron). One 6" x 8" pump, one inlet. One 4" x 20' pre-connect suction. One 10" x 8" cylinder. Wheels 16 x ?? front, 16 x ?? rear. Weight 5500#.

1-17-1873: Tested & accepted by Galion, Ohio. Also one Ahrens hose reel with 500' of hose. Water in 4:00. Threw one $1^{1}/_{8}$" stream 243'; one stream 225' through 600' of hose.

3-13-1878: Joe Terbeck installed new boiler coils at Galion.

8-17-1891: Joe Terbeck installed new boiler coils at Galion.

9- -1902: Sent to American Fire Engine Co., Seneca Falls, N.Y., for repairs. Sat at Seneca Falls (and later American-LaFrance, Elmira, N.Y.) for 3 years because Galion couldn't decide whether to repair or replace it.

10-16-1905: Ahrens Fire Engine Co. sold Continental steamer #106 to Galion, Ohio (delivered in 1906). At Galion's request, American-LaFrance turned over the 1873 Ahrens to Ahrens F.E. Co. as a trade-in.

26 2 SSF/HD/S Upright. $35^{1}/_{2}$" x 67" round coil boiler (iron). One 6" x 8" pump. One 4" x 20' pre-connect suction. One 10" x 8" cylinder.

3-15-1873: Tested & accepted by Fire Co. #1, Owensboro, Ky. Name *Ben Bransford*. Also one Ahrens hose reel with 1000' of hose.

12-21-1874: At a fire, had water in 5:00.

6-12-1876: Joe Terbeck installed new boiler coils at Owensboro.

8-3-1887: New boiler coils.

2-3-1888: Still in service at Owensboro.

1900: A photo of men and apparatus shows that Owensboro then used only chemical engines, no longer had any steamers.

27 2 SSF/HD/I Click. Square 42" x ?? coil boiler. One $5^{1}/_{2}$" x 14" pump. One 4" x 20' pre-connect suction. One 10" x 14" cylinder.

4-18-1873: Tested & accepted by Union Fire Co. #9, Cincinnati, Ohio.

Even today, few would wish to tangle with this tough-looking bunch of firemen from Fire Co. #15, St. Louis, Mo. They are posing in front of an 1873 Ahrens 2nd Class Click steamer, Reg. #28. Sadly, this iron-wheeled gem was scrapped in 1898.
(Harold S. Walker collection)

3-16-1874: Replaced by Ahrens #34. To Fire Co. #3 in place of 1854 Latta #3.

3-27-1877: New boiler coils.

12-15-1879: Replaced by Ahrens #228.

28 2 SSF/HD/I Click. Round 42" x ?? coil boiler (brass). One 6" x 14" pump, signal lamp atop pump. One 4" x 20' pre-connect suction. One $10^{1}/_{2}$" x 14" cylinder.

6-19-1873: Delivered by Chris Ahrens to St. Louis, Mo. Threw one $1^{1}/_{4}$" stream 258' horizontally, 190' vertically; two streams 211' each horizontally, 180' each vertically.

6-24-1873: Tested & accepted by Engine 15, St. Louis, Mo.

1898: Replaced by Metropolitan #2585

C. Ahrens & Co.

22, 24, and 26 Webster St., Cincinnati, 1873-1877
Chris Ahrens and G. Fred Hawekotte, Proprietors

29 2 SSF/HD/I Click. Round 42" x ?? coil boiler (brass). 6" x 14" pump. 4" x 20' pre-connect suction. $10^{1}/_{2}$" x 14" cylinder.

8-9-1873: Tested & accepted by Engine 16, St. Louis, Mo.

1906: Replaced by Metropolitan #3148.

30 2 SSF/HD/I Click. 42" x ?? round coil boiler (brass). One 6" x 14" pump. One 4" x 20' pre-connect suction. One $10^{1}/_{2}$" x 14" cylinder.

10-1-1873: Tested & accepted by Engine 17, St. Louis, Mo. Delivered by Fred Hawekotte.

1906: Replaced by Metropolitan #3149.

31 2 SSF/HD/S Upright. $35^{1}/_{2}$" x 67" round coil boiler (iron). One 6" x 8" pump, three inlets. One 4" x 20' pre-connect suction. One 10" x 8" cylinder. $56^{1}/_{2}$" track.

10-30-1873: Demonstrated at Indianapolis, Ind. Threw one stream 220' through 750' of hose, vs. 196'6" for a new Silsby and 195' for a new Jeffers. Competing with an old Silsby, the Ahrens had water in 6:23 vs. 9:10 for old Silsby; threw one $1^{1}/_{16}$" stream 272'10" vs. 247'10" for old Silsby. With hose length reduced to 50', Ahrens threw one $1^{1}/_{16}$" stream 285', vs. 234' for old Silsby.

11- -1873: 4" x 20' pre-connect suction replaced with two 4" x 10' suctions. Delivered by Chris Ahrens to Chicago, Ill.

11-10-1873: Tested & accepted by Engine 25,

Make	Length of hose, feet.	Highest pressure steam.	Highest pressure water.	Size of Nozzle, inches.	Number of streams	Distance thrown, feet.
Amoskeag	500	100	145	$1^1/4$	1	178
Amoskeag	300	115	110	$1^1/8$	2	197
Amoskeag	300	106	130	$1^1/2$	1	154.6
Silsby	500	110	147	$1^1/4$	1	197
Silsby	300	107	140	$1^1/8$	2*	154.4
Silsby	300	90	110	$1^1/2$	1	127.3
Gould	500	115	170	$1^1/4$	1	209.7
Gould	300			$1^1/8$	2	@
Gould	300	130	175	$1^1/2$	1	213.6
Ahrens	500	130	163	$1^1/4$	1	217
Ahrens	300	124	140	$1^1/8$	2	169
Ahrens	300	110	135	$1^1/2$	1	172.5

* The Silsby broke down on the double stream, distance shown is for a single stream.

@ The Gould did not attempt pumping two streams.

CHART #1

Results of Steam Fire Engine Competition
Chicago, Ill. Sept. 5, 1874

Chicago, Ill. Through 250' of hose, water in 5:43, vs. 11:07 for Silsby; threw one $1^1/_2$" stream 277', vs. Silsby's 1" stream 229'9".

9-5-1874: Competed against an 1872 2nd Class Amoskeag named *William James #3*, an 1868 2nd Class Silsby named *Webansia #2* (Waubansia?), and a new Gould steamer. Results are shown in Chart #1.

10-6-1874: Seven-hour test of Chicago steamers and water system (hydrants). Results are shown in Chart #2. Engine 25 was Ahrens. Engines 9, 10, 11, 12, and 14 were Amoskeags. Engines 1, 2, 5, 18, and 27 were Silsbys. Engines 21 and 22 were of unknown make (possibly Gould?). Retired by 1906.

32 2 SSF/HD/S Upright. $35^1/_2$" x 67" round coil boiler (brass). One 6" x 8" pump, signal lamp atop brass air chamber, two brass vacuum chambers with coach lamps atop, all steam pipes of brass. One 4" x 20' pre-connect suction. One 10" x 8" cylinder (brass casing). Wheels 16 x ?? front, 16 x ?? rear. Double seat, open-railing style.

12-4-1873: Delivered by Chris Ahrens and Fred Hawekotte. Tested and accepted by Hamilton, Ohio. Name *Neptune #1.*

6-30-1881: Joe Terbeck installed new boiler coils at Hamilton.

8-20-1985: Now in storage (not on display) at Hamilton Museum. Complete but unrestored. Boiler tubes reportedly rusted out (unconfirmed), and rig needs cleanup, but otherwise in excellent original condition. Last exhibited in July 4, 1976, parade at Hamilton. This steamer may soon be restored to operating condition.

33 2 SSF/HD/I Click. 42" x ?? round coil boiler (brass). 6" x 14" pump. $4^1/_2$" x 20' pre-connect suction. $10^1/_2$" x 14" cyl. Weight 7575#.

1-1-1874: Chris Ahrens & Fred Hawekotte delivered to Eng. 10, Louisville, Ky. ($5500). Name *John M. Letterle.* Water in 3:48. Threw a $1^3/_4$" stream 261'6", $1^3/_8$" stream 268'.

5-29-1884: Replaced by Ahrens #400.

Test Number	Engine Company	Water Pressure	Steam Pressure	Comments
1 (45 minutes)	2	125	--	
	5	125	--	
	9	125	--	
	10	140	--	
	11	150	--	
	12	50	--	
	14	40	--	
	18	--	--	Fire Box out of order
	21	125	--	
	22	60	--	
	25	200	--	
	27	--	--	Valve out of order
2 (15 minutes)	1	115	--	
	5	110	--	
	9	100	--	
	10	120	--	
	11	140	--	
	12	60	--	
	14	90	--	
	18	--	--	Disabled
	21	80	--	Broke down
	22	70	--	
	25	222	--	Best performance of the day
	27	70	--	Broke down
3 (90 minutes)	5	90	73	
	9	40	100	Stream thrown only 40'
	10	50	75	Pressure dropped to 20#
	11	95	75	
	12	70	100	
	14	70	85	
	21	60	65	
	22	69	80	
	25	110	60	
	27	30	65	Stream thrown only 65'

CHART #2

Results of Steam Fire Engine Test
Chicago, Ill. Oct. 6, 1874

(Continued on next page)

34 2 SSF/HD/I Click. 42" x ?? round coil boiler (brass). One 6½" x 14" pump. One 4½" x 20'6" pre-connect suction. One 10½" x 14" cylinder. Last Click built.

3-16-1874: Tested & accepted by Union Fire Co. #9, Cincinnati, Ohio. Replaced Ahrens #27. Threw one 2¼" stream 243'.

3-29-1877: Replaced by Ahrens #13. To Relief Fire Co. #2.

8- -1877: New 4" x 20'6" suction.

5-26-1881: Sold to Crystal City Plate Glass Works, Crystal City, Mo.

35 2 SSF/HD/I Upright. 35½" x 67" round coil boiler (brass). One 6" x 8" pump. One 4" x 20' pre-connect suction. One 10" x 8" cyl.

4-27-1874: Delivered by Fred Hawekotte to Quincy, Ill. Tested & accepted by Liberty

Test Number	Engine Company	Water Pressure	Steam Pressure	Comments
4	5	40	15	
	9	80	55	
	10	80	100	
	11	90	100	
	12	70	100	
	14	90	100	
	18	40	50	Broke down
	21	45	90	
	22	35	65	
	25	100	100	
	27	80	50	
5	5	--	--	Broke down
	11	145	85	
	14	105	95	
	25	160	100	

CHART #2 (Continued from previous page)

Results of Steam Fire Engine Test
Chicago, Ill. Oct. 6, 1874

Fire Co. #3. Name *Al Bebee.* Water in 4:45 (fastest time of Quincy's five steamers). Through 100' of hose, threw a $1^1/_4$" stream 255'6". Through two 100' hose lines with $1^1/_4$" nozzles, threw one stream 196'6" and second stream 198'7".

8-18-1887: New boiler coils.

36 2 SSF/HD/S Upright. $35^1/_2$" x 67" round coil boiler (brass). $6^1/_4$" x 8" pump, air pump. 4" x 20' pre-connect suction. 10" x 8" cylinder.

5-30-1874: Tested & accepted by Jefferson Fire Co. #5, Cincinnati, Ohio. Name *C.W. Johnston.* Replaced Ahrens #10. Price $4800. Water in 3:58. Threw one $1^1/_4$" stream 274', two 1" streams 220' each. Original engineer on this steamer, 1874-1875, was J. Peter Ahrens, brother of the builder.

5-20-1875: Demonstrated at 5th & Main, Cincinnati, Ohio. Water in 4:40, vs. 4:15 for Ahrens #17 and 3:50 for Ahrens #43.

3-18-1884: Boiler ruptured at Shomaker Lumber & Sash Factory fire, West 9th St.

4-29-1884: New boiler coils.

6-21-1884: New boiler coils. Became a Spare.

37 2 SSF/HD/S Upright. $35^1/_2$" x 67" round coil boiler (brass). One 6" x 8" pump. One 4" x 20' pre-connect suction. One 10" x 8" cylinder. Weight 6730# with water and fuel.

7-8-1874: Delivered by Chris Ahrens and Fred Hawekotte to Indianapolis, Ind. Tested & accepted by Fire Co. #5. Name *Suaaks.* Through 500' of hose, threw one stream 247' in 4:30. Through 100' of hose, threw one $1^1/_4$" stream 286'9". Through two 100' hose lines, threw two $^7/_8$" streams 226' each.

7-30-1874: On test at Indianapolis, through 100' of hose, had water in 5:10. Threw one $1^1/_4$" stream 271'.

6-17-1879: Joe Terbeck installed new boiler coils at Indianapolis.

2-29-1892: Replaced by Ahrens 670. Junked.

38 2 SSF/HD/S Upright. $35^1/_2$" x 67" round coil boiler (brass). One 6" x 8" pump, two inlets. One 4" x 20' pre-connect suction. One 10" x 8" cyl. Wheels 16 x ?? front, 16 x ?? rear.

9-17-1874: Delivered to Perry & Dumont, Nashville agents of C. Ahrens & Co., for the Exposition in Nashville, Tenn. Water in 3:30.

12-12-1874: Tested & accepted by Fire Co. #2, Nashville, Tenn. Water in 4:15. Threw one $1^3/_{16}$" stream 259'6", two $^7/_8$" streams 221'6" each.

The oldest Ahrens steamer so far found surviving, and the only known survivor built by C. Ahrens & Co., is this 1873 2nd Class Upright #32, now in storage at the Hamilton Museum, Hamilton, Ohio. Above: Remarkably-original condition includes 20-foot squirrel-tail suction hose, brass-cased boiler, and primitive tubular seat. Below left: Close-up of pump shows painted flywheel and pump-support columns (later Ahrens engines had these plated). Note the two red-lensed coach lamps atop the twin vacuum chambers. Below right: The etched glass whale-oil lamp atop the air chamber bears the fire company name "Neptune 1" and depicts King Neptune with his trident, rising out of the sea.

(Photos by Jim Van Lieu)

The "Bash Wilson #1" of Greensburg, Ind. was this 1874 Ahrens 2nd Class single-pump, straight-frame Upright #39. The deer antlers, attached to the signal lamp atop the air chamber, signified Greensburg's victory over another steamer in a competition. A steamer was called a "virgin" until it lost a contest, and the antlers were worn by the steamer that had thus "deflowered" a virgin engine. This Ahrens was junked in 1921.

(Courtesy Captain Don Minning, Greensburg Fire Dept.)

10-2-1895: New boiler coils.

39 2 SSF/HD/A Upright. Round $35^1/_2$" x 67" coil boiler (iron). One 6" x 8" pump, signal lamp atop air chamber. One 4" x 20' pre-connect suction. One 10" x 8" cylinder. Wheels 16 x ?? front, 16 x ?? rear.

10-20-1874: Delivered by Fred Hawekotte to Greensburg, Ind. Tested and accepted by Greensburg at Shelbyville, Ind. (this test resulted in Shelbyville ordering Ahrens #42). Also one Ahrens hose reel and 1000' of hose. Price (steamer, reel & hose) $6000.

10-23-1874: Bash Wilson Fire Co. #1 organized at Greensburg to operate this Ahrens.

5-8-1882: Joe Terbeck installed new boiler coils at Greensburg.

8-1-1887: Joe Terbeck installed new boiler coils at Greensburg.

8-2-1893: George Zink installed new boiler coils at Greensburg.

1921: Retired. Probably junked.

40 2 SSF/HD/S Upright. $35^1/_2$" x 67" round coil boiler (iron). One 6" x 8" pump, three inlets, air pump. One 4" x 20' pre-connect suction. One 10" x 8" cylinder.

12-10-1874: Tested & accepted by Fire Co. #3, Memphis, Tenn. Water in 4:00. Drafted at 21' lift. Threw one $1^1/_4$ stream 263' horizontal, 175' vertically. Threw one $1^3/_{16}$" stream 267', two $^7/_8$" streams 234' each.

4-18-1878: New rear wheels.

12-8-1882: New boiler coils.

1-4-1888: New wheels, 16 x 52 front, 18 x 58 rear.

12- -1900: New rear wheels.

1906: Nott boiler.

41 1 SSF/HD/A Upright. Round 38" x 67" coil boiler (nickel). One $6^5/_8$" x 8" pump, two inlets, air pump. One $4^1/_2$" x 21' pre-connect suction. One 11" x 8" cylinder. Wheels 16 x ?? front, 16 x ?? rear. Weight 7300#.

11-7-1874: Tested at Cincinnati, Ohio. Water in 3:32. Through 500' of hose, threw one $1^1/_2$" stream 309'. Threw one $1^1/_2$" stream 273', two $1^1/_8$" streams 268' each.

11-9-1874: Tested at Oxford, Ohio. Steam in 3:15. Threw one stream 200' in 5:00. This test resulted in Oxford ordering Ahrens #45.

11-10-1874: Delivered by Chris Ahrens & Fred Hawekotte to Indianapolis, Ind. (for Engine 6?) Name *Thomas D. Kingan.* Water in 4:53, vs. 5:17 for Amoskeag and 7:02 for Clapp & Jones. Through 500' of hose, threw one $1^{3}/_{8}$" stream 268'. Threw one $1^{1}/_{2}$" stream 273', vs. 257' for Amoskeag and 232' for Clapp & Jones.

11-12-1874: Tested at Indianapolis. Through 100' of hose, threw one $1^{3}/_{8}$" stream 100' in 3:52. Threw one $1^{1}/_{2}$" stream 299'$9^{1}/_{2}$". Threw two $1^{1}/_{8}$" streams, one through 100' of hose to a 165' height, the other through 500' of hose but distance not measured.

7-21-1875: Tested at Indianapolis. Water in 3:40. Through 100' of hose, threw one $1^{3}/_{8}$" stream 302'.

8-27-1878: Rebuilt at factory. New boiler coils, pump valves, front and rear springs. Painted and varnished springs and axles.

2-8-1888: Traded-in on Ahrens #530. Ahrens installed new boiler, pump, springs, wheels, and horse hitch, preliminary to reselling it.

10-18-1888: Shipped to Engine 8, Memphis, Tenn. Name *James Lee, Jr.*

12-31-1894: New $4^{1}/_{2}$" x 20'6" suction.

42 2 SSF/MD-HD/A Upright. Round coil boiler $35^{1}/_{2}$" x 67" (iron). One 6" x 8" pump, one inlet, air pump. One 4" x 20' pre-connect suction. One 10" x 8" cylinder.

1-22-1875: Delivered by Fred Hawekotte to Fire Co. #3, Shelbyville, Ind. Water in 4:10. Threw a $1^{1}/_{16}$" stream 259', two $^{7}/_{8}$" streams 210' each. Through 700' of hose, threw one $1^{1}/_{4}$" stream 248'6". Also one manually-drawn Ahrens hose reel and 1000' of hose.

10-12-1878: New boiler coils.

12-2-1885: Rebuilt by Ahrens in Cincinnati.

5-7-1928: Supplemented by motor Ahrens-Fox Model N-S-4 #3333. The steamer had previously been Shelbyville's only pumper.

1936: This steamer was retired and junked.

1940: The Ahrens hose reel bought with this steamer was junked. Since 1935, the reel had been towed to fires by an automobile.

43 2 SSF/HD/A Upright. Round coil boiler $35^{1}/_{2}$" x 67" (iron & nickel). 6" x 8" pump, three inlets, air pump, signal lamp atop air chamber. 4" x 20' pre-connect suction, brass suction brackets. 10" x 8" cylinder (nickel).

5-17-1875: Tested and accepted by Wooden Shoe Fire Co. #7, Webster & Sycamore, Cincinnati, Ohio. Threw one $1^{3}/_{16}$" stream 287'. Replaced Ahrens #10. The #7 fire-

Ahrens #43, an 1875 2nd Class Upright, is seen in front of Wooden Shoe Fire Co. #7, 17 Webster St., Cincinnati, Ohio. The man at far left, leaning against the boiler, is believed to be Peter Ahrens, younger brother of Chris Ahrens. Peter Ahrens was stoker on this engine from 1875 to 1877, and engineer from 1883 to 1896.
(Courtesy New York Fire Dept. Library)

house was directly across the street from the Ahrens factory. Stoker on this steamer from 1875 to 1877, and engineer from 1883 to 1896, was Peter Ahrens, younger brother of the manufacturer Chris Ahrens.

5-20-1875: Demonstrated at 5th & Main Streets, Cincinnati. Water in 3:50, vs. 4:15 for Ahrens #17 and 4:40 for Ahrens #36.

5-8-1878: Valves destroyed at $75,000 Henshaw Furniture Factory fire, Elm at Canal.

5-17-1879: Experimental set of bronze pump valves installed.

1880: New Engineer-of-Steamer on this rig (Fire Co. #7) was Charles H. Fox, who in 1892 married Chris Ahrens' daughter Lillie.

10-29-1884: Experimental set of B.F. Goodrich rubber pump valves installed. Since the firehouse was across the street from the factory, Ahrens logically used this steamer for field-testing experimental new parts.

7-27-1895: New boiler coils.

1896: New Engineer on this steamer, John P. Ahrens, youngest son of Chris Ahrens and nephew of the engineer (J. Peter Ahrens) he replaced. He was the third relative of the builder to have charge of this steamer.

44 1 SSF/HD/I Upright. Round 38" x 67" coil boiler (nickel). One $6^{5}/_{8}$" x 8" pump, three inlets. One $4^{1}/_{2}$" x 20' pre-connect suction. One 11" x 8" cylinder. Wheels 16 x ?? front, 16 x ?? rear. Springs steel front, rubber rear. $57^{1}/_{2}$" track. Weight 7695#.

4-5-1875: Delivered by Chris Ahrens to St. Louis, Mo. First of an order of seven 1st Class Uprights for St. Louis. Tested & accepted by Engine Co. #18, St. Louis. Name *Future City*. Water in 3:30. Threw one $1^{1}/_{2}$" stream 296', two $1^{1}/_{8}$" streams 252' each.

1901: Replaced by Metropolitan #2799.

45 3 SSF/MD-HD/W Upright. Round 32" x 62" coil boiler (iron). One $5^{1}/_{4}$" x 8" pump, one inlet, automatic air pump. One $3^{1}/_{2}$" x 20' pre-connect suction. One 9" x 8" cylinder. Wheels 16 x 56 front, 16 x 62 rear. Ratchet brake. Springs steel front, gum rear. Driver's seat. Three lanterns, two coach lamps.

4-24-1875: Tested at Webster & Sycamore, Cincinnati. Through 100' of hose, threw one 1" stream 200' in 4:49, one $1^{1}/_{8}$" stream 249'.

5-13-1875: Delivered by Fred Hawekotte to Oxford, Ohio. Name *Pap Ringwood #1*. Water in 3:40. Through 100' of hose, threw one $1^{1}/_{8}$" stream 238', one 1" stream 188'.

6-14-1882: New Ahrens coil boiler.

10-12-1890: George Zink installed new boiler

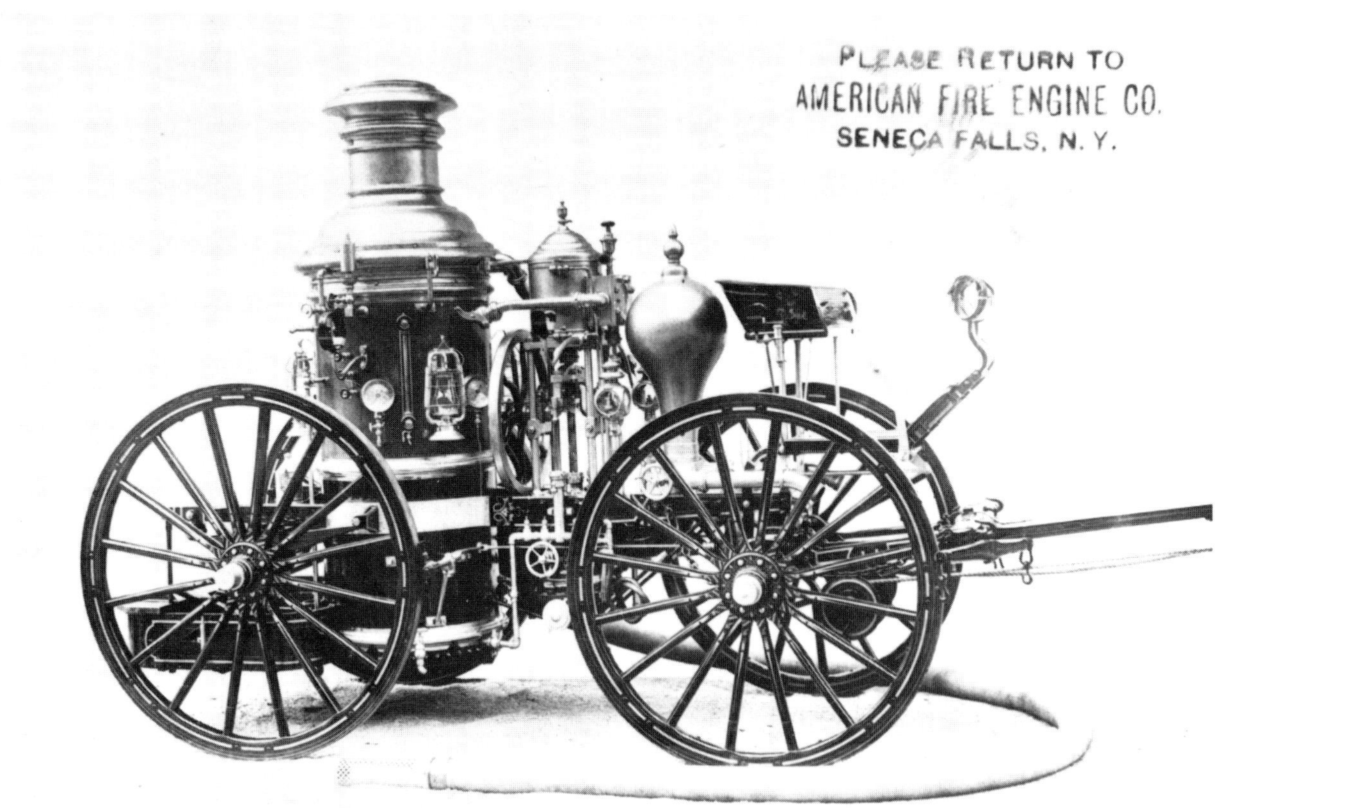

This 1875 Ahrens 3rd Class single-pump, straight-frame Upright #45 was named "Pap Ringwood #1" and served Oxford, Ohio, for 20 years. In this 1898 ad, it had just been rebuilt with a Clapp & Jones boiler, and was offered for sale. According to Ahrens records, it was then sold to Cochran, Indiana (near Aurora).

(Courtesy John Peckham)

coils at Oxford.

7-11-1895: Traded-in on 5th Class #2386.

6-18-1898: New 30" x 62" Clapp & Jones boiler. Sold to Cochran, Ind. Threw one $1^{1}/_{4}$" stream 250'; one $1^{1}/_{8}$" stream 260'; two $^{7}/_{8}$" streams 200' each. Through 1000' of hose, threw one 1" stream 290'. [NOTE: As of 1984, U.S. Postal Service says there is no such town as Cochran, Ind. Old atlas shows that Cochran was just west of Aurora, Ind.]

At this point, Ahrens serial numbers inexplicably jump by 100 numbers. A "1" was added to the front of what would have been No. 46, to make it No. 146. At first, the author thought that the change in numbering coincided with the incorporation of Ahrens Mfg. Co., but it now appears that the corporate name change did not occur until 1877, two years after the changed numbering system. So the reason for padding the numbers by 100 remains a mystery. Perhaps it was done so that the higher serial numbers would make customers think that Ahrens had built more steamers than they actually had.

146 3 SSF/MD/W Upright. Round 32" x 62" coil boiler (iron). One $5^{1}/_{4}$" x 8" pump, piston air pump. One $3^{1}/_{2}$" x 20' pre-connect suction. One 9" x 8" cylinder. Wheels 16 x 56 front, 16 x 62 rear. Springs steel front, gum rear. Driver's seat. Weight 5100#.

6-10-1875: Tested at Nashville, Tenn. Water in 3:30. Threw one $1^{1}/_{16}$" stream 219', two $^{7}/_{8}$" streams 188' each.

6-14-1875: Delivered by Fred Hawekotte to Lookout Fire Co. #1, Chattanooga, Tenn. Name *Harry Wilcox*. Water in 3:18. Through 50' of hose, threw one $1^{1}/_{8}$" stream 270'. Threw four $^{3}/_{4}$" streams 120' each. Two $^{3}/_{4}$" streams 128' each through two 450' hose lines, third stream 175' through 50' of hose.

6-16-1875: Test at Chattanooga. Steam 2:47.

11-10-1888: Rebuilt as #570. New boiler coils, boiler casing, safety rails on rear of boiler, floorboard of coal bunker, and heater. New brass frame casing around boiler. Pump overhauled. New gauges, flywheel, and base for air chamber (complete with shut-off discharge gates). New signal lamp atop air chamber. New steam cylinder. New rear axle, front and rear springs, and 18 x 58 rear wheels. Gong repaired. New drop pole, driver's foot board, and seat bottom. New whip socket and torch, three new lanterns. Surviving records don't say why this extensive rebuilding was needed, but the nature of the repairs suggests that the steamer had been consumed by flames.

147 3 SSF/HD/I Upright. Round 32" x 62" coil boiler (nickel). One $5^{1}/_{4}$" x 8" pump. One 4" x 20' pre-connect suction. One 9" x 8" cylinder. Springs steel front, gum rear. $57^{1}/_{2}$" track. Weight 5575#.

6-25-1875: Delivered by Chris Ahrens to Engine 1, St. Louis, Mo. Replaced 1856 Latta #14. Steam in 2:40. Threw one $1^{1}/_{8}$" stream 259', two $^{7}/_{8}$" streams 215' each.

1909: Was a Spare at Engine 2, and Engine 1 was using 1894 Ahrens #2285.

148 1 SSF/HD/I Upright. 38" x $67^{1}/_{2}$" round coil boiler (nickel). One $6^{5}/_{8}$" x 8" pump, three inlets. One 4" x 20' pre-connect suction. One 11" x 8" cylinder. Springs steel front, rubber rear. $57^{1}/_{2}$" track. Weight 7000#.

7-30-1875: Delivered by Chris Ahrens to Engine 6, St. Louis, Mo. Price $5250. Threw one $1^{1}/_{2}$" stream 285', two $1^{1}/_{8}$" streams 225' each. Replaced 1857 Latta #18.

5- -1883: New pump.

1909: Replaced by 1898 Metropolitan #2584.

149 1 SSF/HD/I Upright. 38" x $67^{1}/_{2}$" round coil boiler (nickel). One $6^{5}/_{8}$" x 8" pump, three inlets. One 4" x 20' pre-connect suction. One 11" x 8" cylinder. Springs steel front, rubber rear. $57^{1}/_{2}$" track. Weight 7000#.

8-20-1875: Delivered by Fred Hawekotte to Engine 4, St. Louis, Mo. Threw one $1^{3}/_{8}$" stream 292', two $1^{1}/_{8}$" streams 275' each. Replaced 1857 Latta #16.

7-12-1887: 38" x $67^{1}/_{2}$" Ahrens coil boiler.

2-16-1899: Sold to Lafayette & Brothers, Checotah, Indian Territory (Oklahoma).

3-8-1899: Boiler repaired at factory, $128.30.

8-3-1903: New 38" x $67^{1}/_{2}$" Ahrens coil boiler, $325.

150 1 SSF/HD/I Upright. 38" x $67^{1}/_{2}$" round coil boiler (nickel). One $6^{5}/_{8}$" x 8" pump, three inlets. One 4" x 20' pre-connect suction. One 11" x 8" cylinder. Springs steel front, rubber rear. $57^{1}/_{2}$" track. Weight 7000#.

9- -1875: Tested at Cincinnati, Ohio. Threw one $1^{5}/_{16}$" stream 312', one $1^{3}/_{4}$" stream 244'.

10-1-1875: Delivered by Fred Hawekotte to Engine 7, St. Louis, Mo. Accepted without a test. Replaced 1858 Latta #19.

1902: Replaced by Metropolitan #2868.

151 3 SSF/HD/W Upright. Round 32" x 62" coil boiler (nickel). $5^{1}/_{4}$" x 8" pump, piston air pump. $3^{1}/_{2}$" x 20' pre-connect suction. 9" x 8" cylinder. Wheels 16 x 56 front, 16 x 62 rear.

10- -1875: Tested at the Exposition in Cincinnati. Through 1000' of hose, threw one $1^{1}/_{8}$" stream 210'. Through three 100' hose lines, threw two $^{7}/_{8}$" streams and one $1^{1}/_{8}$" stream, distance not measured.

10-25-1875: Delivered by Chris Ahrens to Hillsboro, Ohio.

7-12-1882: New boiler coils.

2-5-1890: New boiler coils.

7-16-1891: New Ahrens coil boiler.

152 1 SSF/HD/I Upright. 38" x $67^{1}/_{2}$" round coil boiler (nickel). One $6^{5}/_{8}$" x 8" pump, three inlets. One 4" x 20' pre-connect suction. One 11" x 8" cylinder. Springs steel front, rubber rear. $57^{1}/_{2}$" track. Weight 7000#.

11-8-1875: Delivered by Fred Hawekotte to Engine 3, St. Louis, Mo. Accepted without a test. Replaced 1857 Latta #15.

1909: No longer on St. Louis F.D. roster. Engine 3 was using 1884 Ahrens #399.

153 3 SSF/HD/W Upright. Round 32" x 62" coil boiler (iron & silver). $5^{1}/_{4}$" x 8" pump, three inlets. $3^{1}/_{2}$" x 20' pre-connect suction. 9" x 8" cylinder. Wheels 16 x 56 front, 16 x 62 rear.

11-16-1875: Delivered by Chris Ahrens to Waverly, Ohio.

12-13-1878: At a fire, had water in 4:30.

9-16-1882: Joe Terbeck installed new boiler coils at Waverly.

5-11-1886: Joe Terbeck installed new boiler coils at Waverly.

1-11-1896: New boiler coils.

4-13-1899: New boiler coils.

8-29-1910: New rubber pump valves.

154 1 SSF/HD/I Upright. 38" x $67^{1}/_{2}$" round coil boiler (nickel). One $6^{5}/_{8}$" x 8" pump, three inlets. One 4" x 20' pre-connect suction. One 11" x 8" cylinder. Springs steel front, rubber rear. $57^{1}/_{2}$" track. Weight 7000#.

11-22-1875: Shipped to St. Louis, Mo.

12-1-1875: Accepted by Engine 9, St. Louis, Mo., without a test. Replaced Ahrens #19.

2-7-1885: New front and rear wheels.

7- -1885: New Ahrens coil boiler. Entire steamer repainted. General overhaul.

1909: Replaced by 1891 Ahrens #651.

155 1 SSF/HD/W Upright. Round 38" x 67" coil boiler (nickel). One $6^{5}/_{8}$" x 8" pump, piston air pump. One $4^{1}/_{2}$" x 20' pre-connect suction. One 11" x 8" cylinder. Wheels 16 x 51 front, 16 x 58 rear. Steel spiral springs.

1-7-1876: Delivered by Chris Ahrens and Fred Hawekotte to Eng. 7, Indianapolis, Ind.

2-8-1888: Traded-in on Ahrens #530. New boiler, wheels, springs, and $4^{1}/_{2}$" x 21' suction installed at factory.

10-17-1888: Shipped to Memphis, Tenn. Name *John E. Randle #7*.

1899: New boiler by LaFrance Fire Engine Co., Elmira, N.Y.

156 1 SSF/HD/I Upright. 38" x $67^{1}/_{2}$" round coil boiler (brass). One $6^{5}/_{8}$" x 8" pump. One 4" x 20' pre-connect suction. One 11" x 8" cylinder. Steel springs. $57^{1}/_{2}$" track.

2-7-1876: Delivered by Chris Ahrens to St. Louis, Mo. Last of order of seven 1st Class Uprights for St. Louis.

2-8-1876: Tested & accepted by Engine 5, St. Louis. Replaced 1856 Shawk *Missouri #5*.

1904: Replaced by Metropolitan #2961.

157 3 SSF/HD/W Upright. Round 32" x 62" coil boiler (iron & silver). One $5^{1}/_{4}$" x 8" pump. One $3^{1}/_{2}$" x 20' pre-connect suction. One 9" x 8" cylinder. Wheels 16 x 56 front, 16 x 62 rear. Gum springs. Drop pole.

12-17-1875: Delivered by Fred Hawekotte to Leavenworth, Kansas. Name *Kansas #3*. Through 100' of hose, threw one $1^{1}/_{8}$" stream 240'. Through two 100' hose lines, threw two $^{7}/_{8}$" streams 197' each. Through 1000' of hose, threw one $1^{1}/_{8}$" stream 211'.

5-1-1882: New boiler coils.

6- -1893: Traded-in to Fire Extinguisher Manufacturing Co., Chicago, Ill. FEMCO manufactured Babcock chemical engines, and Champion and Hale water towers.

11-27-1894: FEMCO sold this Ahrens, plus a chemical engine, to Hamilton, Ill. for $2750.

1-10-1898: Still in service at Hamilton, Ill.

158 3 SSF/HD/A Upright. Round 32" x 62" coil boiler (iron & silver). One $5^{1}/_{4}$" x 8" pump, piston air pump. One $3^{1}/_{2}$" x 20' pre-connect suction. One 9" x 8" cylinder. Wheels 16 x 56 front, 16 x 62 rear. Drop pole.

12-30-1875: Delivered by Chris Ahrens to Nashville, Tenn. Name *Citizens Gift #3*.

1-29-1879: New wheels.

12-14-1900: Traded-in for $150 on Metropolitan #2765. Rebuilt as #458.

4-17-1901: Sold to Engine Co. #1, Lawrenceburg, Indiana. Replaced Ahrens #348.

5-24-1901: New 30" x 60" Fox boiler.

159 3 SSF/MD-HD/A Upright. Round 32" x 62" coil boiler (iron & silver). $5^1/4$" x 8" pump, automatic air pump. $3^1/2$" x 20' pre-connect suction. 9" x 8" cylinder. Wheels 16 x 56 front, 16 x 62 rear. Gum springs.

2-17-1876: Delivered by Fred Hawekotte to Maiden City Vol. Fire Dept., Cynthiana, Ky. Name *Cynthiana #1*. Steam in 3:48. Through 1000' of hose, threw a $1^1/16$" stream 196'.

1-28-1886: Joe Terbeck installed new boiler coils at Cynthiana.

160 2 SSF/HD/W Upright. $35^1/2$" coil boiler (nickel). 6" x 8" pump, 2 inlets, automatic air pump. 4" x 20' pre-connect suction. 10" x 8" cyl. Wheels 16 x 58, 16 x 62. Steel springs.

3-8-1876: Delivered by Fred Hawekotte to Indianapolis, Ind. Tested & accepted by Fire Co. #2, Indianapolis. Threw one $1^1/4$" stream 292'6", two 7/8" streams 230' each.

9-30-1890: Joe Terbeck installed new boiler coils at Indianapolis.

6-30-1892: New Ahrens coil boiler. Steamer completely repainted

1905: Replaced by new Ahrens Continental.

161 2 SSF/HD/A Upright. Round $35^1/2$" coil boiler (nickel). One 6" x 8" pump, three inlets, piston air pump. One 4" x 20' pre-connect suction. One 10" x 8" cylinder. Wheels 16 x 58 front, 16 x 62 rear. Steel eliptic springs.

5- -1876: Tested, Milwaukee, Wis. Through 500' of hose, threw one stream 292'.

5-29-1876: Delivered by Fred Hawekotte and Chris Ahrens to Watertown, Wis. Tested at Watertown against a Silsby rotary. Water in 4:30 vs. 5:30 for Silsby, but Ahrens broke a valve. Ahrens distanced Silsby on horizontal stream, but Silsby threw vertical stream higher. Ahrens won $^2/3$rds of the contests, but it broke down twice, so neither emerged the clear winner. The City Council had a tough decision: the east side of town, mostly German, wanted to buy the Ahrens; the west side, mostly Irish, rooted for the Silsby.

6-7-1876: Watertown bought the Ahrens for $5000, also one Ahrens hose reel and 1000' of hose. The Ahrens steamer and reel were placed in service at Watertown's only existing fire company, Pioneer Fire Co. #1 in the German east side. Watertown *also* bought the Silsby, organized the Phoenix Fire Co. #2 in the Irish west side to man it.

6-15-1893: New boiler coils.

162 3 SSF/HD/I Upright. 32" x 62" coil boiler (nickel). $5^1/4$" x 8" pump, piston air pump. 4" x 20' pre-connect suction. 9" x 8" cylinder. Gum springs. $57^1/2$" track. Drop pole.

5-4-1876: Delivered by Fred Hawekotte to Engine 10, St. Louis, Mo. Water in 3:19. Threw one $1^1/8$" stream 268'. Replaced Ahrens #23.

1888: Replaced by Ahrens #564.

1909: Was a Spare at Engine 6.

163 3 SSF/MD-HD/W Upright. Round 32" x 62" coil boiler (nickel). One $5^1/4$" x 8" pump, automatic air pump. One $3^1/2$" x 20' pre-connect suction. One 9" x 8" cylinder. Wheels 16 x 56 front, 16 x 62 rear. Gum springs.

5-19-1876: Delivered by Fred Hawekotte to Cambridge City, Ind., with two Ahrens hose reels, 1000' of hose. Name *General Meredith*. Water in 3:17. Through 1000' of hose, threw one $1^1/8$" stream 198'. Through 650' of hose, threw one $1^1/8$" stream 228'. Through 100' of hose, threw one $1^1/8$" stream 275'.

11-17-1881: Joe Terbeck installed new boiler coils at Cambridge City.

1-23-1882: Boiler rebuilt at factory.

9-14-1894: New boiler coils.

164 3 SSF/HD/W Upright. Round 32" x 62" coil boiler (nickel). One $5^1/4$" x 8" pump, automatic air pump. One $3^1/2$" x 20' pre-connect suction. One 9" x 8" cylinder. Wheels 16 x 56 front, 16 x 62 rear. Gum springs. Drop pole.

11-2-1876: Delivered by Fred Hawekotte to Engine 4, Indianapolis, Ind. Through 100' of hose, threw one stream 244'. Through 750' of hose, threw one stream 260'.

5-27-1889: New 32" Ahrens tubular boiler, $5^1/2$" x 8" pump. Shipped back 7-2-1889.

6-13-1904: Traded-in on Metropolitan #2925.

1-3-1906: Loaned to Oklahoma City, Okla. It was Oklahoma City's first steamer.

1-28-1906: New front springs and axle.

6-21-1906: Sold to Oklahoma City.

1925: All surviving steamers in Oklahoma City were scrapped. The Ahrens, if not already scrapped, was scrapped at that time.

165 1 SSF/HD/W Upright. 38" x $67^1/2$" coil boiler (brass). $6^5/8$" x 8" pump. $4^1/2$" x 20' pre-connect suction. 11" x 8" cyl. Wheels 16 x 51 front, 16 x 58 rear. 59" track. Weight 7365#.

12-26-1876: Loaned to Louisville, Ky., in primer (not painted). Tested against Louis-

After 28 years of service in Indianapolis, this 1876 Ahrens 3rd Class Upright #164 began its second career serving Oklahoma City, Oklahoma. Here it is seen in front of fire headquarters in Oklahoma City, about 1907. (Collection of the late Ben Dancy)

ville's Engine 8 (1872 Ahrens #22), a new Amoskeag, and a new Silsby. Ahrens #165 won, with Ahrens #22 a close second.

2- -1877: Returned to factory for painting.

3-9-1877: Delivered by Chris Ahrens to Engine 1, Louisville, Ky. Name *Charles D. Jacobs*. Replaced 1857 Latta #17.

9-11-1900: Replaced by Ahrens #384 and to Engine 3. This was a four-way trade: #165 from Eng. 1 to Eng. 3, #276 from Eng. 3 to Eng. 16, #220 from Eng. 16 to Eng. 2, #384 from Eng. 2 to Eng. 1.

1908: Replaced by new Ahrens Continental.

166 3 SSF/MD-HD/W Upright. Round 32" x 62" coil boiler (nickel). One $5\frac{1}{4}$" x 8" pump, automatic air pump. One $3\frac{1}{2}$" x 20' pre-connect suction. One 9" x 8" cylinder. Wheels 16 x 56 front, 16 x 62 rear. Weight 5640#.

9-11-1876: Delivered by Fred Hawekotte to Aurora, Ind. Name *Thomas Gaff*. Water in 3:40. Threw one stream 258'. Through 500' of hose, threw one $1\frac{1}{8}$" stream 228'.

Date?: Pumped for 72 hours at a fire in Thomas Gaff's "Crescent Brewery".

7-26-1885: Joe Terbeck installed new boiler coils at Aurora.

5-16-1894: George Zink installed new boiler coils at Aurora.

1918: Still in service at Aurora. Probably not retired until Ahrens-Fox pumper #871 (ex-Chester, Pa.) was purchased in 1939.

c-1943: Sold to local junkyard for $50 for the World War II scrap drive. The steamer was melted down, but Aurora Fire Dept. still has its nameplate.

167 3 SSF/MD-HD/W Upright. 32" x 62" coil boiler (silver). $5\frac{1}{4}$" x 8" pump, automatic air pump. $3\frac{1}{2}$" x 20' pre-connect suction. 9" x 8" cylinder. Wheels 16 x 56 front, 16 x 62 rear. Half-eliptic steel springs. Driver's seat.

10-13-1876: Demonstrated at Firemen's Tournament, Decatur, Ill.

11- -1876: Loaned to Covington, Ky., while #169 built. Returned to Ahrens 12-23-1876.

5-15-1877: Demonstrated at Fostoria, Ohio. Had water on a fire in less than 5:00 after alarm was received. Through 1450' of hose, threw one stream 224'6".

The "Thomas Gaff" of Aurora, Ind., was an 1876 Ahrens 3rd Class single-pump, straight-frame Upright #166. It is seen here in a junkyard, c-1943, just before it was melted-down for the World War II scrap drive. Only its nameplate survives today. This was the first of three Ahrens pumpers Aurora bought, and the only one bought new; a steamer purchased in 1884 and a motorized pumper in 1939 were both bought second-hand.
(Courtesy Fire Chief Hafenbridle, Aurora, Ind.)

This 1907 photo shows an 1876 3rd Class Upright #167 of Van Wert, Ohio, shortly before Ahrens Fire Engine Co. installed a new Continental boiler and a new air pump. The occasion was the arrival of Van Wert's 1907 Continental #139, which is still in Van Wert today; fate of this 1876 engine, however, is unknown.
(Deal Studios photo, courtesy Bill Purmort)

5-26-1877: Demonstrated at Van Wert, Ohio.

5-31-1877: Sold to Van Wert, Ohio. Name *Van Wert #1*.

12- -1878: Raised 70 pounds steam from cold water in 4:00, temperature -10° F.

8-3-1887: Joe Terbeck installed new boiler coils at Van Wert.

12-27-1907: 32" x 64" Radial Inclined Water Tube boiler by Ahrens Fire Engine Co.

1-3-1908: New air pump by Ahrens Fire Engine Co.

168 3 SSF/HD/W Upright. 32" x 62" coil boiler (nickel). $5^1/4$" x 8" pump, piston air pump. $3^1/2$" x 20' pre-connect suction. 9" x 8" cylinder. Wheels 16 x 56 front, 16 x 62 rear. Gum springs. Drop pole. Weight 5685#.

3-1-1877: Tested & accepted by Cleveland, Ohio. Delivered by Chris Ahrens.

6-25-1888: Was at Engine 10, Cleveland.

7-20-1896: Was in service at Engine 22.

169 2 SSF/HD/W Upright. Round $35^1/2$" x 67" coil boiler (nickel). One 6" x 8" pump, piston air pump. One 4" x 20' pre-connect suction. One 10" x 8" cylinder. Wheels 16 x 58 front, 16 x 62 rear. Gum springs. Drop pole.

12-23-1876: Delivered by Fred Hawekotte to Covington, Ky. Name *Ashbrook*.

1- -1884: During flood at Covington, pumped 9 hours non-stop.

10-1-1887: New boiler coils.

12-30-1902: New $35^1/2$" Fox water-tube boiler, miscellaneous repairs, replating. Price $1550. Paid $1300 and trade-in Ahrens #346.

170 4 SSF/MD/W Upright. 30" x 60" coil boiler. $4^3/4$" x 7" pump. $3^1/4$" x 20' pre-connect suction. 8" x 7" cylinder. No seat. Wheels 16 x 54 front, 16 x 60 rear. Steel eliptic springs. Rear brake lever. Weight 4143#.

11-21-1876: Test at Cincinnati, Ohio. Water in 4:40, distance 150' in 5:00. Threw one 1" stream 215', one $1^1/8$" stream 200', one $1^1/4$" stream 178', two $^3/4$" streams 160' each.

4- -1877: Rebuilt from manually-drawn to horse-drawn, seat added.

6-22-1877: Fred Hawekotte del'd to Georgetown, Ky., with Ahrens hose reel, 1000' of hose. Through 700' hose up 80' embankment, two streams over tallest building in town.

8-25-1883: Joe Terbeck installed new boiler coils at Georgetown.

2- -1893: Sold to Midway, Ky.

7- -1894: Boiler and pump rebuilt. Midway borrowed Ahrens #11 while repairs made.

7-1-1899: George Zink installed new boiler coils at Midway.

171 2 SSF/HD/W Upright. $35^1/2$" x 67" coil boiler (nickel). 6" x 8" pump, piston air pump. 4" x 20' pre-connect suction. 10" x 8" cyl. Wheels 16 x 58, 16 x 62. Drop pole.

4-25-1877: Tested & accepted by Covington, Ky. Name *Robert Howe*. Water in 3:40. Threw one $1^3/16$" stream 264'6", one $1^1/8$" stream 256', one $1^3/8$" stream 238', two $^7/8$" streams 200' each. Through 1000' of hose, threw one $1^1/8$" stream 197', one 1" stream 200'.

1- -1884: During flood at Covington, pumped $191^1/2$ hours non-stop.

3-21-1888: New boiler coils.

5-7-1908: New 34" x 64" Inclined Radial water tube boiler by Ahrens Fire Engine Co.

172 3 SSF/HD/W Upright. 32" x 62" coil boiler (nickel). $5^1/4$" x 8" pump, piston air pump. $3^1/2$" x 20' pre-connect suction. One 9" x 8" cylinder. Wheels 16 x 56, 16 x 62. Drop pole.

3-29-1877: Delivered by Fred Hawekotte to Reliable Fire Co. #18, Cincinnati, Ohio. Name *J.L. Thompson*. Replaced Ahrens #13.

3-13-1899: New boiler coils.

7- -1902: Replaced by Ahrens 503. To Spare.

1981: *J.L. Thompson #18* nameplate from this steamer, formerly at Ohio Mechanics Institute, donated to Cincinnati Fire Museum. This nameplate now displayed at museum.

173 3 SSF/HD/W Upright. Round 32" x 62" coil boiler (nickel). One $5^1/4$" x 8" pump. One 4" x 20' pre-connect suction. One 9" x 8" cylinder. Wheels 16 x 56 front, 16 x 62 rear.

4- -1877: Delivered to Fire Co. #4, Nashville, Tenn. Name *William Stockell*.

5-25-1898: New 4" x 20'6" suction.

1903: Still in service per annual report of Nashville Fire Dept.

Ahrens Manufacturing Co., Inc.

22, 24, 26 Webster St., Cincinnati (1877-79)
Chris Ahrens, Pres./G. Fred Hawekotte, Vice Pres.

Note that this company used only even serial numbers. All boilers were round, not square.

174 3 SSF/HD/W Upright. 32" x 62" coil boiler (nickel). 5^1/4" x 8" pump, piston air pump. Non-corrosive pump and boiler, oversized water chambers to work salt water. 3^1/2" x 20' pre-connect suction. 9" x 8" cylinder. Wheels 16 x 56, 16 x 62. Drop pole.

7-4-1877: Delivered by Chris Ahrens to Mechanics Fire Co. #6, Galveston, Texas. Tested at 4th of July picnic in Galveston, against a Silsby and a Jeffers. Threw one stream 180' before Silsby had water; threw two streams farther than Jeffers threw one.

1878: At Texas Cotton Press fire, pumped a brackish mixture, salt water and quicksand, for 39 hours; briefly shut down to clean out boiler, then pumped another 17 hours.

1878: Raised steam faster than 3rd Class [Silsby?] steamer from Waco, Texas. Threw one stream 150' before Waco's had water.

11-15-1879: New boiler coils.

6-28-1883: New boiler coils.

6-13-1885: New Ahrens tubular boiler with two gauges and clock.

176 4 SSF/MD-HD/A Upright. 30" x 60" coil boiler (silver). 4^3/4" x 7" pump, automatic air pump. 3^1/4" x 20' pre-connect suction. 8" x 7" cyl. Wheels 16 x 54, 16 x 60. Driver's seat.

6- -1877: Demonstrated at Pittsburgh, Pa.

6-9-1877: Chris Ahrens delivered to Uniontown, Pa., with hose reel, 750' of hose.

3-29-1892: New boiler coils.

7-22-1910: Pump overhauled at Uniontown.

178 4 SSF/HD/W Upright. 30" x 60" coil boiler (nickel). 4^3/4" x 7" pump, automatic air pump. 3^1/4" x 5' suction. 8" x 7" cylinder. Wheels 16 x 54, 16 x 60. Driver's seat.

7-6-1877: Delivered by Fred Hawekotte to Franklin, Ohio, with hose cart, 600' of hose, and manual hook-and-ladder. Tested at 17' lift on bridge over canal. Through 100' of hose, threw a 1" stream 234'. Through 1000' of hose, threw one 1" stream 184'.

8-26-1877: Franklin's city hall caught fire at 5 a.m. The Ahrens saved the building.

9-28-1886: Rebuilt at factory. New boiler coils, 3^1/4" x 19'6" suction, brakes, gong. Paint & varnish touch-up. Loaned Ahrens #461 to Franklin while repairs were made.

11-29-1887: Pumped for 7 hours non-stop at a fire in Franklin.

180 4 SSF/MD/W Upright. 30" x 60" coil boiler (nickel). 4^3/4" x 7" pump, automatic air pump, air chamber ornament. 3^1/4" x 20' suction. 8" x 7" cyl. Wheels 16 x 54, 16 x 60.

9-5-1877: Demonstrated with Ahrens #182 and #196, at the Exposition in Montreal, Quebec (Canada).

5-5-1879: Delivered by Peter Ahrens and Van Dyne to Marysville, Ohio.

12-19-1884: Joe Terbeck installed new boiler coils at Marysville.

9-19-1888: Joe Terbeck installed new boiler coils at Marysville.

182 3 SSF/HD/W Upright. 32" x 62" coil boiler (nickel). 5^1/4" x 8" pump (nickel-plated air chamber & flywheel), automatic air pump. 3^1/2" x 20' pre-connect suction. 9" x 8" cyl. (nickel). Wheels 16 x 56, 16 x 62, nickel springs & axles. Gum springs. Drop pole.

9-5-1877: Demonstrated with Ahrens #180 and #196, at Exposition, Montreal, Quebec.

11-13-1877: Tested at St. Marys, Ontario.

5-29-1878: Delivered by Chris Ahrens and Dick Raymond to Parkersburg, W. Va.

7-9-1883: New boiler coils.

12-11-1895: New boiler coils.

8- -1911: Rebuilt by American-LaFrance, Elmira, N.Y.

184 3 SSF/HD/W Upright. Round 32" x 62" coil boiler (silver/brass). 5^1/4" x 8" pump, automatic air pump. One 3^1/2" x 20' pre-connect suction. One 9" x 8" cylinder. Wheels 16 x 56 front, 16 x 62 rear. Steel springs. Drop pole.

10-21-1877: Tested at Portage, Wis., against a 3rd size Silsby. Portage bought Silsby, $4200.

12-8-1877: Delivered by Chris Ahrens to Gallipolis, Ohio.

2-9-1887: G. Fred Ahrens installed new boiler coils at Gallipolis.

186 3 SSF/HD/W Upright. Round 32" x 62" coil boiler (nickel). One 5^1/4" x 8" pump, piston air pump. One 3^1/2" x 20' pre-connect suction. One 9" x 8" cylinder. Wheels 16 x 56 front, 16 x 62 rear. Gum springs. Drop pole.

11-5-1877: Tested & accepted by Corryville Fire Co. #19, Cincinnati, Ohio. Replaced Ahrens #14.

12-20-1888: New boiler coils.

188 4 SSF/MD-HD/W Upright. Round 30" x 60" coil boiler (silver). One $4^3/_4$" x 7" pump, automatic air pump. One $3^1/_4$" x 20' suction. One 8" x 7" cylinder (nickel-plated). Wheels 16 x 54 front, 16 x 60 rear. Steel springs.

1- -1878: Demonstrated at Maysville, Ky.

1-8-1878: Delivered by Chris Ahrens to Rescue Fire Co., Paris, Ky.

8-14-1885: Joe Terbeck installed new boiler coils at Paris.

2-14-1888: Still in service at Paris, Ky.

190 4 SSF/MD-HD/W Upright. Round 30" x 60" coil boiler (iron). One $4^3/_4$" x 7" pump, automatic air pump. One $3^1/_4$" x 20' suction. One 8" x 7" cylinder (nickel-plated). Wheels 16 x 54 front, 16 x 60 rear. Steel springs.

2- -1878: Built for Brookville, Ind. Not accepted.

6-5-1878: Delivered by Chris Ahrens to Columbia, Tenn., with one Ahrens hose reel and 300' of hose. Water in 3:15.

1- -1895: Sold to Engine 1, Nashville, Tenn.

2-26-1895: New boiler coils. New $3^1/_4$" x 20' suction. Pump overhauled.

192 4 SSF/HD/W Upright. 30" x 60" coil boiler (silver). Brass fuel box, copper trim, copper number "2" on back. $4^3/_4$" x 7" pump, nickel-plated air chamber on gold mountings, automatic air pump. $3^1/_4$" x 20' suction, silver-plated suction brackets. 8" x 7" cyl. Wheels 16 x 54, 16 x 60. Kingbolt axles and springs. Brass-plated frame. Silver and gold mounting brackets for nozzles and lanterns.

3-16-1878: Shipped via steamboat to New Orleans, La. Delivered by Fred Hawekotte.

3-30-1878: Accepted by Protector Fire Co. #2, New Orleans, La. Water in 3:35. Through 100' of hose, threw one 1" stream 239'.

5-12-1883: New boiler coils.

8-9-1889: New boiler coils.

194 2 SSF/HD/W Upright. $35^1/_2$"x67" coil boiler (iron). 6" x 8" pump, 3 inlets, air pump. 4" x 20' pre-connect suction. 10" x 8" cyl. Wheels 16 x 58, 16 x 62. Steel springs. Drop pole.

9- -1877: Delivered by Chris Ahrens to Topeka, Kansas. Went in service as Engine 1.

6-24-1904: Rebuilt at factory. New 35" x 64" Fox boiler (no jacket), rebored pump and steam cylinder, overhauled pump, repaired front and rear wheels and springs, added 13" gong, touched-up paint. Nickel-plated pump support columns, pump and cylinder heads, crank, flywheel, journal boxes, sliding box yokes, and cylinder casing.

196 2 SSF/HD/W Upright. $35^1/_2$" x 67" coil boiler (silver). 6" x 8" pump (nickel-plated air chamber & flywheel). 4" x 20' pre-connect suction. 10" x 8" cylinder (nickel). Wheels 16 x 58 front, 16 x 62 rear, nickel-plated springs and axles. Gum springs. Drop pole.

10-9-1877: Demonstrated with #180 and 182, at Exposition in Montreal, Quebec (Canada).

12-28-1878: Tested & accepted by Engine Co. #19, St. Louis, Mo.

10-16-1888: New rear springs.

9- -1899: One new front spring.

1907: Replaced by Metropolitan #3200.

198 2 DCN/HD/W Upright. $35^1/_2$" x 67" coil boiler (silver). $4^3/_8$" x 8" pumps. 4" x 20' pre-connect suction. 7" x 8" cyls. Wheels 16 x 58, 18 x 62. Springs steel front, gum rear. The first double crane-neck Ahrens built.

4-1-1878: Raised steam for the first time.

4-27-1878: Tested at 5th & Walnut streets, Cincinnati. Threw one $1^1/_4$" stream 288'. After this test, pumps were rebored to $4^1/_2$".

5-1-1878: In service at Franklin Fire Co. #4, Cincinnati, Ohio. Replaced 1867 Lane & Bodley #21.

5-8-1878: Pumped 18 hours 45 minutes non-stop (from 12:30 a.m. to 7:15 p.m.), Henshaw Furniture Factory fire, Elm St. at Canal.

7-18-1884: Collided with Fire Co. #3 (Ahrens #280) at 9th & Elm, Cincinnati. Rebuilt at factory, including new boiler coils, 8-1-1884.

9-30-1893: New front and rear springs, new boiler coils, and general overhaul.

7-8-1902: Replaced by Ahrens #316. To Fire Co. #13.

200 3 SSF/HD/W Upright. Round 32" x 62" coil boiler (silver). $5^1/_2$" x 8" pump, Centennial signal lamp atop air chamber, piston air pump. 4" x 20' pre-connect suction. 9" x 8" cylinder. Wheels 16 x 56 front, 16 x 62 rear. Gum springs. Drop pole. Gong on foot board.

10-1-1878: In service, Eng. 8, Cleveland, O.

7-20-1896: Was in service at Engine 20.

Ahrens #202 is the second-oldest Cincinnati-built fire engine known to survive today. Built in 1878, this lavishly-plated 4th Class gem served Washington, Ind., for 50 years, then sat in storage another 45 years. In this 1973 photo, it awaits the start of a 7-year renovation project by Daviess County Historical Society of Washington, Ind.
(Courtesy Rodger Birchfield)

1878 Ahrens #202 of Washington, Ind., is seen in fully-renovated condition at the firehouse in Washington, Ind., in 1981. The "Elisha Hyatt" is a 4th Class single straight-frame Upright built by Ahrens Manufacturing Co. just one year after that firm's incorporation. The number (202) is deceptive: it was only the 88th Ahrens built.
(Bruce Neal photo)

202 4 SSF/MD-HD/W Upright. 30" x 60" coil boiler (iron). $4^{3}/_{4}$" x 7" pump, automatic air pump (nickel-plated), air chamber ornament. $3^{1}/_{4}$" x 20' pre-connect suction. 8" x 7" cyl. Wheels 16 x 54 front, 16 x 60 rear. Nickel-plated frame, springs, axles, pump & cylinder heads, flywheel, and coal bunker.

8-17-1878: Delivered to Young America Fire Co. #1, Washington, Ind. Name *Elisha Hyatt*. Price $4300.

4-5-1882: New boiler coils.

1928: Retired and placed in storage.

1973: To Daviess County Historical Society, Washington, Ind., for restoration.

1981: Now fully-restored and displayed at the firehouse in Washington, Ind.

204 4 SSF/MD-HD/W Upright. 30" x 60" coil boiler (iron). $4^{3}/_{4}$" x 7" pump, automatic air pump, no lamp or ornament. $3^{1}/_{4}$" x 20' pre-connect suction. 8" x 7" cyl. Plain finish (not plated), no lanterns. Wheels 16 x 56, 16 x 62.

3-1-1879: Fred Hawekotte delivered to Providence Fire Co. #1, Lake Providence, La. Name *Lady Morgan*. Steam in 3:35. Through 300' of hose, threw one 1" stream 227'.

1884: City of Lake Providence took over firefighting duty from the volunteers.

5-26-1886: New boiler coils.

c-1925: Replaced by Ford Model T pumper. Probably junked.

206 4 SSF/HD/W Upright. Round 30" x 60" coil boiler (silver), brass fuel pan. 5" x 7" pump, Centennial signal lamp on air chamber. One $3^{1}/_{2}$" x 20' pre-connect suction. 8" x 7" cylinder. Wheels 16 x 56 front, 16 x 62 rear. Nickeled frame, springs, axles. Drop pole.

3-17-1879: Delivered by Fred Hawekotte to Young America Fire Co. #11, New Orleans. Through 50' of hose, threw a 1" stream 258'.

5-10-1884: New boiler coils.

3-17-1891: New boiler coils.

12-31-1896: New rear springs.

208 3 SSF/HD/A Upright. 32" x 62" coil boiler (silver). $5^{1}/_{2}$" x 8" pump, piston air pump, nickel plating. 4" x 20' pre-connect suction. 9" x 8" cylinder (nickel). Wheels 16 x 56, 16 x 62, brass hubs, nickel-plated springs & axles. Gum springs. $57^{1}/_{2}$" track. Drop pole.

11-26-1878: Delivered by Fred Hawekotte to Engine 20, St. Louis, Mo.

1909: Was a Spare at Engine 3. Engine 20 was using 1896 Ahrens #2421.

Ahrens Manufacturing Co., Inc.

22, 24, 26 Webster St., Cincinnati (1879-1891)
Chris Ahrens, Pres./G. Fred Hawekotte, Vice Pres.
Hiram H. Nieman, Secretary/Treasurer

There seems to be no logic to this company's even/odd numbering system. Unused numbers may have been other apparatus (hose reels or ladder trucks) or steamers ordered but never built.

210 3 SSF/HD/A Upright. 32" x 62" coil boiler (silver). $5^{1}/_{2}$" x 8" pump, air pump, nickel plating. 4" x 20' pre-connect suction. 9" x 8" cyl (nickel). Wheels 16 x 56, 16 x 62, nickel-plated springs & axles. Drop pole.

3-26-1879: Delivered by Hiram Nieman and Dick Raymond to Eng. 4, Omaha, Nebraska.

3-31-1890: New boiler coils.

1908: American-LaFrance rebuilt at Elmira.

212 3 SSF/HD/A Upright. 32" x 62" coil boiler (silver). $5^{1}/_{2}$" x 8" pump, nickel plating, air pump. 4" x 20' pre-connect suction. 4" x 5' suction. 9" x 8" cyl. (nickel). Wheels 16 x 56, 16 x 62, nickeled springs & axles. Stiff pole.

6-4-1880: Delivered by Fred Hawekotte and Dick Raymond to Rescue Fire Co., Council Bluffs, Iowa.

1892: Traded-in to Fire Extinguisher Mfg. Co. (FEMCO), Chicago, Ill.

1-15-1893: FEMCO sold to Swift & Co. meat-packing plant, Kansas City, Kansas.

12-26-1905: New suction inlet supplied by American-LaFrance, Elmira, N.Y.

214 3 DSF/HD/S Upright. 32" x 62" coil boiler (silver), brass coal bunker. 4" x 7" pumps, Centennial signal lamp on air chamber. Two $3^{1}/_{2}$" x 9'6" suctions. $2^{1}/_{2}$" x 11' pre-connect suction. 8" x 7" cyls. Wheels 16 x 58, 16 x 62, nickel-plated springs and axles. Stiff pole. Two lanterns under seat.

5-19-1879: Demonstrated at Louisville, Ky. Threw one stream 270' in 8:00.

5- -1879: Delivered by Fred Hawekotte and Hiram H. Nieman to Jackson Fire Co. #18, New Orleans, La.

216 2 SSF/HD/S Upright. $35^{1}/_{2}$" x 64" coil boiler (iron), 3" shorter than standard to clear firehouse doors. 6" x 8" pump. 4" x 20' pre-

connect suction. 10" x 8" cylinder. Wheels 16 x 58, 16 x 62. Steel springs. Drop pole.

7-7-1879: Tested & accepted by Newark, Ohio. Delivered by Fred Hawekotte and Dick Raymond. Newark borrowed Ahrens #14 while this was built.

11-11-1885: Joe Terbeck installed new boiler coils at Newark.

12- -1887: Pumped two streams through 900' of hose, for 12 hours non-stop at a fire in Newark, despite temperature of -14° F.

218 4 SSF boiler and frame only. Built as #244.

220 3 DSF/HD/I Upright. 32" x 62" coil boiler (brass). 4" x 7" pumps, Centennial signal lamp on air chamber. 3½" x 20' pre-connect suction. 9" cyls. Polished axles and springs. 59" track. Drop pole. Gong on foot board.

8-20-1879: Delivered by Chris Ahrens to Engine 6, Louisville, Ky. Name *George W. Frantz*. Replaced 1865 Lane & Bodley #9.

1891: Replaced by Ahrens #?. To Engine 16.

9-11-1900: Replaced by Ahrens #276, to Eng. 2 (4-way trade: #220 from Eng. 16 to Eng. 2, #384 from Eng. 2 to Eng. 1, #165 from Eng. 1 to Eng. 3, #276 from Eng. 3 to Eng. 16).

1-2-1901: New front axle.

1904: Replaced by Metropolitan #2992.

222 3 DSF/HD/S Upright. 32" x 62" coil boiler (silver), brass coal bunker. 4" x 7" pumps, Centennial signal lamp on air chamber. Two 3½" x 10' suctions. 8" x 7" cylinders. Wheels 16 x 56, 16 x 60, nickel-plated springs and axles. Stiff pole. Gong on foot board.

11-30-1879: Delivered by Fred Hawekotte to Palmetto Fire Co. #10, New Orleans, La.

224 2 DCN/HD/W Upright. 35½" x 67" coil boiler (nickel). 4½" x 8" pumps. 4" x 20' pre-connect suction. 7" x 8" cylinders. Wheels 16 x 58, 16 x 60, plated springs. Stiff pole.

10-24-1879: Delivered by Fred Hawekotte & Dick Raymond to Engine 1, Milwaukee, Wis.

11-18-1879: One 4" x 10'6" suction, one 4" x 11'6" suction, to replace pre-connect suction.

1892: Replaced by Ahrens #2123. To Eng. 8.

1904: Replaced by Ahrens #401. To Engine 16 in place of Ahrens #451.

1908: New Nott boiler.

1921: Replaced by 1902 Metropolitan #2907.

226 3 DSF/HD/S Upright. 32" x 62" coil boiler (silver), brass coal bunker. 4" x 7" pumps, signal lamp on air chamber. 3½" x 15' pre-connect suction. 3½" x 7' suction. 8" x 7" cyls. Wheels 16 x 58, 16 x 60, nickel-plated springs & axles. Stiff pole. Gong on front.

12-11-1879: Delivered by Dick Raymond to Dexter Fire Co. #1, Montgomery, Alabama.

7-1-1891: New boiler coils.

1-16-1892: New 3½" x 16' suction.

9-1-1899: New boiler coils.

11-19-1903: 32" x 62" Fox boiler, rebored steam cylinder, overhauled pump ($1728).

1916: Was a Spare at Fire Co. #2.

228 3 DSF/HD/S Upright. 32" x 62" coil boiler (iron & brass). 4" x 7" pumps. 3½" x 20' pre-connect suction. 8" x 7" cylinders (brass). Wheels 16 x 56, 16 x 60. Drop pole.

12-15-1879: In service at Citizens Gift Fire Co. #3, Cincinnati, Ohio. Name *L.C. Weir*. Replaced Ahrens #27.

5-24-1881: Replaced by Ahrens #280. To Deluge Fire Co. #10 in place of Ahrens #1.

1-1-1882: To Garfield Fire Co. #21.

5-4-1889: Replaced by Ahrens #573.

12-4-1893: Collided with Fire Co. #3 (Ahrens #280) at 3rd & Elm, Cincinnati. Repaired, including new boiler coils.

12-23-1893: To Fire Co. #27.

6-30-1898: New horse pole ($110).

230 3 DSF/HD/S Upright. 32" x 62" coil boiler (silver), brass coal bunker. 4" x 7" pumps, silver-plated flywheel. One 3½" x 10' (pre-connect), one 3½" x 10' suction. 8" x 7" cyls. Wheels 16 x 56, 16 x 60. Silver-plated frame, springs, and axles. 13" gong on foot board.

4-11-1880: Delivered by Chris Ahrens to Fire Co. #7, New Orleans, La.

232 3 DSF/HD/S Upright. 32" x 62" coil boiler (silver), brass coal bunker. 4" x 7" pumps, nickel-plated air chamber with Centennial signal lamp atop. 3½" x 20' pre-connect suction. 8" x 7" cyls. Wheels 16 x 56, 16 x 62. Silver-plated frame, springs, and axles. Drop pole. 13" gong on foot board.

2-7-1880: Delivered by Chris Ahrens and Peter Ahrens to Chillicothe, Ohio. Name *G.W.A. Clough #1*. Price $4400.

8-27-1898: New boiler coils.

234 3 DSF/HD/A Upright. 32" x 62" coil boiler

(iron), brass coal bunker. 4" x 7" pumps, air chamber ornament. $3^1/_2$" x 20' pre-connect suction. 8" x 7" cylinders (silver). Wheels 16 x 56 front, 16 x 62 rear. Silver-plated frame. Drop pole. Gong on foot board.

3-5-1880: Fred Hawekotte & Dick Raymond delivered (with hose reel) to Vincennes, Ind.

2-27-1883: Gas explosion hurled steamer into cistern. New wheel, boiler coils, seat, flywheel, repaired front axle. Fred Hawekotte delivered to Vincennes 3-2-1883.

7-12-1889: New boiler coils.

9-12-1889: Steamer sold to Kansas City, Mo. In service as Engine 6.

4-14-1898: New boiler coils.

1-8-1904: New boiler coils.

1910: New Continental boiler. To Engine 25.

2-18-1986: John Gambs reports that Vincennes still has their 1880 Ahrens hose reel.

236 3 DSF/HD/A Upright. 32" x 62" coil boiler (silver), brass coal bunker. 4" x 7" pumps, air chamber ornament. 4" x 20' pre-connect suction. 8" x 7" cyls. (silver). Wheels 16 x 56, 16 x 60. Steel springs. $57^1/_2$" track. Drop pole. Silver-plated frame. Gong on foot board.

8-9-1880: Delivered by Fred Hawekotte to Engine 21, St. Louis, Mo.

6-9-1888: New boiler coils.

8-11-1905: New 32" x 62" Fox boiler.

1909: Was a Spare at Engine 1. Engine 21 was using 1887 Ahrens #508.

238 4 DSF/HD/W Upright. 30" x 60" flue boiler (silver), 2 coach lamps on boiler, silver-plated coal bunker with metal number "5" on back. $3^5/_8$" x ? pumps, signal lamps atop pump and air chamber. One $3^1/_2$" x 10' (pre-connect), one $3^1/_2$" x 10' suction. 7" x ? cyls. (silver). Wheels 16 x 54, 16 x 60. Silver-plated frame, springs, axles. Stiff pole. Single seat, Star lamps both sides.

4-25-1880: Delivered by Chris Ahrens to Columbia Fire Co. #5, New Orleans, La. Threw one stream 265'8", two streams 231' each.

7-26-1897: One of 12 New Orleans steamers sold for $100 apiece to Schwartz Foundry Co. as scrap.

240 4 DSF/HD/W Upright. 30" x 60" flue boiler (silver), 2 coach lamps on boiler, silver-plated coal bunker with metal number "21" on back. $3^5/_8$" x ? pumps, signal lamp atop pump and air chamber. One $3^1/_2$" x 10' (pre-connect), one $3^1/_2$" x 10' suction. Two 7" x ? cyls. (silver). Wheels 16 x 54, 16 x 60. Silver-plated frame, springs, and axles. Stiff pole. Single seat with Star lamps on both sides.

4-25-1880: Delivered by Chris Ahrens to Fire Co. #21, New Orleans, La. Threw one stream 265'10", two streams 231' each.

242 4 SSF/MD-HD/W Upright. 30" x 60" coil boiler (iron & silver). 5" x 7" pump. $3^1/_2$" x 20' pre-connect suction. 8" x 7" cyl. Wheels 16 x 54, 16 x 60. Four lanterns. Rear brake lever.

9-29-1880: Shipped to U.S. Government, General Department of Missouri.

10-8-1880: Tested & accepted by U.S. Government at Chicago, Ill. Through 300' of hose, threw one stream 270'. Went in service at Camp Carlin, Mo.

12- -1880: Transferred to U.S. Army Depot, Cheyenne, Wyoming. 50 pounds of steam in 4:00. Pumped muddy water through 600' of hose, up a 70' elevation, 24 hours non-stop.

1-20-1883: Front wheels repaired.

11-22-1897: Moved to Fort Duchesne, Utah.

12-1-1897: New air chamber, discharge gates.

5-26-1902: Destroyed by fire, Fort Duchesne.

244 4 SSF/MD/A Upright. 30" x 60" coil boiler (nickel). 5" x 7" pump. $3^1/_4$" x 20' pre-connect suction. 8" x 7" cyl. Wheels 16 x 54, 16 x 60. Driver's seat. Rear brake lever. Weight 4300#. Boiler/frame of Ahrens #218.

3-28-1881: Delivered by Fred Hawekotte and Dick Raymond to Lena, Ill. Name *Lena #1*. Price $3700. Through 100' of hose, threw one 1" stream in 4:12. Had 100# nozzle pressure at end of 850' hose line.

4-16-1907: Destroyed in $8000 city hall fire at Lena, Ill.

246 2 DCN/HD/W Upright. $35^1/_2$" x 67" coil boiler (iron). $4^1/_2$" x 8" pumps, 2 flywheels. 4" x 10' suctions. 7" x 8" cyls. Wheels 16 x 58, 16 x 60. Stiff pole. Plain finish (no plating).

6-24-1880: In service Eng. 4, Milwaukee, Wis.

5-28-1883: Two pump elbows for pre-connect suction.

9-14-1883: Replaced by Ahrens #372. To Engine 6.

1893: Replaced by Ahrens #2173. To Engine 9 in place of Ahrens #451.

10- -1897: New front springs.

1902: Replaced by Metropolitan #2906.

9-1-1904: To Engine 27.

1910: New Nott boiler.

1916: To Engine 6.

1920: Replaced by 1902 Metropolitan #2905.

248 2 DCN/HD/S Upright. $35^1/_2$" x 67" coil boiler (iron). Two $4^1/_2$" x 8" pumps. Two 4" x 10' suctions. Two $7^1/_4$" x 8" cylinders. Wheels 16 x 58 front, 16 x 62 rear. Steel springs. $56^1/_2$" track. Stiff pole. Front brake lever.

7-24-1880: Delivered by Chris Ahrens and Dick Raymond to Engine 7, Chicago, Ill. Later Engine 2, then Engine 56.

1901: New boiler.

1906: At Engine 52.

1917: At Engine 86.

250 2 DCN/HD/A Upright. $35^1/_2$" x 67" coil boiler (iron). $4^1/_2$" x 8" pumps. 4" x 20' preconnect suction. 7" x 8" cyls. Wheels 16 x 58, 16 x 62. Springs steel front, gum rear. 59" track. Stiff pole. Railing atop seat. Painting of Major Hughes, Louisville Fire Chief, on front of foot board; gong under foot board.

9-6-1880: Delivered by Fred Hawekotte and Dick Raymond to Engine 5, Louisville, Ky. Name *Major Hughes*. Replaced 1859 Latta #24. Should have been #252, but Registered No. plates of #250 and #252 were switched at the factory and the mistake wasn't discovered until after both were delivered.

1895: Replaced by Ahrens 2405. To Eng. 18.

1-29-1898: New $35^1/_2$" x 67" Ahrens coil boiler ($361).

8-11-1903: New rear springs.

1910: New $5^1/_2$" x 8" pump.

1915: Still in service at Engine 18.

1924: Wrecked in collision and junked.

252 2 DCN/HD/A Upright. $35^1/_2$"x67" coil boiler (iron). $4^1/_2$" x 8" pumps. One $4^1/_2$" x 10'6", one $4^1/_2$" x 11'6" suctions. $7^1/_4$" x 8" cyls. Wheels 16 x 58, 16 x 62. Steel springs. Stiff pole. Gong. Railing atop seat. Weight 7575#.

9-4-1880: Delivered by Chris Ahrens and Peter Ahrens to New York, N.Y. In service at Engine 13-1 (Manhattan). This should have been #250, but Registered No. plates of #250 and #252 were switched at the factory and the mistake wasn't discovered until after both were delivered.

11-30-1880: Competed against Amoskeag of FDNY Eng. 20, Clapp & Jones of Engine 33. All three drafted salt water from Hudson river, each pumping 10 hours through 200' of hose with $1^3/_8$" ring nozzle. Performances of each are shown in Chart #3.

1890: Became a Spare.

3-4-1892: New boiler coils. To Engine 34.

1-23-1896: Sold for $120 to Andrew Jersey, sales agent for American Fire Engine Co.

254 2 DCN/HD/S Upright. $35^1/_2$" x 67" coil boiler (iron). Two $4^1/_2$" x 8" pumps. Two 4" x

Make	Steam Pressure	Pump Pressure	Nozzle Pressure	Stream Distance	Lbs. coal per Hour
Ahrens	120.33	166.70	88.78	210'7"	496.000
Amoskeag	101.64	143.14	74.54	201'3"	993.500
C & J	110.83	173.55	93.03	228'1"	776.875

Note: Only the Ahrens operated for the entire 10 hours. The Clapp & Jones worked only $8^1/_2$ hours, and the Amoskeag only $5^1/_2$ hours. The best throw of the Ahrens was 240', vs. 235' for the Clapp & Jones and 220' for the Amoskeag. Note also that the Ahrens was by far the most fuel-efficient of the three steamers.

CHART #3

Results of Steam Fire Engine Test
New York, N.Y. Nov. 30, 1880

10' suctions. Two $7^1/_4$" x 8" cylinders. Wheels 16 x 58 front, 16 x 62 rear. Steel springs. $56^1/_2$" track. Stiff pole. Front brake lever.

9-25-1880: Delivered by Hiram H. Nieman to Engine Co. #5, Chicago, Ill.

1898: New boiler.

1906: At Engine 9. Retired by 1917.

256 3 SCN/HD/I Upright. 32" x 62" coil boiler (iron). $5^1/_2$" x 8" pump. One 4" x 20' pre-connect suction. 9" x 8" cylinder. Stiff pole.

10- -1880: Delivered by Peter Ahrens to Hot Springs, Arkansas.

Date?: New springs. Gone by 1904.

258 4 SSF/HD/W Upright. 30" x 60" flue boiler (silver), silver-plated coal bunker, two coach lamps on boiler. 5" x 7" pump, signal lamp atop pump, copper outlets, brass flywheel, air pump. One $3^1/_2$" x 20' (pre-connect), one $2^1/_2$" x 12' suction. 8" x 7" cyl. Wheels 16 x 56, 16 x 62 (standard 4th Class were 16 x 54 and 16 x 60). Steel springs. Stiff pole. Two Star lamps on seat. Foot board had railing on top, handles on sides, and gong on front.

11-14-1880: Delivered by Fred Hawekotte to Creole Fire Co. #9, New Orleans, La.

260 3 DCN/HD/S Upright. 32" x 62" coil boiler (iron). 4" x 7" pumps, air chamber ornament. $3^1/_2$" x 10' suctions. $6^3/_4$" x 7" cyls. Wheels 16 x 56, 16 x 62. $56^1/_2$" track. Stiff pole.

11-10-1880: Delivered by Chris Ahrens & Hiram H. Nieman to Engine 30, Chicago, Ill.

1906: At Engine 43.

1912: Replaced by Continental 191. Retired.

262 3 DCN/HD/S Upright. 32" x 62" coil boiler (iron). 4" x 7" pumps, air chamber ornament. Two $3^1/_2$" x 10' suctions. $6^3/_4$" x 7" cyls. 16 x 56, 16 x 62 wheels. $56^1/_2$" track. Stiff pole.

11-10-1880: Delivered by Chris Ahrens & Hiram H. Nieman to Engine 31, Chicago, Ill.

1906: Was at Engine 100. Still there 1917.

264 3 DCN/HD/S Upright. 32" x 62" coil boiler (iron). 4" x 7" pumps, air chamber ornament. Two $3^1/_2$" x 10' suctions. $6^3/_4$" x 7" cyls. 16 x 56, 16 x 62 wheels. $56^1/_2$" track. Stiff pole.

11-15-1880: Accepted by Engine 2, Chicago, Ill., without a test.

1906: Was at Engine 38.

1917: Was at Engine 113.

266 3 SCN/MD-HD/S Upright. 32" x 62" coil boiler (iron). $5^1/_2$" x 8" pump, air chamber ornament. 4" x 20' pre-connect suction. 9" x 8" cylinder. Wheels 16 x 56, 16 x 62. Two front springs. Rear brake lever. 4 lanterns.

11-13-1880: Delivered by Foote and Dick Raymond to Wausau, Wis.

12-21-1901: New 32" x 62" Fox boiler.

1909: Replaced by new Nott steamer (which Wausau still owns, 1985). Fate of the Ahrens unknown, probably traded-in on the Nott.

268 1 DSF/HD/S Upright. 38" x 67" coil boiler (iron & silver). 5" x 8" pumps, air chamber ornament. One 5" x 9'6", one 5" x 10'6" suctions. 8" x 8" cyls. Wheels 16 x 60, 18 x 64. $56^1/_2$" track. Stiff pole. (Retired 1907).

2-19-1881: In service, Engine 13, Chicago, Ill.

11-17-1900: 36"x66" Fox boiler, to Engine 59.

270 3 DSF/HD/I Upright. 32" x 62" coil boiler (iron). 4" x 7" pumps, Princess lamp atop air chamber. $3^1/_2$" x 20' pre-connect suction. $6^3/_4$" x 7" cylinders. Wheels 16 x 56, 16 x 62. 59" track. Stiff pole with brace underneath.

12-24-1880: Delivered by Chris Ahrens and Dick Raymond to Engine 3, Louisville, Ky. Name *J.L. Mason.* Replaced 1858 Latta #22.

10-29-1891: Replaced by Ahrens #276. To Eng. 12 in place of Ahrens #384. Renamed *Jacob Thome #12.* (This was a 3-way trade: #270 from Eng. 3 to Eng. 12, #384 from Eng. 12 to E. 2, #276 from Eng. 2 to Eng. 3).

7-11-1892: New front axle.

1895: Replaced by Metropolitan #2406.

3-28-1902: New 32" x 62" Fox boiler ($850).

1923: Retired.

272 3 SSF/HD/W Upright. 32" x 62" coil boiler (iron). One $5^1/_2$" x 8" pump. One 4" x 20' pre-connect suction. 9" x 8" cylinder. Wheels 16 x 56 front, 16 x 62 rear. Stiff pole.

1-28-1881: Delivered by Fred Hawekotte and Dick Raymond to Washington Fire Co. #1, Delphos, Ohio. Name *Harold G. Hayes.* Also manually-drawn hose cart, 500' of hose.

10-3-1883: Traded-in on Ahrens #376.

4-11-1884: Loaned to Eaton, Ohio, while Ahrens #18 was repaired.

274 2 DCN/HD/W Upright. $35^1/_2$" x 67" coil boiler (brass). $4^1/_2$" x 8" pumps. 4" x 22' pre-connect suction. $7^1/_4$" x 8" cylinders. Wheels 16 x 58, 16 x 62. No brake. Drop pole.

3-10-1881: Delivered by Fred Hawekotte and

Dick Raymond to Oshkosh, Wis., snowbound enroute. Name *Brooklyn.* Water in 4:09.

10-30-1887: Drained the fire hydrant faster than the water works could replenish it.

12-24-1908: New 35" x 64" Fox boiler.

276 2 DSF/HD/A Upright. 35$^{1}/_{2}$" x 67" coil boiler (iron). 4$^{1}/_{2}$" x 8" pumps, one flywheel, signal lamp on air chamber. 4" x 20' pre-connect suction. 7$^{1}/_{4}$" x 8" cylinders. Wheels 16 x 58, 16 x 62. 59" track. Stiff pole.

4-2-1881: Delivered by Chris Ahrens and Dick Raymond to Engine 2, Louisville, Ky. Name *B.F. Bache*. Replaced 1858 Latta #21.

5-19-1881: New drop pole.

10-29-1891: Replaced by Ahrens #384. To Engine 3 in place of Ahrens #270. Renamed *J.L. Mason #3*. (This was a 3-way trade: #276 from Eng. 2 to Eng. 3, #270 from Eng. 3 to Eng. 12, #384 from Eng. 12 to Eng. 2).

11-20-1893: Renamed *Paul C. Barth #3.*

9-11-1900: Replaced by Ahrens #165. To E. 16 in place of Ahrens #220. (A 4-way trade: 276 from E. 3 to E. 16; 220 from E. 16 to E. 2; 384 E. 2 to E. 1; 165 from E. 1 to E. 3).

1906: Replaced by Ahrens Continental #111. To Spare at Engine 22.

1920: Retired.

278 2 DSF/HD/A Upright. 35$^{1}/_{2}$" x 67" coil boiler (iron). 4$^{1}/_{2}$" x 8" pumps, one flywheel, signal lamp on air chamber. 4" x 20' pre-connect suction. 7$^{1}/_{4}$" x 8" cylinders. Wheels 16 x 58, 16 x 62. 59" track. Stiff pole.

4-2-1881: Delivered by Chris Ahrens & Dick Raymond to Engine 4, Louisville, Ky. Name *Frank Roberts*. Replaced 1859 Latta #23.

7-25-1893: Renamed *John Doerhoefer #4.*

1904: Replaced by Metropolitan #2993.

280 1 DCN/HD/H Upright. 38" x 67" coil boiler (iron & silver). 5" x 8" pumps, air chamber ornament. 4$^{1}/_{2}$" x 23' pre-connect suction. 8" x 8" cyls. Wheels 16 x 60, 18 x 64. Drop pole.

5-24-1881: Accepted by Fire Co. #3, Cincinnati, Ohio. Delivered by Fred Hawekotte and Peter Ahrens. Replaced Ahrens #228.

7-18-1884: Collided with Franklin Fire Co. #4 (Ahrens #198) at 9th & Elm streets. Rebuilt and to Fire Co. #25, according to Ahrens records (but 1886 Cincinnati City Directory lists only up to Fire Co. #24; Fire Co. 25 apparently didn't exist yet).

11-18-1890: New boiler coils.

1-29-1892: New coil boiler. Replaced by Ahrens #534. To Fire Co. 3.

12-4-1893: Collided with Fire Co. 21 (Ahrens #228) at 3rd & Elm streets.

3-1-1902: Repaired at factory. To Co. #12 in place of Ahrens 17. Renamed *J.A. Archibald.*

7-6-1908: Front springs rebuilt by Ahrens Fire Engine Co.

282 4 SSF/MD-HD/A Upright. 30x60" coil boiler (iron). 5" x 7" pump, lamp on air chamber. 3$^{1}/_{4}$" x 20' pre-connect suction. 8" x 7" cyl. Wheels 16 x 54, 16 x 60. Rear brake lever.

4-19-1881: Delivered by Chris Ahrens to Rushville, Ind. Name *Rushville #1.*

8-25-1887: G. Fred Ahrens repaired boiler coils at Rushville.

3-18-1889: Joe Terbeck installed new boiler coils at Rushville.

284 4 SSF/HD/A Upright. 30" x 60" coil boiler (iron). 5" x 7" pump. 3$^{1}/_{4}$" x 20' pre-connect suction. 8" x 7" cylinder. Wheels 16 x 54, 16 x 62. Stiff pole. Rear brake lever.

6-24-1881: Delivered by Peter Ahrens and Fennessy to Paducah, Ky.

11-10-1886: New boiler coils.

286 4 SSF/HD/A Upright. Round 30" x 60" coil boiler (iron). 5" x 7" pump. One 3$^{1}/_{2}$" x 20' pre-connect suction. 8" x 7" cyl. Wheels 16 x 54 front, 16 x 60 rear. Rear brake lever.

5-25-1882: Delivered by Fred Hawekotte and Foote to Wisconsin Industrial School, Waukesha, Wis.

4-28-1900: Trade-in to A.E. Foote, Milwaukee, Wis. He planned to repair it and sell it. (Not clear whether this was the same "Foote" who helped deliver it 18 years earlier).

288 4 DSF/HD/S Upright. 30"x60" tubular boiler (nickel), silver-plated coal bunker with metal number "13" on back, two coach lamps on boiler. 3$^{5}/_{8}$" x ? pumps, one nickel-plated inlet, signal lamps atop pump and air chamber. 3$^{1}/_{2}$" x 20' pre-connect suction. 7" x ? cylinders (silver). Wheels 16 x 54, 16 x 60. Silver-plated frame, springs, axles. Stiff pole. Single seat, Star lamps on both sides.

6-20-1881: Tested & accepted by Fire Co. #13, New Orleans, La., without anyone from Ahrens Co. present. Threw one stream 275'.

290 3 DCN/HD/S Upright. 32" x 62" coil boiler (iron). 4" x 7" pumps, air chamber ornament. Two 3$^{1}/_{2}$" x 10' pre-connect suctions. Two 6$^{3}/_{4}$" x 7" cylinders. Wheels 16 x 56 front, 16 x 62 rear. Drop pole. Gong on foot board.

8-3-1881: Delivered by Peter Ahrens to Battle Creek, Mich.

10-21-1884: New suction coupling.

10-21-1904: New boiler coils.

1914: Traded-in for $500 on new Robinson pumper. Was in extremely poor condition, so Robinson probably junked it.

292 3 SSF/HD/S Upright. 32" x 62" coil boiler (iron & silver). $5^{1}/_{2}$" x 8" pump, automatic air pump. 4" x 20' pre-connect suction. 9" x 8" cyl. Wheels 16 x 56, 16 x 62. Stiff pole.

11-5-1881: Delivered by Fred Hawekotte and Peter Ahrens to Frankfort, Ky.

1- -1884: Boiler coils repaired ($51.96).

12-27-1889: New boiler coils.

294 3 DSF/HD/W Upright. 32" x 62" tubular boiler (silver), 2 coach lamps on boiler, painted coal bunker with metal number "3" on back. 4" x 7" pumps, signal lamps on pump and air chamber. $3^{1}/_{2}$" x 20' pre-connect suction. $6^{3}/_{4}$" x 7" cyls. (silver). Wheels 16 x 56, 16 x 62. Two front springs. Silver-plated frame, springs, axles. No brake. Star lamps on seat.

8-20-1881: Tested & accepted by Fire Co. #3, New Orleans, La., without anyone present from Ahrens Mfg. Co.

296 1 DSF/HD/H Upright. 38" x 67" coil boiler (iron & silver). 5" x 8" pumps, nickel-plated air chamber (ornament atop), nickel-plated flywheel, pump heads, support columns. Two 5" x 10' suctions. Two 8" x 8" cyls. (nickel-plated). Wheels 16 x 60, 18 x 64. Stiff pole.

8- -1881: Displayed at the Exposition, Cincinnati, Ohio.

9-3-1881: Accepted by Eng. 32, Chicago, Ill., without anyone present from Ahrens Co.

11-11-1903: New 38" x 67" Ahrens coil boiler.

1906: At Engine 3. Retired by 1917.

298 2 DCN/HD/H Upright. $35^{1}/_{2}$" x 67" coil boiler (iron). $4^{1}/_{2}$" x 8" pumps. Two 4" x 10' suctions. $7^{1}/_{4}$" x 8" cyls. Wheels 16 x 58, 16 x 62. Steel springs. $56^{1}/_{2}$" track. Stiff pole.

11-23-1881: In service, Engine 8, Chicago, Ill.

1906: At Eng. 26. 1917: At Engine 79.

300 2 DCN/HD/H Upright. $35^{1}/_{2}$" x 67" coil boiler (iron). Two $4^{1}/_{2}$" x 8" pumps. Two 4" x 10' suctions. Two $7^{1}/_{4}$" x 8" cyls. Wheels 16 x 58 front, 16 x 62 rear. Steel springs. $56^{1}/_{2}$" track. Stiff pole. Plain finish (no plating).

11-23-1881: In service, Eng. 31, Chicago, Ill.

1906: At Engine 65.

302 3 SSF/HD/S Upright. 32" x 62" coil boiler (iron). One $5^{1}/_{2}$" x 8" pump. One 4" x 20' pre-connect suction. One 9" x 8" cyl. Wheels 16 x 56 front, 16 x 62 rear. Drop pole.

4-6-1882: Delivered by John McMahon and Fred Hawekotte to Greencastle, Ind.

1902: Traded-in to Seagrave, Columbus, O.

4-8-1903: Seagrave sold this for $200 to American Fire Engine, Seneca Falls, N.Y.

11-22-1906: After removing pump and all brass parts, American-LaFrance sold what was left for $20 to a scrap dealer.

304 5 SSF/MD/H Upright. 28" x $56^{1}/_{2}$" flue boiler (iron/silver). One $4^{3}/_{8}$" x 7" pump, air chamber ornament. $3^{1}/_{4}$" x 20' pre-connect suction. One 7" x 7" cyl. Wheels 16 x 52 front, 16 x 60 rear. Driver's seat.

10-8-1881: Tested & accepted by West Belleville, Ill., on 10th anniversary of Great Chicago Fire. Threw a $^{15}/_{16}$" stream 210'.

2-20-1888: Sold to Belleville, Ill.

306 1 DSF/HD/H Upright. 38" x 67" coil boiler (iron & silver). Two 5" x 8" pumps. Two 5" x 10' suctions. Two 8" x 8" cyl. Wheels 16 x 60, 18 x 64. Stiff pole. (Retired by 1906).

10-12-1881: In service, Eng. 11, Chicago, Ill.

308 2 DCN/HD/H Upright. $35^{1}/_{2}$" x 67" coil boiler (iron). $4^{1}/_{2}$" x 8" pumps. Two 4" x 10' suctions. $7^{1}/_{4}$" x 8" cyls. Wheels 16 x 58, 16 x 62. Steel springs. $56^{1}/_{2}$" track. Stiff pole.

12-22-1881: Shipped to Eng. 14, Chicago, Ill.

1906: At Engine 50. Retired by 1917.

310 4 DSF/HD/S Upright. 30" x 60" tubular boiler (nickel), coach lamp on boiler, nickel-plated coal bunker with metal number "2" on back. $3^{5}/_{8}$" x ?? pumps, signal lamp on air chamber. $3^{1}/_{2}$" x 20' pre-connect suction. 7" x ? cyls. (nickel-plated). Wheels 16 x 54, 16 x 60. Nickel-plated frame, springs, axles. Stiff pole. Single seat with one Star lamp on side.

12-19-1881: Shipped to Florida Fire Co. #2, Pensacola, Florida.

5-21-1885: Fire Co. renamed *Washington #2.*

6- -1888: Sold to Mobile, Alabama.

5- -1891: Sold to Dantzler Lumber Co., Moss Point, Mississippi. Mounted on a flat boat to protect the saw mills along the river.

1902: Boat and steamer sank in the river.

1903: Dantzler Co. dredged the Ahrens out of the river. After several months under water, no longer deemed fit for service. Sat in a field in Moss Point for 3 years.

1906: Sold to Mr. Torres of Moss Point for scrap (brass and copper).

312 2 DCN/HD/H Upright. $35^1/2$" x 67" coil boiler (iron). Two $4^1/2$" x 8" pumps. Two 4" x 10' suctions. Two $7^1/4$" x 8" cyls. Wheels 16 x 58 front, 16 x 62 rear. Stiff pole.

1-19-1882: Shipped to Engine Co. #5, Town of Lake, Ill.

1984: U.S. Post Office says there is no such place as Lake or Town-of-Lake, Ill., so this steamer's further history could not be traced. May have been in Lake County, Ill., north of Chicago. Probably annexed to another town or the town was renamed.

314 1 DSF/HD/H Upright. 38" x 67" coil boiler (iron). Two 5" x 8" pumps. One $4^1/2$" x 23' pre-connect suction. Two 8" x 8" cyls. Wheels 16 x 60 front, 18 x 64 rear. Weight 8190#.

4-1-1882: In service at Marion Fire Co. #8, Cincinnati, Ohio.

8-12-1885: Several coils in the boiler burst from overheating while working at a fire at 3:30 a.m. Returned to factory for repair.

11-5-1885: Replaced by Ahrens #443.

1-28-1887: New boiler coils.

10-5-1894: New boiler coils, rear section of boiler only.

12-6-1894: The new rear-section boiler coils of two months earlier were allowed to overheat and sprang a leak. New coils installed to replace the leaking coils.

2-18-1896: New boiler coils.

6-30-1902: New 38" x 64" Fox boiler, and repainted ($1808.25). To Fire Co. #14.

316 1 DSF/HD/H Upright. 38" x 67" coil boiler (iron). 5" x 8" pumps. $4^1/2$" x 22' pre-connect suction. 8" x 8" cyls. Wheels 16 x 60, 18 x 64.

3-1-1882: In service, Relief Fire Co. #2, Cincinnati, Ohio. Probably replaced Ahrens #1.

5-21-1895: New boiler coils.

7-8-1902: New 38" x 64" Fox vertical water tube boiler, and repainted. To Fire Co. #4.

3-8-1906: Pump overhauled. Rubber tires installed by Bode Wagon Co., Cincinnati. Repainted by a shop on Spring Grove Ave.

318 2 DCN/HD/H Upright. $35^1/2$" x 67" coil boiler (iron). $4^1/2$" x 8" pumps. One 4" x 22' pre-connect suction. Two $7^1/4$" x 8" cyls. Wheels 16 x 58 front, 16 x 62 rear. Drop pole.

7-12-1882: In service at Engine Co. #1, Memphis, Tenn. Delivered by Fred Hawekotte. Threw one stream 287'.

11-20-1901: New rear wheels ($121).

12-27-1901: New 4" x 22' suction.

320 2 DCN/HD/H Upright. $35^1/2$" x 67" coil boiler (iron). $4^1/2$" x 8" pumps. One 4" x 22' pre-connect suction. Two $7^1/4$" x 8" cyls. Wheels 16 x 58 front, 16 x 62 rear. Drop pole.

8-12-1882: In service at Engine 2, Memphis, Tenn. Accepted without anyone present from Ahrens Co. Threw one stream 286'.

5-24-1891: New 4" x 22' pre-connect suction. To Engine 4. This was a three-way trade: #320 from Eng. 2 to Eng. 4, #414 from Eng. 4 to Eng. 6, #406 from Eng. 6 to Eng. 2.

7-3-1895: New boiler coils.

1-21-1901: New rear wheels.

322 2 DSF/HD/S Upright. $35^1/2$" x 67" coil boiler (iron). $4^1/2$" x 8" pumps. One 4" x 20' pre-connect suction. Two $7^1/4$" x 8" cyls. Wheels 16 x 58 front, 16 x 62 rear. Drop pole.

5-1-1883: Tested & accepted at Webster and Sycamore, Cincinnati. In service at Washington Fire Co. #1, Cincinnati, Ohio.

11-26-1887: Boiler overheated, damaged several boiler coils, and the boiler shell bulged outward just above the fire door. Repaired at Ahrens factory.

7- -1894: To Co. #17. Replaced Ahrens #11.

3-17-1903: New boiler coils.

324 3 DSF/HD/W Upright. 32" x 62" tubular boiler (silver), silver-plated coal bunker. 4" x 7" pumps, nickel-plated crankshaft & columns, signal lamp on air chamber. $3^1/2$" x 20' pre-connect suction. $6^3/4$" x 7" cyls. Two Star and two Globe lamps. Wheels 16 x 56, 16 x 62. One front spring. Stiff pole. Single seat.

10-12-1882: Shipped to Volunteer Fire Co. #1, New Orleans, La.

4-12-1883: Ahrens #338 to St. Bernard Fire Co. #1, New Orleans, La. Not clear whether that steamer replaced this one, or they were two separate fire companies.

326 5 SSF/HD/W Upright. 28" x $56^1/2$" tubular boiler (silver), silver-plated coal bunker. One $4^1/2$" x 7" pump, nickel-plated columns and crankshaft, signal lamp on air chamber. One $3^1/4$" x 20' pre-connect suction. 7" x 7" cyl. 2

Ahrens #320 was a 2nd Class Double Crane Neck built in 1882 for Engine 2, Memphis, Tenn. This c-1898 photo shows it in front of Engine Co. 4, where it had been serving since 1891.
(Ed Hass collection)

Star lamps, 2 Globe lamps. Wheels 16 x 52 front, 16 x 60 rear. Stiff pole. Single seat.

10-12-1882: Shipped to Philadelphia Fire Co. #14, New Orleans, La.

328 3 SSF/MD-HD/A Upright. 32" x 62" coil boiler (iron). 5 1/2" x 8" pump. One 4" x 20' pre-connect suction. 9" x 8" cylinder. Wheels 16 x 56 front, 16 x 62 rear. Front brake lever.

10- -1882: Delivered by G. Fred Ahrens to Huntington, West Virginia. Pumped for 6 days, through 1500' of hose, up a 60' grade.

2-27-1885: Joe Terbeck installed new boiler coils at Huntington.

330 1 DCN/HD/H Upright. 38" x 67" coil boiler (iron). 5" x 8" pumps, one flywheel, air chamber ornament. 4" x 20' pre-connect suction. Two 8" x 8" cyls. Wheels 16 x 60 front, 18 x 64 rear. 57 1/2" track. Plain finish.

10-13-1882: Shipped to St. Louis, Mo. Delivered by G. Fred Hawekotte and Hiram H. Nieman to Engine 12. Replaced Ahrens #19.

9-20-1898: New 36" x 62" Fox boiler.

1909: Was a Spare at Engine 7. Engine 12 was using 1897 Metropolitan #2563.

332 3 DCN/HD/S Upright. 32" x 62" coil boiler (silver), silver-plated coal bunker, raised letters *Washington #3* on back. 4" x 7" pumps, nickel-plated air chamber with signal lamp atop. 3 1/2" x 20' pre-connect suction. 6 3/4" x 7" cyls. Gong on foot board. Wheels 16 x 56, 16 x 62. 58" track. Stiff pole. Double seat with railing atop. Ratchet brake.

10-10-1882: Shipped to Washington Fire Co. #3, Vicksburg, Mississippi.

3-19-1896: New boiler coils.

334 3 DCN/HD/H Upright. 32" x 62" coil boiler (iron). 4" x 7" pumps. One 2 1/2" x 15' plug suction. 6 3/4" x 7" cyls. Wheels 16 x 52 front, 18 x 62 rear. 56 1/2" track. Plain finish.

11-8-1882: Shipped to Chicago, Ill. Harold Walker lists this and #366 as Engine 15.

12-6-1900: New 32" x 62" Fox boiler for #334 of Engine 78.

9-1-1903: 32" x 66 3/4" Ahrens coil boiler.

1906: At Engine 48. 1917: At Engine 95. Later Engine 119.

336 3 DCN/HD/H Upright. 32" x 62" coil boiler (iron). 4" x 7" pumps. One 2 1/2" x 15' plug suction. 6 3/4" x 7" cyls. Wheels 16 x 52 front, 18 x 62 rear. 56 1/2" track. Plain finish.

11-8-1882: Shipped to Chicago, Ill. In service at Engine Co. #56. (New boiler 1902).

1917: Still at Engine 56.

338 5 SSF/HD/W Upright. 25" x $56^{5}/_{8}$" tubular boiler (silver), silver-plated coal bunker. One $4^{1}/_{2}$" x 7" pump, nickel-plated columns and crankshaft, signal lamp atop air chamber. $3^{1}/_{4}$" x 20' pre-connect suction, attached to rear inlet above coal bunker. 7" x 7" cylinder. 2 Star lamps, 2 Globe lamps. Wheels 16 x 52, 16 x 60. Stiff pole. Single seat.

3-27-1882: Shipped to New Orleans, La.

4-12-1882: Tested & accepted by St. Bernard Fire Co. #1, New Orleans, La.

6-12-1911: Sold to Robinson Land & Lumber Co., Chicora, Miss. American-LaFrance supplied new relief valve, new grates for bottom of boiler.

340 3 SSF/MD-HD/A Upright. 32" x 62" coil boiler (iron). $5^{1}/_{2}$" x 8" pump, air pump. $3^{1}/_{2}$" x 20' pre-connect suction. 9" x 8" cylinder. Wheels 16 x 56, 16 x 62. Rear brake lever.

1-24-1883: Delivered by John P. Ahrens (10-year-old son of Chris Ahrens) and John McMahon to Engine 2, Lawrenceburg, Ind.

9-3-1890: Joe Terbeck installed new boiler coils at Lawrenceburg.

6-5-1899: Returned to factory for repairs totaling $1290. New 32" x 64" Fox vertical water tube boiler, new $3^{1}/_{2}$" x 20' pre-connect suction, overhauled pump including replacing all iron pipes on the pump with brass tubing, repaired wheels.

342 3 DCN/HD/W Upright. 32" x 64" coil boiler (iron). Two 4" x 7" pumps. Two $3^{1}/_{2}$" x 10' suctions. Two $6^{3}/_{4}$" x 7" cylinders. Plain finish. Wheels 16 x 56 front, 16 x 62 rear.

12-2-1882: In service, Engine 26, Chicago, Ill.

9- -1900: Shipped new 32" x 64" Fox boiler for this steamer, was then at Engine 49.

1906: Replaced by Metropolitan #3150. To Engine 68. Still at 68 in 1917.

344 3 DSF/HD/H Upright. 32" x 62" coil boiler (iron). 4" x 7" pumps, nickel-plated air chamber with nickel-plated Globe signal lamp atop. $3^{1}/_{2}$" x 20' pre-connect suction. $6^{3}/_{4}$" x 7" cylinders. Wheels 16 x 56, 16 x 62. Plated springs & axles. 59" track. Drop pole.

12-22-1882: Delivered by Chris Ahrens and John McMahon to Engine 11, Louisville, Ky. Name *Vincent Bradas*. Threw a stream 278'.

1906: Rebuilt by Ahrens Fire Engine Co. To Engine 20, and replaced at Engine 11 by 1872 Ahrens #24.

1919: Motorized with 1919 Fordson tractor.

1923: Retired.

346 1 DSF/HD/H Upright. 38" x 67" coil boiler (iron). Two 5" x 8" pumps. One $4^{1}/_{2}$" x 20' pre-connect suction. Two 8" x 8" cyls. Wheels 16 x 60 front, 18 x 64 rear. Drop pole.

12-28-1882: Delivered to Covington, Ky. Name *Samuel Cofflin*. Threw a stream 310'.

1- -1884: Pumped $189^{1}/_{2}$ hours non-stop at Covington during flood of the Ohio River.

12-30-1902: Traded-in for $250 on repairs made to boiler of Ahrens #169.

348 3 SSF/MD-HD/A Upright. 32" x 62" coil boiler (iron). One $5^{1}/_{2}$" x 8" pump, automatic air pump. One $3^{1}/_{2}$" x 20' pre-connect suction. One 9" x 8" cylinder. Wheels 16 x 56 front, 16 x 62 rear. Rear brake lever.

1-24-1883: Delivered by John P. Ahrens and John McMahon to Eng 1, Lawrenceburg, Ind.

12-26-1893: George Zink installed new boiler coils at Lawrenceburg.

4-17-1901: Traded-in on ex-Nashville 1875 Ahrens #158, which Lawrenceburg bought after it was rebuilt as #458.

12-26-1902: Loaned to Wilmington, Ohio, while their Ahrens #605 was rebuilt. Later loaned to Cardington, Ohio, while American Fire Engine Co. (Seneca Falls, N.Y.) rebuilt their 1874 Silsby.

350 3 DCN/HD/S Upright. 32" x 62" coil boiler (iron). Two 4" x 7" pumps. Two 4" x 10' suctions. Two $6^{3}/_{4}$" x 7" cylinders. Wheels 16 x 56 front, 16 x 62 rear. Stiff pole.

3-14-1883: Delivered by John P. Ahrens to Sheboygan, Wis.

2-9-1888: Its record at Sheboygan to date included pumping two streams for $14^{1}/_{2}$ hours despite -10° F weather, and pumping two streams for 16 hours non-stop in less-severe weather.

352 3 DCN/HD/S Upright. 32" x 62" coil boiler (iron). Two 4" x 7" pumps. Two 4" x 10' suctions. Two $6^{3}/_{4}$" x 7" cyls. Wheels 16 x 56 front, 16 x 62 rear. Stiff pole. Weight 6525#.

4-12-1883: Tested & accepted by Minneapolis, Minn. Ahrens records say this went in service as Engine 3, but Minneapolis records show it originally served Engine 4.

1886: To Engine 9.

1888: Moved from Engine 9 to Engine ??.

1893: To Engine 15.

1904: Rebuilt with Nott boiler.

1913: To Engine 18.

1921: Retired.

354 4 SSF/HD/A Upright. 30" x 60" coil boiler (iron). One 5" x 7" pump, air pump. One $3^1/2$" x 20' pre-connect suction. One 8" x 7" cylinder. Wheels 16 x 54 front, 16 x 60 rear. Drop pole. Front brake lever. Weight 4650#.

7-9-1883: Delivered by Chris Ahrens and Peter Ahrens to New York, N.Y. Ahrens records say this was Engine 42 and Ahrens #356 was Engine 45, but FDNY shows this was Engine 45 (on Division St. near Boston Road) and Ahrens #356 was Engine 42.

4-24-1889: New boiler coils.

11-6-1894: New boiler coils, wheels repaired.

12- -1898: Fox boiler installed by American Fire Engine Co., Seneca Falls, N.Y.

1902: Became a Spare.

1909: Was a Spare at 18th Battalion.

356 4 SSF/HD/A Upright. 30" x 60" coil boiler (iron). One 5" x 7" pump, air pump. One $3^1/2$" x 20' pre-connect suction. One 8" x 7" cylinder. Wheels 16 x 54 front, 16 x 60 rear. Drop pole. Front brake lever. Weight 4650#.

7-9-1883: Delivered by Chris Ahrens and Peter Ahrens to New York, N.Y. Ahrens records say this was Engine 45 and Ahrens #354 was Engine 42, but FDNY shows this was Engine 42 (on Fulton St. between 167th and 168th) and Ahrens #354 was Engine 45.

1902: Had been scrapped.

358 3 DCN/HD/A Upright. 32" x 62" coil boiler (iron). $4^1/4$" x 7" pumps. One 4" x 9', one 4" x 10', one $2^1/2$" x 9' suctions. Two $6^3/4$" x 7" cylinders. Wheels 16 x 56 front, 18 x 62 rear. $56^1/2$" track. Stiff pole. Weight 6075#.

6-26-1883: Delivered by Chris Ahrens and Peter Ahrens to New York, N.Y. In service at Engine 13-2, 99 Wooster St., Manhattan.

2-23-1888: New boiler coils.

7- -1892: New front and rear wheels.

1893: To FDNY Engine 28, Manhattan.

8- -1894: Rebuilt by LaFrance Fire Engine Co., Elmira, N.Y.

1902: Became a Spare.

1909: Was a Spare at 10th Battalion.

360 3 DCN/HD/A Upright. 32" x 62" coil boiler (iron). $4^1/4$" x 7" pumps. One 4" x 9', one 4" x 10', one $2^1/2$" x 9' suctions. Two $6^3/4$" x 7" cyls. Wheels 16 x 56 front, 18 x 62 rear. $56^1/2$" track. Stiff pole. Weight 6075#.

6-26-1883: Delivered by Chris Ahrens and Peter Ahrens to New York, N.Y. In service at Engine 20, 47 Marion St., Manhattan.

8- -1893: Rebuilt by LaFrance Fire Engine Co., Elmira, N.Y. To Engine 11, Manhattan.

1902: To Engine 45, Manhattan.

1909: Was in service at Engine 208.

362 3 DCN/HD/S Upright. 32" x 62" coil boiler (iron). 4" x 7" pumps. One $3^1/2$" x 20' pre-connect suction. Two $6^3/4$" x 7" cyls. Wheels 16 x 56 front, 16 x 62 rear. Drop pole.

6-30-1883: Delivered by Chris Ahrens and Hiram H. Nieman to Portsmouth, Ohio.

6-20-1892: George Zink installed new boiler coils at Portsmouth.

4-24-1897: George Zink installed new boiler coils at Portsmouth.

10-20-1905: Rebuilt by American-LaFrance, Elmira, N.Y. New 32" x 62" Fox boiler, pump overhauled, entire steamer repainted.

364 3 DCN/HD/A Upright. 32" x 62" coil boiler (iron). 4" x 7" pumps. One 4" x 9'6", one 4" x 10'6", one $2^1/2$" x 8'6" suctions. Two $6^3/4$" x 7" cylinders. Wheels 16 x 56 front, 16 x 62 rear. Stiff pole. Plain finish.

8-23-1883: Delivered by G. Fred Hawekotte and John McMahon to Peoria, Ill.

1-1-1888: Joe Terbeck installed new boiler coils at Peoria. Boiler had been allowed to get so hot that one of the coils had split open along 12 inches of its length.

10-24-1902: New 32" x 62" Fox boiler, miscellaneous repairs ($1250).

366 3 DCN/HD/A Upright. 32" x 67" coil boiler (iron), heater pipes right. 4" x 7" pumps, one flywheel, nickel-plated valves, relief valve left. Two 4" x 10'6", one $2^1/2$" x 9' suctions. $6^3/4$" x 7" cylinders. Wheels 16 x 56 front, 16 x 62 rear. $56^1/2$" track. Plain finish.

12-19-1883: Shipped to Chicago, Ill. (Harold Walker lists #366, #334, & #655 as Eng. 15).

9-1-1903: 32" x $66^3/4$" Ahrens coil boiler.

1906: At Engine 39. Retired 1907.

367 3 DCN/HD/A Upright. 32" x 67" coil boiler (iron), heater pipes right. 4" x 7" pumps, one flywheel, nickel-plated valves, relief valve right. Two 4" x 10'6", one $2^1/2$" x 9' suctions. $6^3/4$" x 7" cylinders. Wheels 16 x 56 front, 16 x 62 rear. $56^1/2$" track. Plain finish.

12-19-1883: Shipped to Chicago, Ill. (Harold Walker lists this as Engine 32. Ahrens records show #236 as Engine 32).

1904: New boiler.

1906: At Engine 67. Retired by 1917.

370 4 DSF/HD/W Upright. 30" x 60" tubular boiler (silver), silver-plated coal bunker. $3^5/8$" x ?? pumps, nickel-plated air chamber, signal lamp atop. One $3^1/2$" x 20' pre-connect suction. Two 7" x ?? cyls. Wheels 16 x 52 front, 16 x 62 rear. One Star and one Globe lamp.

6-6-1883: In service at Jefferson Fire Co. #23, New Orleans, La.

372 1 DCN/HD/S Upright. 38" x 67" coil boiler (iron). 5" x 8" pumps, relief valve, ornament atop air chamber. Two 5" x 10' suctions, two $2^1/2$" x 10' suctions. Two 8" x 8" cyls. Wheels 16 x 60 front, 18 x 64 rear. Weight 9140#.

9-14-1883: In service, Engine 4, Milwaukee, Wis. Delivered by Fred Hawekotte. Replaced Ahrens #246.

9-1-1903: Replaced by Metropolitan #2989. To Engine 18 in place of Ahrens #2177.

1911: Rebuilt by Ahrens Fire Engine Co., including new Continental boiler.

1918: Replaced by Seagrave pumper. Retired.

1919: Returned to service at Engine 18.

1921: Replaced by Ahrens #2123. Retired.

374 2 DCN/HD/A Upright. $35^1/2$" x 67" coil boiler (iron). Two $4^1/2$" x 8" pumps, relief valve, one nickel-plated flywheel; nickel-plated air chamber, crankshaft and pump support columns. Two 4" x 9' suctions. Two $7^1/4$" x 8" nickel-plated cylinders. Wheels 16 x 58 front, 16 x 62 rear. $56^1/2$" track.

11-12-1883: Shipped to Chicago, Ill.

10-23-1900: 35" x 64" Fox vertical water tube boiler for #374 of Engine 3.

12-23-1903: Shipped new $35^1/2$" x 67" Ahrens coil boiler for #374.

1906: At Engine 95. 1917: At Engine 117.

376 4 SSF/MD-HD/A Upright. 30" x 60" tubular boiler (iron). One 5" x 7" pump, automatic air pump. One $3^1/4$" x 20' pre-connect suction. One 8" x 7" cylinder. Wheels 16 x 54 front, 16 x 60 rear. Rear brake lever.

10-3-1883: Delivered by John P. Ahrens and G. Fred Ahrens to Washington Fire Co. #1, Delphos, Ohio. Took Ahrens #272 in trade.

2-29-1892: New boiler coils.

7- -1895: $3^1/4$" x 20' pre-connect suction.

1920: Junked.

Engine 18 in Milwaukee, Wis., used this 1883 Ahrens 1st Class Double Crane Neck #372 from 1903 to 1918, and again (after a brief retirement) from 1919 to 1921. It is seen here while assigned to Engine 18, sometime after it received a new Continental boiler in 1911. Original assignment (1883-1903) was Engine 4.

(Wayne Sorenson collection)

The "Henry Young #1" is an 1883 3rd Class single straight-frame Ahrens #378 that served Sidney, Ohio until 1928. The Fox boiler was installed in 1901. Photo above shows how it looks today, following a 1955 restoration by firefighters of Dayton, Ohio. Sketch below shows how this engine (in fact,all Ahrens engines) operated. (Courtesy Carillon Park, Dayton, Ohio)

378 3 SSF/MD-HD/A Upright. 32" x 62" tubular boiler (iron). One $5^1/2$" x 8" pump, piston air pump. One 4" x 20' pre-connect suction. One 9" x 8" cylinder. Wheels 16 x 56 front, 16 x 62 rear. Rear brake lever. Weight 5800#.

9-27-1883: Delivered by G. Fred Ahrens and G. Fred Hawekotte to Sidney, Ohio. Name *Henry Young #1*. Price $4000.

1-26-1888: Pumped $6^1/2$ hours non-stop at Anderson & Frazier Wheel Works fire (3-story, 80' x 100'). From 8:30 a.m. to 9:30 a.m., had 4 streams from the water works and 3 from the Ahrens, until the water works broke down. From 9:30 a.m. to 3:00 p.m., the Ahrens extinguished the blaze alone.

10-15-1901: 30" x 60" Fox boiler ($1000).

1916: To reserve.

1928: Retired and to storage.

1953: Restored for Sidney by members of the Dayton Fire Dept.

1955: Donated for display in a permanent pavilion at Carillon Park, Dayton, Ohio. It is still there as of 1984.

380 2 DCN/HD/S Upright. $35^1/2$" x 67" coil boiler (iron), safety rails above coal bunker, heater connections. $4^1/4$" x 8" pumps, relief valve, ornament atop air chamber. One 4" x 22' pre-connect suction. 8" x 8" cyls. Wheels 16 x 58 front, 16 x 62 rear. Drop pole.

11- -1883: Delivered by Chris Ahrens and G. Fred Ahrens to Nashville, Tenn. Name *John M. Lea #5*.

5-6-1889: New Ahrens coil boiler and gauges.

11-23-1894: Pump overhaul, converted from one 4" x 22' pre-connect to two 4" x 10' suction hoses. Suction bracket added, left side of boiler. Foot rail for driver added on foot board. Repairs made by Nashville Fire Dept., using parts supplied by Ahrens.

382 2 DCN/HD/S Upright. $35^1/2$" x 67" coil boiler (iron), safety rails above coal bunker, heater connections. $4^1/4$" x 8" pumps, relief valve, ornament atop air chamber. One 4" x 22' pre-connect suction. 8" x 8" cyls. Wheels 16 x 58 front, 16 x 62 rear. Drop pole.

11- -1883: Delivered by Chris Ahrens and G. Fred Ahrens to Nashville, Tenn. Name *James Geddes #6*.

11-17-1890: New Ahrens coil boiler. Pump bored-out and overhauled.

12-17-1894: Pump overhauled, and converted from one 4" x 22' pre-connect to two 4" x 10' suction hoses. Suction bracket added to left side of boiler. Foot rail for driver added on foot board. Repairs made by Nashville Fire Dept., using parts supplied by Ahrens.

6-1-1903: New Archibald rear wheels.

7-13-1904: New boiler coils.

384 1 DSF/HD/H Upright. Round 38" x 67" coil boiler (iron). Two 5" x 8" pumps, lamp atop air chamber, 2 front discharge gates. One $4^1/2$" x 21' pre-connect suction. Two 8" x 8" cylinders. Wheels 16 x 60 front, 18 x 64 rear. Two front springs. 59" track. Drop pole.

8-30-1883: Delivered by Chris Ahrens to Eng. 12, Louisville, Ky. Name *Jacob Thome*.

10-29-1891: Replaced by Ahrens #270. To Engine 2 in place of Ahrens #276. Renamed *B.F. Bache*. (This was a 3-way trade: #384 from Eng. 12 to Eng. 2, #276 from Eng. 2 to Eng. 3, #270 from Eng. 3 to Eng. 12).

8- -1900: New 36" x 66" Fox sectional water tube boiler. Pump rebuilt. Wheels repaired. Entire steamer repainted, and nickel parts replated. An old Ahrens was loaned to Louisville, no charge, while repairs made.

9-11-1900: Returned to Louisville and in service as Engine 1 in place of Ahrens #165. Renamed *Charles D. Jacobs*. (This was a 4-way trade: #384 from Eng. 2 to Eng. 1, #165 from Eng. 1 to Eng. 3, #276 from Eng. 3 to Eng. 16, #220 from Eng. 16 to Eng. 2).

3-4-1907: New fifth wheel and new brakes by Ahrens Fire Engine Co.

1915: Still active at Engine 1, Louisville.

386 3 DCN/HD/A Upright. 32" x $62^1/2$" coil boiler (iron), heater pipes left. $4^1/4$" x 7" pumps, air chamber ornament. 4" x 21' pre-connect suction. $6^3/4$" x 7" cyls. Wheels 16x56, 18x62. Drop pole. Gong on foot board. Plain finish.

12-4-1883: Delivered by John P. Ahrens to Ottawa, Kansas. Pumped two streams at 120 pounds steam pressure, fire-box door open.

3-1-1893: New boiler coils. They refused to pay the bill for this because it was still covered by the 10-year warranty.

388 4 SSF/MD-HD/A Upright. 30" x 60" coil boiler (iron). 5" x 7" pump, automatic air pump, ornament atop air chamber. One $3^1/2$" x 20' pre-connect suction. 8" x 7" cyl. Wheels 16 x 54 front, 16 x 60 rear. Rear brake lever.

11-27-1883: Chris, Peter, & Fred Ahrens delivered to Harrison, O. Name *Brackenridge*.

390 4 SSF Upright. Listed in Ahrens records as Celina, Ohio, 1884, but no details.

391 1 DCN/HD/A Upright. 38" x 67" coil boiler (iron), heater pipes left. 5" x 8" pumps, air

chamber ornament, relief valve left. $4^{1}/_{2}$" x 20' pre-connect suction. 8"x8" cylinders. Wheels 16 x 60 front, 18 x 64 rear. Drop pole. Gong on foot board. Plain finish.

2-29-1884: Delivered by Fred Hawekotte and Peter Ahrens to Engine 2, St. Paul, Minn.

10-20-1896: New boiler coils.

1906: Nott boiler.

392 3 DCN/HD/S Upright. 32" x 62" coil boiler (iron), heater pipes left. $4^{1}/_{4}$" x 7" pumps, relief valve left, nickel-plated air chamber with nickel-plated Globe signal lamp atop. 4" x 22' pre-connect suction. Two $6^{3}/_{4}$" x 7" cylinders. Wheels 16 x 56 front, 18 x 62 rear. Stiff pole. Gong on foot board. Plain finish.

2-8-1884: Delivered by Peter Ahrens to Engine 1, Dallas, Texas. Name *Old Tige* (for Dallas mayor W.A. "Old Tige" Cabell).

5-9-1893: New boiler coils.

2-5-1894: New circulator.

6-14-1894: New pump bottom without suction pre-connect. Two new 4" x 10' suctions, with strainer. All 3 Ahrens in Dallas alike, carrying suctions disconnected on each side.

6-11-1898: New 32" x 62" Ahrens coil boiler, installed by W.C. Kahn at Dallas.

1- -1905: One new 4" x 10'6" suction.

7-7-1911: American-LaFrance rblt. To Eng. 6.

1921: Retired.

1984: On display, Dallas Fire Museum.

393 3 DCN/HD/A Upright. 32" x 62" coil boiler (silver). Two $4^{1}/_{4}$" x 7" pumps, nickel-plated air chamber, Globe signal lamp atop. One 4" x 22' pre-connect suction. Two $6^{3}/_{4}$" x 7" cyls. Wheels 16 x 56 front, 18 x 62 rear. Drop pole. No gong. Portrait of H. Clay Sexton (Fire Chief, St. Louis) on foot board.

2-5-1884: Delivered by Fred Hawekotte to Missouri State Penetentiary, Jefferson City, Mo. Name *H. Clay Sexton.*

394 3 DSF/HD/A Upright. 32" x 62" coil boiler (iron), safety rails on coal bunker, heater connections left. 4" x 7" pumps, relief valve right, ornament on air chamber. $3^{1}/_{2}$" x 20' pre-connect suction. $6^{3}/_{4}$" x 7" cyls. Wheels 16 x 56 front, 16 x 62 rear. $57^{1}/_{2}$" track. Drop pole. Gong on foot board. Plain finish.

3-25-1884: Shipped via O&M Railroad to St. Louis, Mo., for Engine 11.

3-27-1884: Wrecked enroute to St. Louis; O&M train carrying it derailed at Aurora, Ind. St. Louis bought Ahrens #398 instead.

"Old Tige" is an 1884 3rd Class double crane-neck Ahrens #392 that served Dallas, Texas, until 1921. It was named for Dallas mayor W.A. "Old Tige" Cabell. In this c-1960 photo, Old Tige still had its original wooden wheels. As currently displayed at the Dallas Firefighters Museum, #392 has iron wheels.

(Wayne Sorenson collection)

5-5-1884: Sold ($4600) to Ohio & Missouri Railroad, Aurora, Ind. Rebuilt in their shop.

2-20-1885: O&M RR sold this for $3000 to Fire Dept., Aurora, Ind. Name *Aurora #1.*

3-4-1892: George Zink installed new boiler coils at Aurora.

4-4-1898: George Zink installed new boiler coils at Aurora.

1911: Rebuilt by Ahrens-Fox Fire Engine Co., including new Continental boiler.

1-25-1937: Pumped several days at Aurora during flood of the Ohio River.

1949: New Howe pumper arrived at Aurora. Possibly the Ahrens was not retired until then, although probaby retired pre-WW II.

1962: Donated by Aurora Fire Dept. to Cincinnati Fire Museum, Cincinnati, Ohio. On permanent display there as of 1982.

395 2 SSF/HD/A Upright. $35^{1}/_{2}$" x 67" coil boiler (iron), safety rails on coal bunker, heater pipes left. 6" x 8" pump, piston air pump, air chamber ornament. 4" x 20' pre-connect suction. 10" x 8" cyl. Wheels 16 x 52, 18 x 58. Drop pole. Gong on foot board. Plain finish.

4-18-1884: Delivered by Peter Ahrens to Engine 5, Memphis, Tenn.

6-11-1889: New rear wheels.

1910: Replaced by Continental steamer #179.

396 1 DCN/HD/A Upright. 38" x 67" coil boiler (iron), heater pipes left. 5"x8" pumps, relief valve left, air chamber ornament. Two 5" x 10'6" suctions. 8" x 8" cyls. Wheels 16 x 60, 18 x 64. Stiff pole. Gong. Plain finish.

5-4-1884: Delivered by Fred Hawekotte to Milwaukee, Wis. Water in 3:50. Through 200' of hose, threw a $1^{3}/_{8}$" stream 325'. Through two siamesed lines of 300' each, threw one $1^{1}/_{4}$" stream 318', then one $1^{1}/_{8}$" stream 276'.

7-2-1884: In service at Engine 9, Milwaukee.

9-21-1886: Replaced by Ahrens #451. To Engine 10. Ahrens made *Engine 10* nameplate for Ahrens #396 on 11-18-1886.

1904: Replaced by Metropolitan #2990. To Engine 14 in place of Ahrens #451.

1911: Nott boiler.

1921: Retired.

397 4 SSF/HD/A Upright. 30" x 60" coil boiler (iron), heater pipes left. 5"x7" pump, automatic air pump, signal lamp on air chamber. $3^{1}/_{4}$" x 20' pre-connect suction. 8" x 7" cyl. Wheels 16 x 54, 18 x 60. 59" track. Drop pole. Gong under foot board. Plain finish.

Originally intended for St. Louis, this 1884 3rd Class double straight-frame Ahrens #394 was wrecked when the train carrying it to Missouri derailed at Aurora, Ind. The freight carrier, O&M Railroad Co., paid Ahrens the full value of the steamer ($4600), rebuilt the wrecked engine at a cost of thousands more, then sold it at a loss (for $3000) to Aurora Fire Dept in 1885. Retired in the 1940s, #394 was donated in 1962 to the Cincinnati Fire Museum, where it is currently displayed. (Ed Hass collection)

This 1912 photo shows an 1884 Ahrens 1st Class double crane-neck #396, which was then assigned to Engine 14, Milwaukee, Wis. A few months earlier, it had been rebuilt with a Nott boiler.
(Bill Hass collection)

Milwaukee's 1884 1st Class double crane-neck Ahrens #401 originally served Engine 3, but was assigned to Engine 22 from 1921 to 1926, as in this circa-1923 photo.
(C.P. Fox photo, courtesy Harold S. Walker)

3-4-1884: Delivered by G. Fred Ahrens to Louisville, Ky. Name *C.B. Blackburn #13.*

9-26-1892: New rear springs.

1908: Rebuilt.

1923: Retired.

398 3 DSF/HD/A Upright. 32" x 62" coil boiler (iron), heater pipes left, safety rails on coal bunker. 4" x 7" pumps, relief valve right. $3^1/_2$" x 20' pre-connect suction. $6^3/_4$" x 7" cyls. Wheels 16 x 56, 16 x 62. $57^1/_2$" track. Drop pole. Gong on foot board. Plain finish.

6-10-1884: Shipped to Eng. 11, St. Louis, Mo., to fulfill contract for wrecked Ahrens #394. Replaced Ahrens #16.

10-11-1892: Replaced by #2125. To Eng. 31.

4-18-1898: New boiler coils.

4-4-1901: New 32" x 62" Fox boiler ($850).

1909: Was a Spare at Engine 10. Engine 11 was using 1897 Ahrens #2562.

399 3 DSF/HD/A Upright. 32" x 62" coil boiler (iron), heater pipes left, safety rails on coal bunker. 4" x 7" pumps, relief valve on right. $3^1/_2$" x 20' pre-connect suction. $6^3/_4$" x 7" cyls. Wheels 16 x 56, 16 x 62. $57^1/_2$" track. Drop pole. Gong on foot board. Plain finish.

8- -1884: Shipped to St. Louis, Mo., for display at the Exposition.

12-22-1884: In service, Eng. 22, St. Louis, Mo.

4-7-1899: New 32" x 62" Fox boiler.

1909: Was in service at Engine 3. Engine 22 was using 1889 Ahrens #600.

400 3 DSF/HD/A Upright. 32" x 62" coil boiler (iron), heater pipes left, safety rails on coal bunker. $4^1/_4$" x 7" pumps, air pump, signal lamp on air chamber. 4"x20' pre-connect suction. $6^3/_4$" x 7" cyls. Wheels 16 x 56, 16 x 62. 59" track. Drop pole. Gong and portrait of John M. Letterle on foot board. Whip socket on seat, torch on frame. Plain finish.

5-29-1884: Delivered by Chris and Fred Ahrens to Eng. 10, Louisville, Ky. Name *John M. Letterle.* Replaced Ahrens #33.

1-22-1904: New 32" x 62" Fox boiler, two discharge gates, frame, front axle, springs.

1921: Retired.

401 1 DCN/HD/A Upright. 38" x 67" coil boiler (iron), heater pipes. 5" x 8" pumps, air pump, relief valve, air chamber ornament, two $2^1/_2$" discharge gates. Two 5" x 10' suctions. 8" x 8" cyls. Wheels 16 x 60, 18 x 64. Stiff pole. Gong on foot board. Plain finish.

6- -1884: Delivered by Chris Ahrens to Engine 3, Milwaukee, Wis. Took 1863 Amoskeag 1st Class steamer #77 in trade, sold it to Chicago 8-21-1885 for use on a fireboat.

1904: Replaced by Metropolitan #2991. To Engine 8 in place of Ahrens #224.

1909: Nott boiler.

1921: Replaced by 1900 Metropolitan #2737. To Engine 22 in place of Ahrens #2445.

1926: Retired.

402 3 DSF/HD/A Upright. 32" x 62" coil boiler (iron), heater pipes left, safety rails on coal bunker. 4" x 7" pumps, relief valve on right. $3^1/_2$" x 20' pre-connect suction. $6^3/_4$" x 7" cyls. Wheels 16 x 56, 16 x 62. $57^1/_2$" track. Drop pole. Gong on foot board. Plain finish.

11-27-1884: Shipped to St. Louis, Mo.

12-22-1884: In service, Eng. 23, St. Louis, Mo.

3-7-1903: New 32" x 62" Fox boiler ($850).

1909: Was in service at Engine 34. Engine 23 was using 1893 Ahrens #2247.

403 2 DCN Upright. No details in Ahrens ledger.

404 2 DCN Upright. No details in Ahrens ledger.

405 4 DSF/MD-HD/A Upright. 30"x60" coil boiler (iron). $3^3/_4$" x ?? pumps. One $3^1/_2$" x 20' pre-connect suction. Two 6" x ?? cyls. Wheels 16 x 54 front, 16 x 60 rear. Plain finish.

8- -1884: Loaned to Wilmington, Ohio, for 29 days while Ahrens installed new Ahrens boiler on their Clapp & Jones steamer.

9-12-1884: Tested & accepted by Pekin, Ill. Delivered by Peter Ahrens.

11-1-1890: Geo. Zink put in new boiler coils.

406 4 SSF/MD-HD/A Upright. 30" x 60" coil boiler (iron), safety rails on coal bunker, heater pipes left. 5"x7" pump, two inlets with lamps atop, piston air pump, nickel-plated air chamber with ornament atop. $3^1/_2$" x 20' pre-connect suction. 8" x 7" cyl. Wheels 16 x 54, 18 x 60. Drop pole. Front brake lever. Gong on foot board. Railing on seat. Plain finish.

12- -1884: Delivered by Peter Ahrens to Memphis, Tenn. Name *C.L. Pullen #6.*

5-24-1891: Replaced by Ahrens 414. To E. 2. (3-way trade: #406 from E. 6 to E. 2, #320 from E. 2 to E. 4, #414 from E. 4 to E. 6).

407 3 DCN/HD/A Upright. 32" x 62" coil boiler (iron), heater pipes right. $4^1/_4$" x 7" pumps,

one flywheel, relief valve left. Two 4" x 10', two $2^{1}/_{2}$" x 9' suctions. $6^{3}/_{4}$" cyls. Wheels 16 x 56, 16 x 62. Platform springs. $56^{1}/_{2}$" track. Stiff pole. (New boiler coils 1904).

9-6-1884: Shipped to Chicago, Ill.

1906: At Engine 31. 1917: At Engine 74.

408 3 DCN/HD/A Upright. 32" x 62" coil boiler (iron), heater pipes right. $4^{1}/_{4}$" x 7" pumps, one flywheel, relief valve left. Two 4" x 10', two $2^{1}/_{2}$" x 9' suctions. Two $6^{3}/_{4}$" cylinders. Wheels 16 x 56 front, 16 x 62 rear. Platform springs. $56^{1}/_{2}$" track. Stiff pole.

9-6-1884: Shipped to Chicago, Ill.

1906: At Engine 54. 1917: At Engine 67.

409 3 DCN/HD/A Upright. 32" x 62" coil boiler (iron), heater connections right. $4^{1}/_{4}$" x 7" pumps, one flywheel, relief valve left. Two 4" x 10', two $2^{1}/_{2}$" x 9' suctions. Two $6^{3}/_{4}$" cyls. Wheels 16 x 56 front, 16 x 62 rear. Platform springs. $56^{1}/_{2}$" track. Stiff pole.

10-15-1884: Shipped to Chicago, Ill.

1906: At Engine 57. 1917: At Engine 96.

410 3 DSF/HD/S Upright. 32" x 62" coil boiler (iron), heater pipes left. $4^{1}/_{4}$" x 7" pumps. $3^{1}/_{2}$" x 20' pre-connect suction. Two $6^{3}/_{4}$" x 7" cylinders. Wheels 16 x 56, 16 x 62. Drop pole. Weight with coal and water 6125#.

9-5-1884: In service, Fire Co. 16, Cincinnati, O. Name *Chris Kiechler*. Replaced Ahrens #6.

8-13-1887: While pumping from a cistern at McMillan and Gilbert Ave., boiler became red hot and began spewing water out top of smokestack. Steam pressure dropped to almost nothing. Safety valve blew off, water leaking like a sieve around fire-box door.

8-29-1887: Boiler rebuilt including new coils.

3-20-1900: New 32" x 62" Fox boiler. First Fox boiler used by Cincinnati Fire Dept.

411 3 DCN/HD/A Upright. 32" x 62" coil boiler (iron), heater pipes left. $4^{1}/_{4}$" x 7" pumps, relief valve left; nickel-plated flywheel, columns, air chamber. Suction & discharge adaptors for Amoskeag steamer. Two 4" x 10', one 3" x 9' suctions. $6^{3}/_{4}$" x 7" cyls. Wheels 16x56, 16x62. Nickel-plated springs (two front springs). Stiff pole. Gong on foot board. Weight with water and coal 6620#.

2-4-1885: Shipped to Oakland, Cal. Name *Alameda #5*. Delivered by Chris Ahrens and Peter Ahrens.

6-21-1889: New boiler coils, no charge.

10-12-1900: R.S. Chapman (Pacific Coast agent, American Fire Engine Co.) installed new boiler coils at Oakland. Replaced by Metropolitan #2715, to (Engine 16?).

1-22-1904: New 32" x 62" Fox boiler by American Fire Engine Co., Seneca Falls, N.Y., installed at Oakland.

12- -1909: American-LaFrance supplied new shell for boiler, and new front springs.

1913: Was assigned to Engine 16. Still pulled by two horses (never motorized).

1918: Replaced by new Seagrave pumper. Probably became a spare.

1922: All remaining horse-drawn steamers in Oakland retired. The Ahrens was probably junked at that time.

412 3 DCN/HD/A Upright. 32" x $62^{1}/_{2}$" coil boiler (iron), heater connections right, clock on steam chest. $4^{1}/_{4}$" x 7" pumps, relief valve left. Two 4" x 10' suctions, one 3" x 9' plug suction. $6^{3}/_{4}$" x 7" cyls. Wheels 16 x 56 front, 16 x 62 rear. Two front springs. Stiff pole.

1- -1885: Delivered by Peter Ahrens and G. Fred Ahrens to Racine, Wis. According to Racine, this was their first steamer, named *Stephan Freeman*, but Racine reputedly had a Latta steamer before this.

7-25-1889: Racine built a new water works, and felt that they no longer needed the Ahrens, so they sold it to Cedarburg, Wis.

11-26-1892: New boiler coils.

8-19-1904: New boiler coils.

1966: *Fire Engineering* magazine reports that the Fire Bell Club of Racine has an 1885 steamer named *Stephan Freeman* that was Racine's first steamer. Article does not say whether this was the Ahrens. It was displayed at the club's museum in old fire station #3. Unable to find out what steamer they had or whether they still have it.

413 3 DSF/HD/S-A Upright. 32"x62" tubular boiler (silver/nickel), silver-plated coal bunker, nickel-plated feed pump, 2 lamps on boiler. 4" x 7" pumps; nickel-plated air chamber, inlets, discharge gates, crankshaft, columns, pump heads. Signal lamp atop air chamber. Two $3^{1}/_{2}$" x 9'6", one $2^{1}/_{2}$" x 12' suctions, nickel-plated couplings. $6^{3}/_{4}$" x 7" cyls. (silver-plated casing, nickel-plated heads). Silver-plated frame, springs, axles. Plain Sarven wheels for service, fancy Archibald wheels for parades, all 16 x 60, 18 x 62. Platform springs. Plain pole for service, plated pole (brass/ nickel) for parades. Gong on foot board. Two Star lamps on seat.

11-22-1884: Shipped to Louisiana Hose Co., New Orleans, La.

Left: Straight-frame chassis of 1885 5th Class #415 in 1975, during early stages of restoration at St. Paris, Ohio. Wheels were by Archibald (Lawrence, Mass.). Note the two suction brackets on the front foot board. Right: This close-up shows the restored 1897 Ahrens tubular boiler and serial-number ID plate on the St. Paris 1885 Ahrens. Ahrens used this style of nameplate from 1870 through 1904. Note the original steam-pressure gauge to the left of the nameplate. Also note the safety-valve and steam-whistle to the right of the nameplate.

(Courtesy Chief Ralph Stradling, Johnson-St. Paris)

1885 Ahrens 5th Class single straight-frame #415 as it looks today, following a 1976 restoration by firefighters at Saint Paris, Ohio. Although the firefighters returned it to fully-operable condition, the Ahrens is considered too valuable to risk damaging it by firing it up and pumping water.

(Courtesy Chief Ralph Stradling, Johnson-St. Paris)

414 4 SSF/HD/A Upright. 30" x 60" coil boiler (iron), silver-plated coal bunker with safety rails above, heater pipes left. 5" x 7" pump, air pump, two inlets, body painted carmine; nickel-plated pump columns, crankshaft, and air chamber with eagle atop. $3^{1}/_{2}$" x 20' pre-connect suction. 8" x 7" cyl. Wheels painted white, 16 x 54 front, 16 x 60 rear, transfer pictures on wheels and pole. Silver-plated frame. Nickel-plated front brake lever and braces. Gong on foot board. Railing on seat.

12- -1884: Delivered by Peter Ahrens to Memphis, Tenn. Name *James Speed #4.*

5-24-1891: Replaced by Ahrens 320. To Eng. 6 in place of Ahrens 406. (A 3-way trade: #414 from Eng. 4 to Eng. 6, #406 from Eng. 6 to Eng. 2, #320 from Eng. 2 to Eng. 4).

6-1-1899: New 30" x 54" Fox vertical water tube boiler, new wheels, pump overhauled, entire steamer repainted ($1800).

415 5 SSF/MD-HD/A Upright. 25" x $56^{1}/_{2}$" tubular boiler (iron). $4^{1}/_{2}$" x 7" pump, automatic air pump, air chamber ornament. $3^{1}/_{4}$" x 20' pre-connect suction. 7" x 7" cylinder. Wheels 16 x 52, 16 x 60. Rear brake lever.

1-7-1885: Delivered by G. Fred Hawekotte and G. Fred Ahrens to Saint Paris, Ohio. Name *Saint Paris.* Also two Preston hose carts and 1000' of hose.

9- -1897: New 25" x $56^{1}/_{2}$" tubular boiler, and minor repairs ($800).

1976: Restored by the Johnson-Saint Paris Fire Dept., Saint Paris, Ohio.

416 5 SSF/MD/A Upright. 28" x $56^{1}/_{2}$" tubular boiler (iron). $4^{1}/_{2}$" x 7" pump, automatic air pump, bell atop air chamber. One $3^{1}/_{4}$" x 20' pre-connect suction. 7" x 7" cyl. Wheels 16 x 52 front, 16 x 60 rear. Rear brake lever.

7-6-1885: Delivered to Auburn, Ind. Name *Auburn.* Also one Ahrens manually-drawn hose cart and 500' of hose.

11-7-1885: Hoosier Boys Fire Co. organized in Auburn to put this steamer in service.

Date?: Converted to horse-drawn.

417 4 SSF/MD-HD/A Upright. 30" x 60" coil boiler (iron), extension stack. 5" x 7" pump, automatic air pump, ornament on air chamber. One $3^{1}/_{4}$" x 20' pre-connect suction, one $3^{1}/_{4}$" x 8' suction. One 8" x 7" cyl. Wheels 16 x 54 front, 16 x 60 rear. Rear brake lever.

11-22-1884: Delivered by G. Fred Hawekotte and Fred Ahrens to Catlettsburg, Ky. Also two Preston hose carts and 1000' of hose.

12-1-1884: Shipped gong, no charge.

7-11-1889: Joe Terbeck installed new boiler

This c-1904 photo shows Hoosier Boys Fire Co. of Auburn, Ind., with their 1885 Ahrens 5th Class single straight-frame #416. Usually, Ahrens engines came with gongs, but in this case, the bell was factory-original and the gong added later. Originally manually-drawn, #416 had obviously been "modernized" with horses. Fate of this Ahrens is unknown; if it survives, which is doubtful, Auburn seeks to buy it back and restore it.

(Courtesy Clarence Boger, Auburn Fire Dept.)

coils at Catlettsburg.

8-20-1891: Returned to factory for repairs. The suction pre-connect inlet was broken and the 20' suction missing, male coupling missing from the 8' suction. Rear steam gauge broken. Missing oil cans, lanterns, scraper, shovel, rear foot board, play pipes, nozzles, Doctor Shield, and drag ropes.

3-1-1899: Repairs made at Catlettsburg ($51).

8-20-1901: Returned to Ahrens factory for repairs. Ahrens Mfg. Co. kept the steamer for non-payment of the 3-1-1899 repair bill.

8-27-1904: Shipped back to Catlettsburg by boat, still unrepaired. Catlettsburg firemen were to make the needed repairs themselves. (One can only surmise that the management of the newly-created American-LaFrance did not want the headaches of trying to collect on a 5-year-old repair bill and trying to dispose of a 20-year-old steamer, deciding that the lowest-cost alternative was to return the Ahrens to Catlettsburg).

418 4 SSF/HD/A Upright. 30" x 60" coil boiler (iron), lamps on boiler, safety rails on coal bunker. 5" x 7" pump, two inlets with lamps, air pump, air chamber ornament. $3^1/2$" x 20' pre-connect suction. 8" x 7" cyl. Wheels 16 x 56, 16 x 58. Drop pole. Front brake lever.

3-17-1885: Delivered by Hiram H. Nieman and G. Fred Ahrens to Chattanooga, Tenn. Name *Thomas J. Carlisle #2.*

10-21-1889: New 18 x 58 rear wheels.

10-8-1892: New boiler coils.

11-3-1904: New boiler coils.

12-24-1909: New 30" x 60" Fox boiler.

419 3 DSF/HD/A Upright. 32" x 62" coil boiler (iron), heater pipes left, safety rails on coal bunker. 4" x 7" pumps. $3^1/2$" x 18'6" pre-connect suction. $6^3/4$" x 7" cyls. Wheels 16 x 56, 16 x 62. $57^1/2$" track. Gong on foot board. Plain finish, except brass-plated drop pole.

5-30-1885: Delivered by G. Fred Hawekotte to Engine Co. #24, St. Louis, Mo.

9-5-1902: New 32" x 62" nickel-plated Fox vertical water tube boiler ($850).

1909: Was in service at Engine 44. Engine 24 was using 1902 Metropolitan #2881.

420 3 DSF/HD/A Upright. 32" x 62" coil boiler (iron), safety rails on coal bunker, heater connections left. 4" x 7" pumps. $3^1/2$" x 18'6" pre-connect suction. $6^3/4$" x 7" cyls. Wheels 16 x 56 front, 16 x 62 rear. $57^1/2$" track. Gong on foot board. Plain finish, brass pole.

5-30-1885: Delivered by G. Fred Hawekotte

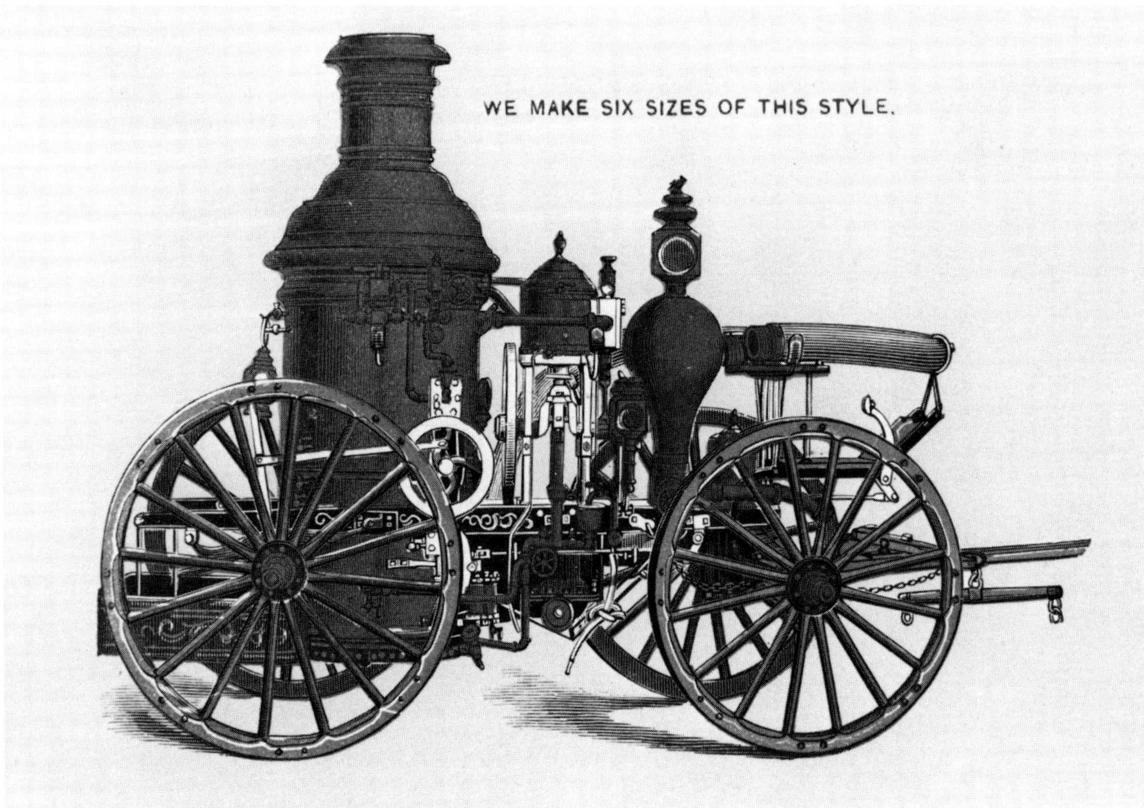

"Phoenix #2" was an 1885 4th Class Ahrens single straight-frame #422 that served Vicksburg, Mississippi. Note the signal lamp on top of the air chamber, and the smaller coach lamp on top of the vacuum chamber. Also note the ratchet-type foot brake on the driver's foot board.

(Ed Hass collection)

to Engine Co. #25, St. Louis, Mo.

5-13-1887: New boiler coils.

4-26-1901: New boiler coils.

3-7-1903: New 32" x 62" Fox boiler ($850).

1909: Was a Spare at Engine 4. Engine 25 was using 1904 Metropolitan #2962.

422 4 SSF/HD/A Upright. 30" x 60" coil boiler (silver), silver-plated coal bunker, raised letters *Phoenix #2* on back, safety rails above. 5" x 7" pump, two nickel-plated inlets (lamps atop), nickel-plated air chamber with signal lamp atop, brass flywheel. Nickel-plated crankshaft, columns, heads. $3^{1}/_{2}$" x 20' pre-connect suction. 8" x 7" cyl., nickel-plated heads. Wheels 16 x 56, 16 x 62. Drop pole. Ratchet brake. Gong on foot board. Two lanterns under seat. Weight 5015#.

8-1-1885: Delivered by Fred Hawekotte to Phoenix Fire Co. #2, Vicksburg, Mississippi.

12-27-1904: Vicksburg inquired about a new boiler. Ahrens records don't say whether they actually put on a new boiler.

423 4 SSF/MD-HD/A Upright. 30" x 60" coil boiler (iron). 5" x 7" pump, automatic air pump, ornament atop air chamber. One $3^{1}/_{4}$" x 20' pre-connect suction. 8" x 7" cyl. Wheels 16 x 54 front, 18 x 60 rear. Rear brake lever.

1885: Built as a demonstrator.

2-22-1887: Delivered by Chris Ahrens and G. Fred Ahrens to Lebanon, Ohio. Price $2500. Beat Silsby in competition.

11-16-1894: George Zink installed new boiler coils at Lebanon.

11-13-1907: To Ahrens Fire Engine Co. as a relief engine (for loan to customers while their steamers were repaired). New $30^{1}/_{2}$" x 62" Nest Tube boiler (Ahrens' improved version of LaFrance nest tube boiler).

425 4 SSF/MD/A Upright. 28" x $56^{1}/_{2}$" tubular boiler (iron). 5" x 7" pump, automatic air pump, air chamber ornament. $3^{1}/_{4}$" x 20' pre-connect suction. 8" x 7" cyl. Wheels 16 x 54, 18 x 60. Rear brake lever. Weight 4500#.

4-2-1885: Delivered by Fred Hawekotte and Fred Ahrens to Arcanum, Ohio. Price $4250. Name *Arcanum*. Through 100' of hose, threw 1" stream 210' in 4:45; a $^{3}/_{4}$" and a $^{7}/_{8}$" stream 200' each. Through 850' of hose into two 75' lines (total 1000' hose), threw $^{3}/_{4}$" stream and $^{7}/_{8}$" stream 165' each. Through 450' of hose, threw $1^{1}/_{8}$" stream 245'.

1-3-1893: $28^{1}/_{2}$" x $56^{1}/_{2}$" tube boiler ($737.06)

7-30-1901: New grate bars, bottom of boiler.

5-23-1905: New 28" x 58" Fox boiler and pump overhaul. Loaned Arcanum another Ahrens while repairs were made.

1983: Arcanum still owns this, and maintains it in fully-operable condition.

427 3 DCN/HD/A Upright. 32" x 62" coil boiler (iron), heater pipes right. $4^{1}/_{4}$" x 7" pumps, one flywheel, relief valve left. Two 4" x 10', one $2^{1}/_{2}$" x 9' suctions. $6^{3}/_{4}$" x 7" cyls. Wheels 16x56, 16x62. Platform springs. $56^{1}/_{2}$" track.

6-22-1885: Shipped to Chicago, Ill. New boiler, 1902. At Engine 29 in 1906. At Engine 116 in 1917.

429 3 DCN/HD/A Upright. 32" x 62" coil boiler (iron), heater pipes right. $4^{1}/_{4}$" x 7" pumps, one flywheel, relief valve left. Two 4" x 10', one $2^{1}/_{2}$" x 9' suctions. $6^{3}/_{4}$" x 7" cyls. Wheels 16x56, 16x62. Platform springs. $56^{1}/_{2}$" track.

8-14-1885: Shipped to Chicago. 32" x 64" Fox boiler for #429 of Chicago Eng. 23 in 1900.

1904: Nott boiler. Still at Engine 23 in 1906.

1907: Replaced at Engine 23 by Metropolitan #3197. At Engine 77 in 1917.

431 4 DSF/HD/S Upright. 30" x 60" tubular boiler (silver and brass), two lamps on sides, silver-plated coal bunker. $3^{3}/_{4}$" x ?? pumps; silver-plated air chamber, heads, columns, crankshaft, discharge gates. Two $3^{1}/_{2}$" x 9', one $2^{1}/_{2}$" x 12' suctions. 6" x ?? cyls. (silver). Wheels 18 x 52, 18 x 62. Platform springs. Silver-plated frame. Gong on foot board. Two lamps on seat, two lanterns.

12-30-1885: Delivered to Fire Co. #6, New Orleans, La.

432 4 DCN/MD-HD/A Upright. 30"x60" coil boiler (nickel). $3^{3}/_{4}$" x ?? pumps, one flywheel, air pump, air chamber ornament, pressure gauge on frame. $3^{1}/_{4}$" x 21'6" pre-connect suction. 6" x ?? cylinders. Wheels 16 x 50 front, 16 x 60 rear. Rear brake lever. Two torches on frame. Gong on foot board.

12-8-1885: Delivered by Fred Hawekotte to Pendleton, Oregon. Water in 4:00. Threw one stream 253'. Pumped for 12 hours non-stop.

3-16-1897: New boiler coils.

433 5 SSF/MD-HD/A Upright. 25" x $56^{1}/_{2}$" tubular boiler (iron). $4^{1}/_{2}$" x 7" pump, air pump, air chamber ornament. $3^{1}/_{4}$" x 20' pre-connect suction. 7" x 7" cyl. Wheels 16 x 52, 18 x 60. Gong on foot board. Rear brake lever.

12-28-1885: Tested at Webster & Sycamore, Cincinnati, Ohio. Threw one 1" stream 215'; two $^{3}/_{4}$" streams 164' each; one $^{7}/_{8}$" stream 201'; two $^{3}/_{4}$" streams 149' each.

This sketch from an 1888 Ahrens catalog shows an 1885 4th Class single straight-frame #425 of Arcanum, Ohio. A Fox boiler was installed in 1905, but otherwise its overall appearance has changed little in the 100 years that Arcanum has owned it.

(Ed Hass collection)

1885 Ahrens 4th Class single straight-frame #425 of Arcanum, Ohio prepares to pump water at a 1983 fire engine muster. Although it still has its original manually-drawn pole and rope reel, the Ahrens is sometimes pulled by horses in parades. Note the location of the brake, behind (rather than in front of) the rear wheel.

(Bruce Neal photo)

"Shields #2" is an 1885 5th Class single straight-frame Ahrens #433 that has been carefully preserved by its original owner, the fire department of Seymour, Ind. Note the sign advertising "Oktoberfest". Ahrens engines were popular in communities with large German populations, while Silsby steamers found favor in Irish areas. Some cities even installed Ahrens engines in their German neighborhoods and Silsbys in Irish districts.

(Bruce Neal photo)

12-31-1885: Fred Ahrens delivered to Seymour, Ind. Name *Shields #2*. Water in 6:30.

1-1-1886: Tested & accepted by Seymour, Ind. Water through 300' of hose in 9:00. Threw a 1" stream through 400' of hose over Baptist church, tallest building in Seymour.

1984: Maintained at Seymour in operating condition. Pulled by ponies in parades.

435 5 SSF/MD-HD/A Upright. 25"x56 1/2" tubular boiler (iron). 4 1/2" x 7" pump, automatic air pump, ornament on air chamber. 3 1/4" x 20' pre-connect suction. 7" x 7" cyl. Wheels 16 x 52 front, 18 x 60 rear. Rear brake lever.

1-7-1886: Delivered by Chris Ahrens and G. Fred Ahrens to Sullivan, Ind. Water through 100' of hose in 6:00. Through 1000' of hose, threw one 7/8" stream over the court house.

437 2 DCN/HD/A Upright. 35 1/2" x 67" coil boiler (silver), clock right, heater pipes left. 4 1/2" x 8" pumps; nickel-plated air chamber with Globe signal lamp atop; relief valve left; nickel-plated pump-support columns, crankshaft, flywheel, heads; pressure gauge on frame. 4" x 22' pre-connect suction. 7 1/4" x 8" cyls. Wheels 16 x 58 front, 18 x 62 rear. Drop pole (4 horses). Railing atop seat.

8-13-1885: Delivered by Chris Ahrens to Engine 6, Minneapolis, Minn.

1890: To Engine 12. 1894: To Engine 19.

1911: Rebuilt. 1919: Retired.

439 3 DCN/HD/A Upright. 32" x 62" coil boiler (iron), heater pipes right. Two 4 1/4" x 7" pumps, one flywheel, relief valve on left. Two 4" x 10' suctions, one 2 1/2" x 9' suction. Two 6 3/4" x 7" cyls. Wheels 16 x 56 front, 16 x 62 rear. Platform springs. 56 1/2" track.

9-7-1885: Shipped to Engine 94, Chicago, Ill.

1906: At Engine 36. Retired by 1917.

440 3 DCN/HD/A Upright. 32" x 62" coil boiler (iron), heater pipes right. Two 4 1/4" x 7" pumps, one flywheel, relief valve on left. Two 4" x 10' suctions, one 2 1/2" x 9' suction. Two 6 3/4" x 7" cyls. Wheels 16 x 56 front, 16 x 62 rear. Platform springs. 56 1/2" track.

9-28-1885: Shipped to Chicago, Ill. At Engine 62 in 1906, Engine 99 in 1917.

441 3 DCN/HD/A Upright. 32" x 62" coil boiler (iron), heater pipes right. Two 4 1/4" x 7" pumps, one flywheel, relief valve on left. Two 4" x 10' suctions, one 2 1/2" x 9' suction.

The identity of this double crane-neck Ahrens steamer from Chicago is somewhat uncertain. Steamer expert Harold Walker has identified it as Engine 87 (formerly Engine 91), an 1885 3rd Class #441. But lettered on the face of the photograph, below the rear wheel, it is identified as Engine 27.

(Courtesy Harold S. Walker)

Two 6$^3/_4$" x 7" cyls. Wheels 16 x 56 front, 16 x 62 rear. Platform springs. 56$^1/_2$" track.

9-28-1885: Shipped to Eng. 91, Chicago, Ill.

1901: New boiler.

1906: At Engine 84. 1917: At Engine 87.

443 3 DSF/HD/A Upright. 32" x 62" coil boiler (iron), safety rails on coal bunker, heater pipes left. 4$^1/_4$" x 7" pumps, one flywheel. 3$^1/_2$" x 21' pre-connect suction. 6$^3/_4$" x 7" cyls. Wheels 16 x 56 front, 16 x 62 rear. Two front springs. Drop pole. Railing atop seat.

11-5-1885: Tested, 7th & Walnut, Cincinnati. Sold on spot ($4200) to Co. #8, Cincinnati. Name *Chris Kinsinger*. In service same day.

7-1-1891: New boiler coils.

6-30-1898: New front platform springs, pump and boiler overhauled, all plated parts replated in nickel. To Fire Co. #32.

1-18-1904: Tipped over on the ice on Station street. Rebuilt and returned to service.

445 3 DCN/HD/A Upright. 30$^1/_2$" x 62" tubular boiler (silver & brass), safety rails on coal bunker, heater pipes left. 4$^1/_4$" x 7" pumps, one flywheel. One 4" x 22' pre-connect suction. 6$^3/_4$" x 7" cyls. Wheels 16 x 56 front, 18 x 62 rear. 56$^1/_2$" track. Drop pole.

11-12-1885: Delivered by Chris Ahrens, Fred Ahrens, and Hiram Nieman to Dayton, Ohio. Name *E.F. Pryor #3*.

5-23-1899: New boiler tubes.

1901: Rebuilt by LaFrance Fire Engine Co., Elmira, N.Y.

446 Rebuild of 1873 Silsby, Adrian, Mich.

448 1 DCN/HD/A Upright. 38"x67" tubular boiler (iron), safety rails on coal bunker, heater pipes left. 5" x 8" pumps, one flywheel, relief valve left, air chamber ornament. 4$^1/_2$" x 23'3" pre-connect suction, female couplings both ends. 8" x 8" cyls. Wheels 16 x 60, 18 x 64. Drop pole. Gong on foot board.

12-8-1885: Delivered by Chris Ahrens to Engine 8, Saint Paul, Minn.

8-6-1897: New boiler flues.

4- -1901: New rear wheels.

1906: Nott boiler. To Engine 4.

The "Chris Kinsinger" was an 1885 Ahrens 3rd Class #443 that served Fire Co. #8 in Cincinnati, Ohio. Its combination of double-pump and straight frame made this a rare model. Even rarer was the single flywheel on the right side of the pump; most double-pump Uprights had either two flywheels, or a single flywheel on the left side. The second, smaller flywheel (attached to the boiler) is part of the air-pump mechanism.

(Ed Hass collection)

449 3 DCN/HD/A Upright. 32" x 62$^1/_2$" coil boiler (iron), torch and safety rails on coal bunker, heater pipes on left. Two 4$^1/_4$" x 7" pumps, relief valve left. 4" x 22' pre-connect suction. 6$^3/_4$" x 7" cyls. Wheels 16 x 56 front, 16 x 62 rear. Platform springs. 57$^1/_2$" track. Pole for 3 horses. Gong on foot board. Whip socket on seat. Plain finish. Weight 6700#.

1886: Built on prospect for St. Louis, Mo., but they did not buy it.

4-27-1886: Ordered by San Antonio, Texas.

6-14-1886: Delivered by Chris Ahrens to Ben Milam Fire Co. #1, San Antonio, Texas. Name *Sam Houston.*

c-1890: Changed from one 4" x 22' pre-connect suction to two 4" x 10' suctions.

9-1-1891: San Antonio became paid department. Steamers of the volunteers (1886 Ahrens #449 of Co. #1 and 1871 Silsby of Co. #2) to Eng. 1 and 2, respectively.

6-13-1892: Replaced by #2107. To Engine 3.

12-7-1894: New boiler coils.

9-19-1910: New boiler coils.

451 3 DCN/HD/A Upright. 32" x 62" coil boiler (iron), heater pipes left. 4$^1/_4$" x 7" pumps, relief valve left, ornament atop air chamber. Two 4" x 10' suctions. 6$^3/_4$" x 7" cyls. Wheels 16 x 56 front, 18 x 62 rear. Platform springs. 56$^1/_2$" track. Stiff pole. Gong on foot board. Whip socket on seat. Plain finish. Weight with water, coal, tools, and 3 men, 7360#.

6-29-1886: Built on prospect for Chicago, Ill., but they did not buy it.

9-21-1886: In service, Engine 9, Milwaukee, Wis. Delivered by Fred Hawekotte. Replaced Ahrens #396.

1893: Replaced by 1880 Ahrens #246. To Engine 16.

1904: Replaced by Ahrens #224. Retired.

453 5 SSF/MD-HD/A Upright. 25"x56$^1/_2$" tubular boiler (iron). 4$^1/_2$" x 7" pump, air pump, air chamber ornament. One 3$^1/_4$" x 20', one 3$^1/_4$" x 5' suctions. 7" x 7" cylinder. Wheels 16 x 52, 16 x 60. Rear brake lever. Weight 3825#.

12-18-1885: Chris and Fred Ahrens delivered to Bardstown, Ky. Name *Governor Johnson.*

455 5 SSF/MD-HD/A Upright. 25"x56$^1/_2$" tubular boiler (iron). 4$^1/_2$" x 7" pump, automatic air pump, ornament on air chamber. 3$^1/_4$" x 20'

Front and rear views of 1886 Ahrens 4th Class double crane-neck #459 in the basement of the Seagle Building, University of Florida, 1978. Originally used in Birmingham, Alabama, it later served in Jacksonville, Fla. Front view, above, shows that the following parts are missing: driver's foot board, left-panel and backrest of seat, suction and discharge caps, smokestack on the boiler, and 20-foot suction that once wrapped around the front. Rear view, below, reveals additional items missing: coal bunker, top of boiler, and lanterns. The Ahrens has been stored in this condition for 35 years.
(Photos by Bill Hass)

pre-connect suction. 7" x 7" cyl. Wheels 16 x 52 front, 16 x 60 rear. Rear brake lever.

3-6-1886: Delivered by G. Fred Hawekotte to Ashtabula, Ohio. Name *H.J. Topky*.

457 5 SSF/MD-HD/A Upright. 25"x56$^{1}/_{2}$" tubular boiler (iron). 4$^{1}/_{2}$" x 7" pump, automatic air pump, ornament on air chamber. 3$^{1}/_{4}$" x 20' pre-connect suction. 7" x 7" cyl. Wheels 16 x 52 front, 16 x 60 rear. Rear brake lever. Adams & Westlake (A&W) lanterns.

3-10-1886: Chris Ahrens, Fred Hawekotte, H.H. Nieman delivered to Waynesville, Ohio.

458 See #158.

459 4 DCN/HD/A Upright. 30" x 60" coil boiler (silver/brass), torch & heater-pipes left. 3$^{3}/_{4}$" x ?? pumps, plated inlets, air chamber ornament. 3$^{1}/_{2}$" x 20' pre-connect suction. 6" x ?? cyls. Wheels 16x54, 18x60. 56$^{1}/_{2}$" track. Stiff pole. Platform springs. Front brake lever. 2 torches on frame. Single seat with railing.

2-16-1886: Shipped to Birmingham, Alabama. Delivered by Chris and Fred Ahrens.

6- -1887: New boiler coils.

10-5-1895: New boiler coils.

4-26-1905: New 30" x 60" Fox boiler, pump overhauled, repainted.

c-1912: Sold to Jacksonville, Florida. Rebuilt with double seat (no railing) and drop pole.

8-21-1951: Postcard from Florida Museum to Jim Fileul, noted collector and restorer of Amoskeag steamers in Manchester, N.H. Card says that Ahrens #459 now belongs to Florida State Museum, Seagle Building, University of Florida, Gainesville, Fla.

1978: Bill Hass photographed Ahrens 459 in basement, Seagle Building, Gainesville, Fla.

461 4 DCN/HD/A Upright. 30" x 60" coil boiler (iron/silver), heater pipes left, safety rails on coal bunker. 3$^{3}/_{4}$" x ?? pumps, air chamber ornament. 3$^{1}/_{2}$"x20' pre-connect suction. 6" x ?? cyls. Wheels 16 x 54, 18 x 60. 56$^{1}/_{2}$" track. Platform springs. Stiff pole. Front brake lever. Two torches on frame. Gong on foot board. Double seat, railing atop.

7- -1886: Delivered by G. Fred Hawekotte to Chattanooga, Tenn. They tried it for about one month, but did not buy it. Returned to the factory in Cincinnati.

9- -1886: Loaned to Franklin, Ohio, while their Ahrens #178 was rebuilt.

10- -1886: Returned to factory, rebuilt as #520. See #520 for its subsequent history.

463 4 DCN/HD/A Upright. Round 30" x 60" coil boiler. Two 3$^{3}/_{4}$" x ??" pumps. One 3$^{1}/_{2}$" x 20' pre-connect suction. Two 6" x ?? cylinders. Wheels 16 x 54 front, 18 x 60 rear.

Listed in Ahrens records, but doesn't say where it was delivered, nor if it was built.

465 3 DCN/HD/A Upright. 32" x 62" coil boiler (silver), heater pipes left, torch on coal bunker. 4$^{1}/_{4}$" x 7" pumps, air chamber ornament. 4" x 20' pre-connect suction, strainer threaded both ends to attach to suction hose and hydrant. 6$^{3}/_{4}$" x 7" cyls. Wheels 16 x 56, 18 x 62, steel tires. Plated platform springs. 57$^{1}/_{2}$" track. Stiff pole. Gong on foot board. Whip socket on seat. Two torches on frame.

1886: Built for St. Louis, Mo., but they didn't buy it.

10-16-1886: Delivered by G. Fred Hawekotte, G. Fred Ahrens, and Dick Raymond to Akron, Ohio. Name *B.F. Manderbach #1*.

12-7-1889: New boiler coils, no charge.

8-15-1903: New 32" x 64" LaFrance boiler installed by American Fire Engine Co., Seneca Falls, N.Y.

467 2 DCN/HD/A Upright. 35$^{1}/_{2}$" x 67" coil boiler (iron), safety rails on coal bunker, heater pipes left. 4$^{1}/_{2}$" x 8" pumps, relief valve left. 4" x 21' pre-connect suction. 7$^{1}/_{4}$" x 8" cyls. Wheels 16 x 58, 16 x 62. Platform springs. Drop pole. Gong on foot board. Plain finish.

3-29-1886: Delivered by Chris Ahrens to Engine 26, St. Louis, Mo.

2-13-1888: Replaced by #532. To Engine 14.

2-23-1897: Two new rear wheels.

8-24-1904: New 35" x 64" Fox boiler ($850).

1909: Fire Chief's report shows Ahrens #467 as a 3rd Class at Engine 45. Engine 14 was using 1900 Metropolitan #2747.

470 4 SSF/MD-HD/A Upright. 28"x56$^{1}/_{2}$" tubular boiler (iron). 5" x 7" pump, automatic air pump, ornament on air chamber. 3$^{1}/_{4}$" x 20' pre-connect suction. 8" x 7" cyl. Wheels 16 x 54 front, 18 x 62 rear. Rear brake lever.

5-4-1886: Chris Ahrens, Fred Ahrens, Hiram Nieman delivered to Miamisburg, Ohio.

8-20-1903: New 28" x 58" Fox boiler. Pump overhauled.

475 5 SSF/MD-HD/A Upright. 25"x56$^{1}/_{2}$" tubular boiler (iron). 4$^{1}/_{2}$" x 7" pump, automatic air pump, ornament on air chamber. 3$^{1}/_{4}$" x 20' pre-connect suction. 7" x 7" cyl. Wheels 16 x 52 front, 16 x 60 rear. Rear brake lever.

5-3-1886: Delivered by Chris Ahrens and G. Fred Ahrens to Germantown, Ohio.

3-19-1904: New 25" x 58" Fox boiler.

477 5 SSF/MD-HD/A Upright. 25"x$56^{1}/_{2}$" tubular boiler (iron). $4^{1}/_{2}$" x 7" pump, automatic air pump, bell on air chamber. $3^{1}/_{4}$" x 20' pre-connect suction. 7" x 7" cylinder. Wheels 16 x 52 front, 16 x 60 rear. Rear brake lever.

11-18-1886: Delivered by G. Fred Hawekotte and G. Fred Ahrens to North Vernon, Ind.

479 5 SSF/MD-HD/A Upright. 25"x$56^{1}/_{2}$" tubular boiler (iron). $4^{1}/_{2}$" x 7" pump, automatic air pump, ornament on air chamber. $3^{1}/_{4}$" x 20' pre-connect suction. 7" x 7" cyl. Wheels 16 x 52 front, 16 x 60 rear. Rear brake lever.

3-1-1887: Leased to Portland, Ind. Delivered by G. Fred Ahrens.

8-5-1887: Returned to Ahrens Mfg. Co.

9-26-1887: Sold to Merchants Cotton Compress Co., Memphis, Tenn., to protect their factory. Manager of factory was H.A. Montgomery, fire commissioner of Memphis.

5- -1900: Sold to Memphis Fire Dept.

481 5 SSF/MD-HD/A Upright. 25"x$56^{1}/_{2}$" tubular boiler (iron). $4^{1}/_{2}$" x 7" pump, automatic air pump, air chamber ornament. $3^{1}/_{4}$" x 20' pre-connect suction. 7" x 7" cyl. Wheels 16 x 52, 16 x 60. Rear brake lever. A&W lanterns.

12-5-1887: Delivered by G. Fred Ahrens to Ripley, Ohio. Name *Archibald Liggett.*

1982: Ripley still owns this Ahrens.

483 1 DCN/HD/A Upright. 38"x67" tubular boiler (iron), safety rails on coal bunker, gasoline-powered heater & gasoline tank on left (first gasoline-power in fire service?). Two 5" x 8" pumps, one flywheel, relief valve left, air chamber ornament. $4^{1}/_{2}$" x 23'3" pre-connect suction, female couplings both ends. 8" x 8" cyls. Wheels 16 x 60, 18 x 64, extra front & rear wheel (ancestor of spare tire?). Gong. Drop pole, extra pole.

8-2-1886: In service, Eng. 3, St. Paul, Minn.

1894: Rebuilt. To Engine 22.

3-26-1902: To Engine 1.

485 1 DCN/HD/A Upright. Round 38" x 67" coil boiler. Two 5" x 8" pumps. Two 8" x 8" cylinders. Wheels 16 x 60 front, 18 x 64 rear.

Listed in Ahrens records, but these records don't show where it was delivered, nor even whether it was actually built.

487 6 SSF/MD/A Upright. 25" x $56^{1}/_{2}$" tubular boiler (iron). 4" x 7" pump, air pump, air chamber ornament, one goose-neck discharge gate. $3^{1}/_{4}$" x 20' pre-connect suction. $6^{1}/_{2}$" x 7" cylinder. Wheels 16 x 50, 16 x 58. Rear brake lever. Two A&W lanterns under seat.

3-26-1887: Chris Ahrens del'd to Deshler, O.

1986: at Allen County Hist. Society, Lima, O.

489 1 DCN/HD/A Upright. Round 38" x 67" coil boiler. Two 5" x 8" pumps. Two 8" x 8" cylinders. Wheels 16 x 60 front, 18 x 64 rear.

Listed in Ahrens records, but these records don't show where it was delivered, nor even whether it was actually built.

491 3 DCN/HD/A Upright. 32" x 62" coil boiler (iron), heater pipes right. $4^{1}/_{4}$" x 7" pumps, one flywheel, relief valve left, air chamber ornament. Two 4" x 10', one $2^{1}/_{2}$" x 9' suctions. $6^{3}/_{4}$"x7" cyls. Wheels 16x56, 18x62. Platform springs. $56^{1}/_{2}$" track. Gong.

12-7-1886: Shipped to Chicago, Ill. Name *Tim Monaghan.* (New boiler 1901).

1906: At Engine 49. 1917: At Engine 91.

493 3 DCN/HD/A Upright. 32" x $62^{1}/_{2}$" coil boiler (iron), heater pipes on right. $4^{1}/_{4}$" x 7" pumps, relief valve left, piston air pump, ornament atop air chamber, gauge on frame. Two 4" x 10', one $2^{1}/_{2}$" x 9' suctions. $6^{3}/_{4}$" x 7" cyls. Wheels 16 x 56 front, 18 x 62 rear. Platform springs. 62" track. Stiff pole.

4-14-1887: Delivered by G. Fred Ahrens to Los Angeles, Cal. Name *Walter S. Moore #4.*

11-21-1892: New boiler coils.

1900: To Engine 8. 1905: To reserve.

7-29-1906: New boiler coils.

1907: To Engine 8.

1918: To reserve. Sold 1924.

499 2 DCN/HD/A Upright. $35^{1}/_{2}$" x 67" coil boiler (iron), heater pipes on left, clock right, lantern at rear. $4^{3}/_{4}$" x 8" pumps; nickel-plated flywheel, crankshaft, columns; relief valve left; nickel-plated headlight atop air chamber. Two $4^{1}/_{2}$" x 10' hard suctions, one $3^{1}/_{2}$" x 15' soft suction. Two $7^{3}/_{4}$" x 8" cyls. Wheels 16 x 58, 18 x 62. Platform springs. Gong on foot board. Two torches on frame.

10-21-1886: Shipped to Eng. 11, Minneapolis, Minn. Delivered by Chris and Fred Ahrens.

1901: To Engine 22.

1914: Retired.

500 2 DCN/HD/A Upright. $35^{1}/_{2}$" x 67" coil boi-

The "Walter S. Moore" was an 1887 Ahrens 3rd Class double crane neck #493, used in Los Angeles, Cal. Originally Engine 4, it subsequently served Engine 8 until retired in 1918. It was sold in 1924. Note the air-pump on the boiler and the pressure-gauge on the crane-neck portion of the frame.

(Ed Hass collection)

Engine 11 in Minneapolis, Minn., used this 1886 Ahrens 2nd Class #499. Note the special headlight on top of the air chamber, and the torch on the frame just forward of the crane-neck.

(Ed Hass collection)

ler (iron), heater pipes on left, clock right, lantern at rear. $4^3/4$" x 8" pumps; nickel-plated flywheel, crankshaft, columns; relief valve left; nickel-plated headlight atop air chamber. Two $4^1/2$" x 10' hard suctions, one $3^1/2$" x 15' soft suction. Two $7^3/4$" x 8" cyls. Wheels 16 x 58, 18 x 62. Platform springs. Gong on foot board. Two torches on frame.

10-21-1886: Shipped to Eng. 3, Minneapolis, Minn. Delivered by Chris and Fred Ahrens.

1891: To Engine 6.

1906: To Engine 5.

1909: To Engine 14.

1912: To Engine 8. 1914: Retired.

501 2 DCN/HD/A Upright. $35^1/2$" x 67" coil boiler (iron), heater pipes left. $4^3/4$" x 8" pumps; nickel-plated flywheel, crankshaft, columns; relief valve left; air chamber ornament. Two $4^1/2$" x 10'6", one $2^1/2$" x 9' suctions. $7^3/4$" x 8" cyls. Wheels 16 x 58, 18 x 62. Platform springs. $56^1/2$" track. Gong on foot board.

12-7-1886: Shipped to Eng. 40, Chicago, Ill.

1906: At Engine 4. Retired by 1917.

502 3 DCN/HD/A Upright. 32" x 62" coil boiler (iron), safety rails above coal bunker, heater pipes left. 4" x 7" pumps. One $3^1/2$" x 20' pre-connect suction. $6^3/4$" x 7" cyls. Wheels 16 x 56 front, 18 x 62 rear. Platform springs. $57^1/2$" track. Stiff pole. Gong on foot board.

4- -1887: Delivered by G. Fred Hawekotte to Engine 27, St. Louis, Mo.

12- -1888: Replaced by Ahrens #564.

8-12-1904: New 32" x 62" Fox boiler.

1909: St. Louis Fire Dept. Annual Report again shows Ahrens #502 at Engine 27.

503 3 DSF/HD/A Upright. 32" x $66^3/4$" coil boiler (iron), safety rail on coal bunker (left), heater pipes left. $4^1/4$" x 7" pumps. 4" x 21' pre-connect suction. $6^3/4$" x 7" cyls. Wheels 16 x 56, 18 x 62. Platform springs. Two stiff poles (for three horses). Gong on foot board. Weight, loaded for service, 6870#.

6-2-1887: In service at Brighton Fire Co. #13, Cincinnati, Ohio. Name *James M. Doherty*. Replaced Ahrens #10.

8-9-1895: New boiler coils.

7- -1902: To Co. #18. Replaced Ahrens 172.

505 2 DCN/HD/A Upright. $35^1/2$" x 67" coil boiler (iron), heater pipes left. $4^3/4$" x 8" pumps, relief valve left, ornament atop air chamber. Two $4^1/2$" x 10'6" suctions. $7^3/4$" x 8" cyls. Wheels 16 x 58 front, 18 x 62 rear. Platform

Nicknamed the "Little Steamer", this 1887 Ahrens 4th Class double crane-neck #506 served Engine 1 in Phoenix, Arizona. This engine may have been "little", but it was quite powerful, as it once proved by drafting water at a 24'6" lift. The photo was apparently taken during a test (note the charged hose behind left front wheel).
(Courtesy Phoenix Fire Dept. Archives)

springs. Gong on foot board. Two torches on frame. Two lanterns. Whip socket on seat.

5-3-1887: Tested & accepted by Engine 7, Milwaukee, Wis. Delivered by Chris Ahrens.

1902: Replaced by Metropolitan #2907.

7-2-1903: To Engine 24.

1918: Sold for $750 to Fond-du-Lac, Wis.

506 4 DCN/HD/A Upright. 28$^{1}/_{2}$" x 60" tubular boiler (iron). 3$^{3}/_{4}$" x ??" pumps, air chamber ornament. One 3$^{1}/_{2}$" x 20' (pre-connect), one 3$^{1}/_{2}$" x 10' suctions. Nathan & Dreyfuss Ejector (syphon) to draft at 22' lift. 6" x ?? cyls. Wheels 16 x 54, 16 x 60. Rear brake lever. Single seat. Two torches on frame.

4-11-1887: Tested, Canal Race on Eggleston Ave., Cincinnati, Ohio. Successfully drafted at 24'6" lift with Ejector, but could not draft at that lift without the Ejector.

5-10-1887: Delivered by Fred Ahrens to Eng. 1, Phoenix, Arizona. Also 2 Ahrens hose carts, 1000' of hose. Name *Little Steamer.*

508 2 DCN/HD/A Upright. 35$^{1}/_{2}$" x 67" coil boiler (iron), safety rails on back, heater pipes left. 4$^{1}/_{2}$" x 8" pumps, relief valve left. 4" x 21' pre-connect suction. 7$^{1}/_{4}$" x 8" cyls. Wheels 16 x 58, 16 x 62. Platform springs. 57$^{1}/_{2}$" track. Gong on foot board.

5-10-1887: Delivered by Chris Ahrens to Saint Louis, Mo.

10-4-1905: 35" x 64" Fox boiler.

1909: Was in service at Engine 21 (in place of Ahrens #296).

509 3 DCN/HD/A Upright. 32" x 62$^{1}/_{2}$" coil boiler (iron), Jungs heater on left, torch on left rear of frame, safety rail on back of boiler. 4$^{1}/_{4}$" x 7" pumps, relief valve on left. One 4" x 22' pre-connect suction, female couplings both ends. 6$^{3}/_{4}$" x 7" cyls. Wheels 16 x 56 front, 16 x 62 rear. 56$^{1}/_{2}$" track. Platform springs. Stiff pole. Gong on foot board. Tool box under seat. Lantern left side of frame.

8-30-1887: Delivered by Chris Ahrens to Engine 10, Saint Paul, Minn.

7-14-1898: New boiler coils.

1912: Fox boiler.

511 3 DCN/HD/A Upright. 32" x 62$^{1}/_{2}$" coil boiler (iron), Jungs heater on left, torch on left rear of frame, safety rail on back of boiler. 4$^{1}/_{4}$" x 7" pumps, relief valve on left. One 4" x 22' pre-connect suction, female couplings both ends. 6$^{3}/_{4}$" x 7" cyls. Wheels 16 x 56 front, 16 x 62 rear. 56$^{1}/_{2}$" track. Platform springs. Stiff pole. Gong on foot board. Tool box under seat. Lantern left side of frame.

8-30-1887: Delivered by Chris Ahrens to Engine 15, Saint Paul, Minn.

3-25-1895: New boiler coils.

1912: Fox boiler. To Engine 19.

512 4 SSF/MD-HD/A Upright. 28"x56$^{1}/_{2}$" tubular boiler. One 5" x 7" pump. One 3$^{1}/_{2}$" x 20' pre-connect suction. One 8" x 7" cylinder. Wheels 16 x 54 front, 16 x 60 rear.

5-18-1888: Loaned to Morrow, Ohio.

6-9-1889: Ahrens #555 delivered to Morrow, and #512 returned to Ahrens Mfg. Co.

1890: Rebuilt as #625 for Elwood, Indiana (see #625 for its subsequent history).

514 6 SSF/MD/A Upright. 25" x 56$^{1}/_{2}$" tubular boiler. One 4" x 7" pump. One 3$^{1}/_{4}$" x 20' pre-connect suction. One 6$^{1}/_{2}$" x 7" cylinder. Wheels 16 x 52 front, 16 x 58 rear.

1888: Built as a demonstrator.

8- -1891: Rebuilt as #661 for Westwood, O.

515 6 SSF/MD/A Upright. 25" x 56$^{1}/_{2}$" tubular boiler. One 4" x 7" pump. One 3$^{1}/_{4}$" x 20' pre-connect suction. One 6$^{1}/_{2}$" x 7" cylinder. Wheels 16 x 52 front, 16 x 58 rear.

Listed in Ahrens records, but doesn't say where it was delivered, nor if it was built.

517 6 SSF/MD/A Upright. 25" x 56$^{1}/_{2}$" tubular boiler. One 4" x 7" pump. One 3$^{1}/_{4}$" x 20' pre-connect suction. One 6$^{1}/_{2}$" x 7" cylinder. Wheels 16 x 52 front, 16 x 58 rear.

Listed in Ahrens records, but doesn't say where it was delivered, nor if it was built.

518 2 DCN/HD/A Upright. 36"x64" tubular boiler (iron), torch on coal bunker; one large lantern at rear, one on left; small lantern on right side of boiler. 4$^{3}/_{4}$" x 8" pumps, relief valve right, air chamber ornament. 7$^{3}/_{4}$" x 8" cyls. Two 4$^{1}/_{2}$" x 10'6" suctions. Wheels 16 x 58, 18 x 62, extra front & rear wheel. Platform springs. Stiff pole. Gong on foot board. Whip socket on seat. Two torches on frame. Height 9'9". Weight 8020#.

10-12-1887: Shipped to Engine 18, Detroit, Mich. Delivered by Chris Ahrens, G. Fred Ahrens, and Hiram H. Nieman. Price $7350. Detroit Fire Dept. records call this a 1st Class, but Ahrens records say 2nd Class.

Later used at Engines 1, 8, 9, 17, and 21.

1911: Retired and scrapped.

520 4 DCN/HD/A Upright. 30" x 60" coil boiler

(iron/silver), safety rails on coal bunker, heater pipes left. $3^3/_4$" x ?? pumps, air chamber ornament. 6" x ?? cyls. $3^1/_2$" x 20' pre-connect suction. Wheels 16 x 54, 18 x 60. Platform springs. $56^1/_2$" track. Stiff pole. Front brake lever. Gong on foot board. Two torches on frame. Double seat, railing atop.

6- -1887: Delivered by Fred Hawekotte to Birmingham, Alabama. Rebuild of #461.

2-13-1899: New boiler coils.

4-26-1905: New 30" x 60" Fox boiler. Repaired and repainted.

521 2 DCN/HD/A Upright. $35^1/_2$" coil boiler (iron), heater pipes right, nickel-plated torch on coal bunker. Two $4^3/_4$" x 8" pumps, relief valve right, one flywheel, ornament on air chamber, water-gauge on frame. Two $4^1/_4$" x 10'6", one $2^1/_2$" x 9' suctions; whip socket on suction bracket. $7^3/_4$" x 8" cyls. Wheels 16 x 58 front, 18 x 62 rear. Platform springs. $56^1/_2$" track. Gong on foot board. Two lanterns. Two nickel torches on frame.

11-8-1887: Shipped to Chicago, Ill. (E. 105?)

10-25-1900: New 35" x 64" Fox boiler for Ahrens #521 of Chicago Engine 27.

1906: At Engine 27. Retired by 1917.

c-1912: Rebuilt with Metropolitan pump.

523 3 DCN/HD/A Upright. 32" x 62" coil boiler (iron), heater pipes right, large lantern on back, nickel-plated torch on coal bunker. $4^1/_4$" x 7" pumps, one flywheel, air chamber ornament, relief valve right, Marsh gauges. Two 4" x 10', one $2^1/_2$" x 9' suctions. $6^3/_4$" x 7" cyls. Wheels 16 x 56, 18 x 62. Platform springs. $56^1/_2$" track. Gong on foot board. Two nickel torches on frame. Whip socket.

11-8-1887: Shipped to Chicago, Ill.

1906: At Engine 22. 1917: At Engine 76.

524 3 DCN/HD/A Upright. 32" x 62" coil boiler (iron), heater pipes right, large lantern on back, nickel-plated torch on coal bunker. $4^1/_4$" x 7" pumps, one flywheel, air chamber ornament, relief valve right. Two 4" x 10', one $2^1/_2$" x 9' suctions. $6^3/_4$" x 7" cyls. Wheels 16 x 56, 18 x 62. Platform springs. $56^1/_2$" track. Gong on foot board. 2 nickel torches on frame. Whip socket. (New boiler 1899).

11-8-1887: Shipped to Chicago, Ill.

1906: At Engine 78. 1917: At Engine 105.

526 4 DCN/MD-HD/A Upright. 30"x60" coil boiler (iron). $3^3/_4$" x ?? pumps, piston air pump, relief valve right, ornament on air chamber, gauge on frame. One $3^1/_2$" x 20' pre-connect suction. Two 6" x ?? cyls. Wheels 16 x 54 front, 16 x 62 rear. Platform springs. $56^1/_2$" track. Rear brake lever. Gong on foot board.

1887 Ahrens 2nd Class double crane-neck #521 at Chicago Engine 105, sometime before 1900. Note that the steamer was then in remarkably-original condition, right down to the buggy-whip beside the driver's seat. Also note the unusual location of the pressure-gauge on the frame.

(Wayne Sorenson collection)

This 1889 photo shows Alert #1 of Fresno, Cal., with their 1887 4th Class double crane-neck Ahrens #526. A Fox boiler was installed in 1911. Fresno had this steamer until 1924, but its fate is unknown. (Courtesy Fresno Firefighters Union)

11-30-1887: Shipped to Fresno, Cal. Name *Alert #1*. Del'd by Vince Chloupek, engineer of #411. Water in 3:25. Threw a stream 194'.

1-25-1911: New 30" x 60" Fox boiler.

10-8-1924: Was in parade at Fresno.

527 4 DCN/HD/A Upright. 30" x 60" coil boiler (iron), heater pipes on left. $3^{3}/4$" x ?? pumps, air chamber ornament. Two $3^{1}/2$" x 10', one $2^{1}/2$" x 9' suctions, expanding ring couplings. 6" x ?? cylinders. Wheels 16 x 54, 16 x 60. Platform springs. Gong on foot board. Brake controlled by driver.

1-14-1888: Delivered by G. Fred Hawekotte to Lake View, Ill.

529 3 DCN/HD/A Upright. 32" x 62" coil boiler (iron), gasoline-powered heater, gasoline tank on left, safety rail on coal bunker. $4^{1}/4$" x 7" pumps, relief valve left, water gauge on frame. 4" x 22' pre-connect suction, female couplings both ends for 4" hydrant. $6^{3}/4$" x 7" cyls. Wheels 16 x 56, 18 x 62, spare rear wheel. Platform springs. $56^{1}/2$" track. Stiff pole. Gong on foot board. Tool box & two 3-pint oil cans under seat. Two torches on frame, lantern left side of frame.

12-29-1887: Shipped to St. Paul, Minn. Accepted without anyone from Ahrens being present. In service at Engine 9, Saint Paul.

Date?: To Engine 16.

1908: Replaced by Continental #143. To Engine 20.

1912: Fox boiler.

530 3 DCN/HD/A Upright. 32"x62" tubular boiler (iron), safety rails on coal bunker. $4^{1}/4$" x 7" pumps, 2 flywheels, air chamber ornament. 4" x 22' pre-connect suction. $6^{3}/4$" x 7" cyls. Wheels 16x56, 18x62. 62" track. Stiff pole. Platform springs. Gong on foot board.

2-8-1888: In service, Engine 6, Indianapolis, Ind., before it was tested. Delivered by G. Fred Hawekotte and G. Fred Ahrens. Ahrens #41 and #155 traded-in.

2-10-1888: Tested & accepted by Indianapolis, two days *after* it went in service.

Date?: To Engine 7.

5- -1896: 32" x $66^{3}/4$" Ahrens coil boiler. Two rear wheels. Repainted. To Engine 9.

532 4 DCN/HD/A Upright. 30" x 60" coil boiler (iron), safety guards & torch on coal bunker,

heater pipes left. Pumps $3^3/4$" x ??, relief valve left, air chamber ornament. $3^1/2$" x 18' pre-connect suction, 4" hydrant coupling one end, 3" coupling other end; one 3" x 2' plug suction. 6" x ?? cyls. Wheels 16 x 54, 16 x 60. Platform springs. $57^1/2$" track. Stiff pole. Gong on foot board. Whip socket on suction bracket. Double seat. Driver-control brake.

2-13-1888: Shipped to Eng. 26, St. Louis, Mo. Accepted with nobody present from Ahrens.

4-26-1898: Tipped over, only minor damage.

2-11-1899: New boiler coils.

9-1-1905: New 30" x 60" Fox boiler.

1909: Was at Engine 35.

533 4 DCN/HD/A Upright. Listed in Ahrens records, without specifications; doesn't say where it was delivered, nor if it was built.

534 3 DSF/HD/A Upright. 32" x $66^3/4$" coil boiler (iron), heater pipes left, safety rails on coal bunker. $4^1/4$" x 7" pumps, air chamber ornament. 4" x 21' pre-connect suction. $6^3/4$" x 7" cylinders. Wheels 16 x 56, 18 x 62. Platform springs. Stiff pole. Gong on foot board. Whip socket and torch.

3-20-1888: Tested, 7th & Walnut, Cincinnati, Ohio. In service at Fire Co. #3, Cincinnati. Name *Abe Furst.*

1-29-1892: Replaced by 1881 Ahrens #280. To Fire Co. #25.

3-2-1895: New boiler coils.

2-18-1905: New 32" x 62" Fox boiler.

535 3 DCN/HD/A Upright. 32" x $66^3/4$" coil boiler (silver/nickel), heater pipes left, safety rails on coal bunker. Safety whistle back of boiler, active when boiler exceeded pre-set pressure (usually 100 psi). $4^1/4$" x 7" pumps, ornament on air chamber, oil-can holder on vacuum chamber. 4" x 22' pre-connect suction. $6^3/4$" x 7" cyls. Wheels 16 x 56 front, 18 x 62 rear. Platform springs. Stiff pole. Gong under foot board, portrait of five brothers on foot board. Whip socket and torch.

2-23-1888: Delivered to Louisville, Ky. In service at Engine 9. Name *Five Brothers* (probably refers to the five Bache brothers, descendants of Benjamin Franklin, all members of the Louisville Fire Dept.).

5-8-1905: New 32" x 62" Fox boiler for Ahrens #535 of Engine 5, Louisville.

537 4 DCN/HD/A Upright. 30"x60" tubular boiler (iron), heater pipes on left, nickel-plated safety valves, torch on coal bunker. $3^3/4$" x ?? pumps, ornament on air chamber. Two $3^1/2$" x 10', one $2^1/2$" x 9' suctions. 6" x ?? cyls. Wheels 16 x 54 front, 16 x 60 rear. Platform springs. 62" track. Stiff pole. Gong on foot board. Brake controlled by driver. Two torches on frame. Two large lanterns.

8-8-1888: Shipped to Los Angeles, Cal. In service at Engine 7.

1905: To Engine 14.

1919: To reserve.

1921: Sold.

1984: One of the four Los Angeles Ahrens steamers (#493, 537, 543, or 544) is now at Calico Ghost Town, Yermo, Cal.

538 4 DCN/HD/A Upright. 30" x 60" coil boiler, lantern on back, heater pipes right, torch on coal bunker. $3^3/4$" x ?? pumps, relief valve right. Two $3^1/2$" x 10' suctions, increaser to fit $4^1/2$" hydrant. 6" x ?? cyls. Wheels 16 x 54, 18 x 60, spare front & rear wheel. $56^1/2$" track. Platform springs. Stiff pole and spare pole. Gong on foot board. Whip socket on seat. Frame painted black, running-gear red.

8- -1888: Delivered by H.H. Nieman & Fred Ahrens to Engine 15, Detroit, Mich.

Date?: To Engine 22.

8-18-1905: Letter from American-LaFrance, Elmira, N.Y. to George W. Krapp. Ahrens #538 received at Elmira 3-28-1905, now used as a relief engine (loaned to customers while their steamers repaired). American-LaFrance apparently took the Ahrens in trade.

1911: Scrapped by Detroit Fire Dept., per Clarence Woodard's Detroit-steamers list, published in the 1965 *Visiting Fireman* magazine. Conflicts with 1905 letter above.

540 5 SSF/MD-HD/A Upright. 25x56 tube boiler (iron). $4^1/2$" x 7" pump, air pump, goose-neck discharge gates, air chamber ornament. 7" x 7" cylinder. $3^1/4$" x 20' pre-connect suction. Wheels 16 x 52, 16 x 60. Rear brake lever.

1-27-1888: Ordered by Versailles, Ky. Also one hose carriage and one hook-and-ladder.

5-10-1888: Delivered by Fred Hawekotte and Fred Ahrens to Versailles, Ky. Price $3200.

11-20-1899: 25" x 56" Fox boiler, pump overhaul, non-corrosive pipes installed in pump.

1940: Retired. To a quarry-Scott County, Ky.

541 5 SSF/MD-HD/A Upright. 25" x 56" tubular boiler (iron). $4^1/2$" x 7" pump, air pump, air chamber ornament, copper discharge pipes. $3^1/4$" x 20' pre-connect suction. 7" x 7" cyl. Wheels 16 x 52, 16 x 60. Rear brake lever.

7-24-1888: Delivered by G. Fred Hawekotte

North Amherst, Ohio, was the original owner of this 1888 5th Class single straight-frame Ahrens #541. It is now displayed at the Frederick C. Crawford Auto-Aviation Museum in Cleveland, Ohio. Note the copper discharge pipe extending forward from the air chamber (visible between the spokes of the front wheel).
(Tim Elder photo)

and G. Fred Ahrens to North Amherst, Ohio.

Date?: To Thompson Products Corporation museum, Cleveland, Ohio.

1971: To Frederick Crawford Auto-Aviation Museum, Cleveland, Ohio.

543 3 DCN/HD/A Upright. 32" x $62^1/_2$" coil boiler (iron), heater pipes left. $4^1/_4$" x 7" pumps, air chamber ornament, relief valve left. Two 4" x 10', one $2^1/_2$" x 9' suctions. $6^3/_4$" x 7" cyls. Wheels 16 x 56, 18 x 62. Platform springs. 62" track. Stiff pole.

5-30-1888: Shipped to Los Angeles, Cal.

6-13-1888: Sold and delivered by Major W.C. Furry, Los Angeles agent for Ahrens Co.

6-16-1888: Tested and accepted by Engine 5, Los Angeles City. Water in 4:10.

7-29-1896: New boiler coils.

1900: To Engine 2.

1907: Rebuilt in L.A.F.D. Repair Shop.

1914: Motorized with Christie tractor.

1917: Tractor put on a 1909 Nott steamer. The Ahrens was then mounted on a fireboat.

1918: To reserve.

1919: To Engine 39.

1921: To reserve.

1922: Sold.

544 3 DCN/HD/A Upright. 30" x $60^1/_2$" coil boiler (iron), heater pipes left. $4^1/_4$" x 7" pumps, air chamber ornament, relief valve left. Two 4" x 10', one $2^1/_2$" x 9' suctions. $6^3/_4$" x 7" cyls. Wheels 16 x 56, 18 x 62. Platform springs. 62" track. Stiff pole.

8-8-1888: Delivered by W.C. Furry to Los Angeles, Cal. In service as Engine 6.

8-20-1901: 32" x $60^1/_2$" tubular boiler from American Fire Engine, Seneca Falls, N.Y.

1907: To Engine 20.

1913: To reserve.

1914: To Engine 6.

1916: To reserve.

1920: To Engine 25.

This ex-Los Angeles, Cal., double crane-neck Ahrens is now at Calico Ghost Town, in Yermo, Cal. It could be any of the four L.A. Ahrens steamers, but is probably 1888 4th Class #537.
(Courtesy Calico Ghost Town)

Fresh from a complete rebuild in the Los Angeles Fire Department Repair Shop was an 1888 Ahrens 3rd Class double crane-neck #543. Originally Engine 5, it was assigned to Engine 2 in this 1907 photo.
(Courtesy Chris Cavette)

1921: To reserve.

1923: Sold.

546 3 DCN/HD/A Upright. 32" x 62" coil boiler (iron), heater pipes left, torch on coal bunker. $4^1/4$" x 7" pumps, oil pump atop vacuum chamber, relief valve left. 4" x 10' suctions, strainer stand rear of frame. $6^3/4$" x 7" cyls. Wheels 16 x 56, 18 x 62. Front springs parallel with frame. $56^1/2$" track. Stiff pole. Gong on foot board. Whip socket on seat.

6-26-1888: Delivered by Fred Ahrens to Milwaukee, Wis.

9-8-1888: In service as Engine 12.

1915: Replaced by 1904 Metropolitan #2990.

547 3 DCN/HD/A Upright. 32" x 62" coil boiler (iron), heater pipes left, torch on coal bunker. $4^1/4$" x 7" pumps, oil pump atop vacuum chamber, relief valve left. 4" x 10' suctions, strainer stand rear of frame. $6^3/4$" x 7" cyls. Wheels 16 x 56, 18 x 62. Front springs parallel with frame. $56^1/2$" track. Stiff pole. Gong on foot board. Whip socket on seat.

6-26-1888: Delivered by G. Fred Ahrens to Milwaukee, Wis.

8-1-1889: In service as Engine 13.

1902: Replaced by 1891 Ahrens #635.

9-1-1903: To Engine 26.

1910: Nott boiler.

1917: Retired.

549 3 DCN/HD/A Upright. 32" x $66^3/4$" coil boiler (iron), heater pipes left, torch and safety rails on coal bunker. $4^1/4$" x 7" pumps, ornament on air chamber, oil pump atop vacuum chamber, relief valve on left. 4" x 22' preconnect suction, expansion ring couplings. $6^3/4$" x 7" cyls. Wheels 16 x 48 front, 18 x 58 rear. Platform springs. 62" track. Stiff pole. Gong on foot board. Whip socket on seat.

1888: Built on prospect for Cincinnati, Ohio.

12-3-1888: Fred Hawekotte and Fred Ahrens delivered to Canton, O. Name *Daniel Worley.*

2-18-1904: New boiler coils.

7-20-1971: Now on display at Stark County Historical Museum, Canton, Ohio.

550 3 DCN/HD/A Upright. 32"x$66^3/4$" coil boiler (iron), heater pipes left, torch and safety rails on coal bunker. $4^1/4$" x 7" pumps, ornament on air chamber. 4" x 22' pre-connect suction, expansion ring couplings. $6^3/4$" x 7" cyls. Wheels 16 x 48, 18 x 58. Platform springs. 62" track. Stiff pole. Gong on foot board. Whip socket on seat.

1888 Ahrens 3rd Class double crane-neck #546 poses in front of Engine 12, Milwaukee, Wis., sometime between 1900 and 1915. A unique feature of this engine was its two independent front springs (a forerunner of today's independent front suspension systems on passenger automobiles).

(Wayne Sorenson collection)

2-25-1889: In service at Fire Co. #22, Cincinnati, Ohio. Name *John Goetz, Jr.*

5-6-1891: New boiler coils.

4-1-1897: Nameplate *John Goetz Jr.* removed and put on Ahrens #2517. To Fire Co. #20 and renamed *Junior A. Caldwell.*

10-5-1897: Two new valve rod guides and new lubricator tops.

552 4 DCN/HD/A Upright. 30" x 60" coil boiler (iron), heater pipes left, torch and safety rails on coal bunker. $3^3/4$" x ?? pumps, ornament on air chamber. $3^1/2$" x 21' pre-connect suction. 6" x ?? cyls. Wheels 16 x 46, 16 x 58. Platform springs. 62" track. Stiff pole. Gong on foot board. Whip socket on seat.

10- -1888: In service at Fire Co. #23, Cincinnati, Ohio. Name *Edwin Stevens.*

9-9-1896: New boiler coils.

553 4 DCN/HD/A Upright. 30" x 60" coil boiler (iron), heater pipes left, torch and safety rails on coal bunker. $3^3/4$" x ?? pumps, ornament on air chamber. $3^1/2$" x 21' pre-connect suction. 6" x ?? cyls. Wheels 16 x 46, 16 x 58. Platform springs. 62" track. Stiff pole. Gong on foot board. Whip socket on seat.

2-19-1889: In service at Fire Co. #26, Cincinnati, Ohio. Name *Charles Fleishman.*

555 5 SSF/MD-HD/A Upright. 25" x 56" tubular boiler (iron), nickel-plated frame around boiler. $4^1/2$" x 7" pump, automatic air pump, ornament on air chamber, nickel flywheel, two inlets, two goose-neck discharge gates under sides of frame. One $3^1/4$" x 20' pre-connect suction, one $3^1/4$" x 5' suction without mounting brackets. Wheels 16 x 52, 16 x 60. Rear brake lever. Driver's seat.

6-9-1889: Delivered by Chris Ahrens to Morrow, Ohio. Name *Hope.* Replaced Ahrens #512, which had been on loan to Morrow since May 18, 1888 and was returned to factory when #555 was delivered.

4-6-1903: 25" x 56" Fox boiler. Pump overhauled. Running gear repaired. Repainted. Price $1365.

1929: Retired. Sold to 18-year-old Lewis H. Brown of Morrow, Ohio.

3-3-1932: Ahrens-Fox supplied valves to L.H. Brown, Morrow, Ohio for Ahrens 555 ($327).

7- -1975: Operated at fire engine muster, Oeder's Lake, Morrow, Ohio.

7-26-1985: Lewis Brown still owns this steamer after 56 years. He had it pumping water during the annual Chili Festival at the Fire Museum, Cincinnati, Ohio.

556 5 SSF/MD-HD/A Upright. 25" x 56" tubular boiler (iron). $4^1/2$" x 7" pump. $3^1/4$" x 20' pre-connect suction. Wheels 16 x 52, 16 x 60.

1888: Built as a demonstrator.

7-31-1891: Rebuilt as #641 and sold to Belle Plaine, Iowa. See #641 for further history.

558 4 DCN/HD/A Upright. 30" x 60" coil boiler (iron), heater pipes left, torch and safety rails on coal bunker. $3^3/4$" x ?? pumps, ornament on air chamber. $3^1/2$" x 21' pre-connect suction. 6" x ?? cyls. Wheels 16 x 54, 16 x 60. Platform springs. Gong. Whip socket on seat.

11-10-1888: Shipped to Chattanooga, Tenn. Delivered by G. Fred Hawekotte and G. Fred Ahrens. Name *W.L. Dugger #3.*

1-14-1897: New boiler coils.

2-20-1905: New boiler coils.

559 Ahrens record book just has a blank page with number "559" at top. No specifications, date, or city, doesn't say if engine was built.

561 3 DCN/HD/A Upright. 32" x $66^3/4$" coil boiler (iron), heater pipes left, torch and safety rails on coal bunker. $4^1/4$" x 7" pumps, ornament on air chamber, Prunty relief valve on left; nickel-plated heads, columns, valves, flywheel. Two 4" x 10', one $2^1/2$" x 9' suctions. $6^3/4$" x 7" cylinders (nickel plated). Wheels 16 x 48, 18 x 58. Platform springs. Stiff pole. Gong on foot board. Whip socket.

9-20-1888: Shipped to Hamilton Fire Co. #2, San Diego, Cal. Delivered by Fred Ahrens.

7-12-1900: New boiler coils.

2-11-1903: New 32" x 62" Fox boiler.

562 3 DCN/HD/A Upright. 32" x $62^1/2$" coil boiler (iron), heater pipes left, torch and safety rails on coal bunker. Pumps 4" x 7", relief valve left, air chamber ornament. $3^1/2$" x 20' (pre-connect) suction. 3" x 9' plug suction, 4" male and 3" female couplings, on forked brackets on boiler. $6^3/4$" x 7" cyls. Wheels 16 x 56, 18 x 62. Platform springs. Stiff pole. Gong on foot board. Whip socket on seat.

11-14-1888: Delivered by G. Fred Hawekotte to Engine 28, St. Louis, Mo.

12-27-1902: New 32" x 62" Fox boiler ($850).

1909: Still in service at Engine 28.

564 2 DCN/HD/A Upright. $35^1/2$" x 67" coil boiler (iron), heater pipes left, torch and safety rails on coal bunker. $4^3/4$" x 8" pumps, relief valve left, air chamber ornament. 4" x 21' pre-connect suction. Plug suction 3" x 2', 4" male and 3" female couplings. $7^3/4$" x 8" cyls. Wheels 16 x 58, 18 x 62. Platform springs.

This 1889 Ahrens 5th Class single straight frame #555 served Morrow, Ohio, for 40 years (1889-1929), and has belonged to Lewis Brown of Morrow since its retirement 57 years ago. In this 1975 photo, Brown had steam up and was preparing to draft water from the lake in background.

(Ed Hass photo)

Close-up details of Lewis Brown's 1889 Ahrens #555. Left: Bright plating, an Ahrens hallmark, includes the gleaming copper air chamber, polished brass deck between the frame rails, and nickel-plated frame encasing the boiler. Right: Close-up of one-cylinder piston pump. S-shaped valve at upper left allowed steam from the boiler to flow into the steam cylinder, forcing the steam cylinder downward, turning the crankshaft, which in turn moved the pump piston. The two gauges just left of flywheel showed the steam pressure and pump (water) pressure.

(Ed Hass photos)

1888 Ahrens 3rd Class double crane-neck #562 early in its career, when it still had original coil boiler. Assigned to Engine 28 in St. Louis, Mo., #562 was rebuilt with a Fox boiler in 1902. (Courtesy Harold S. Walker)

This 1888 Ahrens 2nd Class double crane-neck #564 originally served Engine 27 in St. Louis, Mo., then Engine 10, and finally Engine 8. Retired in 1925, it was sold to Anheuser-Busch brewery in St. Louis. It is now part of the brewery's "Grant's Farm" museum of horse-drawn vehicles, home of the renowned Clydesdale-drawn Budweiser beer wagons. (Courtesy Harold S. Walker)

$57^{1}/_{2}$" track. Stiff pole. Gong on foot board. Whip socket on seat.

12- -1888: Delivered by G. Fred Hawekotte to Engine 27, St. Louis, Mo.

Date?: To Engine 10, replacing Ahrens #162.

11-16-1900: New 35" x 64" Fox boiler. Two new rear wheels.

1909: Listed as Engine 10 on St. Louis Fire Dept. apparatus roster.

1914: To Engine 8. Replaced Ahrens #2337.

1921: Retired.

1925: Sold to August Busch, of Anheuser-Busch brewery (Budweiser), St. Louis, Mo. Restored and displayed at Anheuser-Busch's "Grant's Farm" museum.

1975: Still at Grant's Farm/Anheuser-Busch.

565 2 DCN/HD/A Upright. $35^{1}/_{2}$" x 67" coil boiler (iron), heater pipes left, torch and safety rails on coal bunker. $4^{3}/_{4}$" x 8" pumps, relief valve left, air chamber ornament. 4" x 21' pre-connect suction. Plug suction 3" x 2', 4" male and 3" female couplings. $7^{3}/_{4}$" x 8" cyls. Wheels 16 x 58, 18 x 62. Platform springs. $57^{1}/_{2}$" track. Stiff pole. Gong on foot board. Whip socket on seat.

12- -1888: Delivered by G. Fred Hawekotte to Engine 29, St. Louis, Mo.

10- -1900: New 35" x 64" Fox boiler.

1909: Was a spare at Engine 8.

567 3 DCN/HD/A Upright. 32" x $66^{3}/_{4}$" coil boiler (iron), nickel-plated safety valve, heater pipes right, lamp for stoker, nickel-plated torch and safety rails on coal bunker. Boiler attached to frame with iron brace, not bolts. $4^{1}/_{4}$" x 7" pumps, relief valve right, air chamber ornament, vacuum chamber lamp. $6^{3}/_{4}$" x 7" cyls. Two 4" x 10', one $2^{1}/_{2}$" x 8' suctions. Wheels 16 x 56, 16 x 62. Platform springs. $56^{1}/_{2}$" track. Stiff pole. Gong on foot board. Whip socket. 2 torches on frame.

12-19-1888: Shipped to Chicago, Ill.

1906: At Engine 44. 1917: At Engine 75.

568 3 DCN/HD/A Upright. 32" x $66^{3}/_{4}$" coil boiler (iron), nickel-plated safety valve, heater pipes right, lamp for stoker, nickel-plated torch and safety rails on coal bunker. Boiler attached to frame with iron brace, not bolts. $4^{1}/_{4}$" x 7" pumps, relief valve right, air chamber ornament, vacuum chamber lamp. $6^{3}/_{4}$" x 7" cyls. Two 4" x 10', one $2^{1}/_{2}$" x 8' suctions. Wheels 16 x 56, 16 x 62. Platform springs. $56^{1}/_{2}$" track. Stiff pole. Gong on foot board. Whip socket. 2 torches on frame.

12-19-1888: Shipped to Chicago, Ill. In primer, not painted. (Fox boiler 1901).

1906: At Engine 53. 1917: At Engine 69.

570 Repairs to 1875 3rd Class #146, the *Harry Wilcox* of Lookout Fire Co. #1, Chattanooga, Tenn. Although surviving records don't say why these repairs were made, the nature and extent of repairs suggest that the steamer had been consumed by flames:

New boiler coils and shell, safety rail on back of boiler, boiler valves, heater added to boiler and torch added to coal bunker, brass frame-casing around boiler, stoker's foot board replaced on coal bunker. Pump overhauled (new piston, valves, springs, and leather cups). New flywheel, air-chamber base with shut-off discharge gates, repaired signal lamp atop air chamber. New $3^{1}/_{2}$" x 20' pre-connect suction with Expanding Ring couplings. New steam cylinder. New 18 x 58 rear wheels, $47^{3}/_{4}$" rear axle, front and rear springs, drop pole, single and double trees. Gong repaired. New driver's seat (with whip socket) and foot board. 3 new lanterns.

11-10-1888: Shipped to Chattanooga, Tenn.

5-16-1898: New boiler coils.

571 3 DCN/HD/A Upright. 32" x $66^{3}/_{4}$" coil boiler (iron), safety rail on back, heater pipes on left, torch on coal bunker. Two $4^{1}/_{4}$" x 7" pumps, air chamber ornament. 4" x 22' pre-connect suction, Expanding Ring couplings. $6^{3}/_{4}$" x 7" cyls. Wheels 16 x 48 front, 18 x 58 rear, steel tires. Platform springs. Stiff pole. Gong on foot board. Whip socket on seat.

2-25-1889: Delivered to Cincinnati, Ohio.

3-7-1889: In service at Fire Co. 29, Cincinnati. Name *J.R. Megrue*.

7-20-1900: Boiler and pump overhauled at factory. To Fire Co. #6. Later Fire Co. #4.

2- -1904: 1" stay-bolts in boiler shell. Shell so badly corroded (down to only $^{1}/_{16}$" thick) that this extra reinforcement was needed.

Date?: Wrecked by falling wall at Cincinnati Hotel fire on 9th St.

573 3 DCN/HD/A Upright. 32" x $66^{3}/_{4}$" coil boiler (iron), safety rail on back, heater pipes left, torch on coal bunker. Two $4^{1}/_{4}$" x 7" pumps, air chamber ornament. 4" x 22' pre-connect suction, Expanding Ring couplings. Two $6^{3}/_{4}$" x 7" cyls. Wheels 16 x 48 front, 18 x 58 rear. Platform springs. Stiff pole. Gong on foot board. Whip socket on seat.

5-4-1889: In service at Fire Co. #21, Cincinnati, Ohio. Name *Abe Steinau.*

3-15-1895: New boiler coils.

Falling walls have long been a threat to the safety of firemen and fire engines at fires. Here, a wall had collapsed on top of the "J.R. Megrue #4", Cincinnati's 1889 3rd Class double crane-neck Ahrens #571, during the Cincinnati Hotel fire on 9th street. Note broken left rear wheel, twisted coal bunker, and rubble at left.

(Ed Hass photos)

574 3 DCN/HD/A Upright. 32" x 66$^3/_4$" coil boiler (iron), safety rail on back, heater pipes left, torch on coal bunker. 4$^1/_4$" x 7" pumps, ornament on air chamber. 4" x 22' pre-connect suction. 6$^3/_4$" x 7" cyls. Wheels 16 x 56, 18 x 62. Platform springs. 62" track. Stiff pole. Gong on foot board. Whip socket.

8-12-1889: Ordered by Columbus, Ohio.

12-6-1889: Shipped to Engine 2, Columbus, Ohio. Delivered by Fred Hawekotte, Hiram H. Nieman, John P. Ahrens, and Woehle.

2-11-1893: New boiler coils.

12- -1893: Suction cut in two 4" x 10' pieces, pump bottom without pre-connects.

7-9-1903: New 32" x 62" Fox boiler.

576 3 DCN/HD/A Upright. 32" x 66$^3/_4$" coil boiler (iron), safety rail and lantern on back, heater pipes on right, torch on coal bunker. 4$^1/_4$" x 7" pumps, ornament on air chamber, relief valve right. Two 4" x 10', one 2" x 9' suctions. 6$^3/_4$" x 7" cyls. Wheels 16 x 56, 18 x 62. Platform springs. 56$^1/_2$" track. Stiff pole. Gong on foot board. Whip socket on seat.

10-7-1889: Ordered by Chicago, Ill.

12-2-1889: Shipped to Chicago, Ill.

1899: New boiler.

1906: At Engine 85. Retired by 1917.

577 4 DCN/HD/A Upright. 30" x 60" coil boiler (iron), heater pipes left, torch on coal bunker. Pumps 3$^3/_4$" x ??, air chamber ornament. Two 3$^1/_2$" x 10', one 2$^1/_2$" x 9' suctions. 6" x ?? cyls. Wheels 16 x 54, 18 x 60. Platform springs. 56$^1/_2$" track. Driver controlled brake. Gong on foot board. Whip socket.

9-21-1889: Delivered by G. Fred Hawekotte to Tacoma, Wash. Name *Tacoma #4*. Steam in 3:50. Maintained 120# steam, 260# water pressure for one hour. Threw four streams.

1900: New 3$^1/_2$" x 10' pre-connect suction.

Date?: Nott water-tube boiler. To Engine 5.

Date?: Sold to North Yakima, Wash.

1982: At Yakima Valley Museum, Wash.

579 4 DCN/HD/A Upright. Ahrens records do not show if this was built or who bought it.

580 4 DCN/MD-HD/A Upright. 30"x60" coil boiler (iron), heater pipes left, torch on coal bunker. Pumps 3$^3/_4$" x ??, air chamber ornament. 3$^1/_2$" x 20' pre-connect suction. 6" x ?? cyls. Wheels 16 x 54, 18 x 60. Platform

1889 4th Class double crane-neck #577 originally served Engine 4 in Tacoma, Wash. Later used in North Yakima, it is now displayed at the Yakima Valley Museum in Yakima, Wash. Non-original boiler is a Nott spiral water-tube type.

(Courtesy Yakima Valley Museum)

springs. First straight rear axle, first combination brake for driver and stoker. Gong on foot board.

5-17-1889: Delivered by John P. Ahrens and Holle to Mount Sterling, Ky. Name *J.W. Barnes #2.* Water in 4:00. Threw one $1^{1}/_{8}$" stream 200', two $1^{1}/_{8}$" streams 200' each. Through 500' hose, threw a stream 155' vertically, vs. 135' for Mt. Sterling's Button steamer *R. Apperson Jr. #1*, which Ahrens had recently rebuilt.

2-21-1899: New 28" Clapp & Jones coil boiler installed at Ahrens factory ($910). Also new $3^{1}/_{2}$" x 20' pre-connect suction ($66).

582 4 DCN/HD/A Upright. $28^{1}/_{2}$" x 60" coil boiler (silver), heater pipes left, one lantern on back and one on side of boiler, torch on coal bunker. $3^{3}/_{4}$" x ?? pumps, ornament on air chamber. Two $3^{1}/_{2}$" x 10', one $2^{1}/_{2}$" x 9' suctions. 6" x ?? cyls. Wheels 16 x 54, 18 x 60. Platform springs. Straight rear axle. 62" track. Brake controls for *both* driver and stoker. Gong on foot board. Single seat with railing atop and whip socket on side.

8-21-1889: Delivered by G. Fred Hawekotte to Wilkes-Barre, Pa. Name *T.S. Hillard #4.*

583 3 DCN/HD/A Upright. 32"x62" tubular boiler (iron), heater pipes right, torch on coal bunker. $4^{1}/_{4}$" x 7" pumps, ornament on air chamber. Two 4" x 10', one $2^{1}/_{2}$" x 9' suctions. $6^{3}/_{4}$" x 7" cyls. Wheels 16 x 56, 18 x 62. 62" track. Stiff pole. Gong on foot board. Whip socket on seat. Two torches on sides of frame, stand for strainer on back of frame.

8-7-1889: Delivered by G. Fred Hawekotte to Pueblo, Colo. Name *A.T. Stewart #1.*

10-7-1889: On test at water works, threw one stream 270' horizontally, 180' vertically. Threw four streams simultaneously.

10-1-1898: Pumped 16 hours non-stop at Antlers Hotel fire in Colorado Springs, Colo. (5:30 p.m. Oct. 1 to 9:30 a.m. Oct. 2).

1- -1905: New boiler tubes, general overhaul by Iron City Boiler Works, Pueblo, Colo., parts supplied by American-LaFrance ($738).

585 3 DCN/HD/A Upright. 32"x62" tubular boiler. Two $4^{1}/_{4}$" x 7" pumps. Two $6^{3}/_{4}$" x 7" cylinders. Wheels 16 x 56 front, 18 x 62 rear.

Ahrens records don't show whether this was built or who bought it.

587 2 DCN/HD/A Upright. $35^{1}/_{2}$" x 67" coil boiler (iron), heater pipes left, two lanterns on back, torch on coal bunker. Two $4^{3}/_{4}$" x 8" pumps, air chamber ornament, oil pump atop

The "A.T. Stewart #1" of Pueblo, Colo., was an 1889 3rd Class double crane-neck Ahrens #583. The carefully-matched trio of white fire horses, the gleaming nickel-plated steamer, and the double row of brass buttons on the firemen's uniforms, had typically found admirers in the two boys at right.

(Courtesy Dr. Lester L. Williams)

vacuum chamber, relief valve on right. Two $4^1/2$" x 10'6" suctions. $7^3/4$" x 8" cyls. Wheels 16 x 56, 18 x 62. Platform springs. Stiff pole. Two torches on frame, one lantern on left, stand for suction strainer on back. Gong on foot board. Whip socket. One 3-pint Cylinder Oil can and one 3-pint Lubricating Oil can under seat. Weight 7770#.

6-25-1889: Shipped to Milwaukee, Wis.

8-1-1889: In service, Engine 14, Milwaukee.

1904: Replaced by Ahrens #396.

11-1-1905: To Engine 28.

1908: Replaced by Ahrens #2280.

1913: To Engine 30.

1917: Replaced by American-LaFrance pumper. Retired and probably junked.

589 2 DCN/HD/A Upright. $35^1/2$" x 67" coil boiler (iron), heater pipes left, safety rails and torch on coal bunker. Two $4^3/4$" x 8" pumps, air chamber ornament. $4^1/2$" x 22' pre-connect suction. $7^3/4$" x 8" cyls. Wheels 16 x 56, 18 x 62. Platform springs. $56^1/2$" track. Stiff pole. Gong on foot board. Whip socket. Two torches on frame, strainer at back of frame.

9-18-1889: Hiram Nieman & John Ahrens delivered to Dayton. Name *Charles H. Ware #4*.

9-28-1889: Tested & accepted at Dayton, O. Water in 5:30. Through 200' of hose at 180# pump pressure, threw a $1^1/2$" stream 150' vertically, two $1^1/2$" streams 85' vertically.

591 2 DCN/HD/A Upright. $35^1/2$" x 67" tubular boiler (iron), heater pipes left, torch on coal bunker. $4^3/4$" x 8" pumps, air chamber ornament, oil pump on vacuum chamber, relief valve left. Two $4^1/2$" x 10'6" suctions. $7^3/4$" x 8" cyls. Wheels 16 x 56 front, 18 x 62 rear. Two torches on frame, stand for suction strainer back of frame. Gong on foot board. Whip socket. Two 3-pint oil cans under seat.

10-12-1889: Shipped to Engine 5, Milwaukee, Wis. Accepted without anyone from Ahrens.

9-18-1893: New Ahrens coil boiler. Shipped back to Milwaukee 10-19-1893.

2- -1898: New front platform springs.

1906: Replaced by Nott steamer. To Engine 25 in place of Ahrens #2177.

1912: Nott boiler.

1925: Retired. Probably junked.

This circa-1900 photo shows **Engine 4** of Seattle, Wash., with their 1889 3rd Class double crane-neck Ahrens #597. The rare three-horse hitch was installed in 1898. Teams of fire horses were usually all of a matched color, but this was obviously one exception. Note the large buggy-whip beside the driver.

(Courtesy Harold S. Walker)

Engine 3, Seattle, Wash., operating at draft with their 1889 3rd Class double crane-neck Ahrens #598. The occasion was reputedly a fire chief's convention in Seattle, which explains formal attire of the pump engineer (center) and stoker (right). The background smoke at right was apparently a controlled burn to demonstrate Engine 3's record-setting pumping power (this Ahrens broke the world's record for fastest water in 1903, timed at 3 minutes flat).

(Collection of the late Fire Marshal Crapo, Chicago)

593 3 DCN/HD/A Upright. 32"x62" tubular boiler (iron), heater pipes left, torch on coal bunker. Pumps $4^1/_4$" x 7", air chamber ornament, relief valve left. Two 4" x 10'6" suctions. $6^3/_4$" x 7" cyls. Wheels 16 x 56, 18 x 62. Platform springs. 62" track. Three-horse hitch. Gong on foot board. Whip socket.

7-27-1889: Ordered by Toledo, Ohio.

12-11-1889: Shipped to Toledo, Ohio. Delivered by Fred Hawekotte and Ed Steinway. Price $3500 plus trade-in of old steamer. On test, had steam in 3:00, and water in 5:00. Through 350' of hose, threw a $1^1/_4$" stream 268' horizontally, threw a $1^1/_2$" stream 225' vertically. In service at Engine 2.

1901: Rebuilt by LaFrance Fire Engine Co., Elmira, N.Y.

594 3 DCN/HD/A Upright. 32"x62" tubular boiler (iron), heater pipes left, torch on coal bunker. Pumps $4^1/_4$" x 7", air chamber ornament, relief valve left. Two 4" x 10'6" suctions. $6^3/_4$" x 7" cyls. Wheels 16 x 56, 18 x 62. Platform springs. 62" track. Three-horse hitch. Gong on foot board. Whip socket.

7-27-1889: Ordered by Toledo, Ohio.

12-11-1889: Delivered by Fred Hawekotte and Ed Steinway to Toledo, Ohio. Price $3500 plus trade-in of old steamer. Steam in 3:15, water in 5:15. In service at Engine 5.

1901: Rebuilt by LaFrance Fire Engine Co., Elmira, N.Y.

2-2-1914: New $32^1/_2$" x 64" Nest Tube boiler made and installed by Ahrens-Fox.

596 3 DCN/HD/A Upright. 32"x$66^3/_4$" coil boiler (iron), heater pipes left, safety rail on back, torch on coal bunker. $4^1/_4$" x 7" pumps, ornament on air chamber. One 4" x 22' pre-connect suction. $6^3/_4$" x 7" cyls. Wheels 16 x 56, 18 x 62. Platform springs. 62" track. Stiff pole. Gong on foot board. Whip socket.

8-12-1889: Ordered by Columbus, Ohio.

12-6-1889: Delivered by John Ahrens, Fred Hawekotte, Hiram H. Nieman, and Woehle to Columbus, Ohio. In service as Engine 4. Also delivered one Ahrens stationary heater.

12- 1893: Suction hose cut into two pieces and new couplings added. New pump bottom installed without suction pre-connects.

11-5-1898: New boiler coils.

597 3 DCN/HD/A Upright. 32"x$66^3/_4$" coil boiler (iron), heater pipes left, safety rail on back, two lanterns on sides, torch on coal bunker. $4^1/_4$" x 7" pumps, air chamber ornament. Two 4" x 10' suctions. $6^3/_4$" x 7" cyls. Wheels 16 x 56, 18 x 62. Platform springs. 62" track. Stiff pole. Gong on foot board. Whip socket on seat, tool box under seat, fresh-water tank behind seat. Brake controls for driver and stoker. Two torches on sides of frame.

8-17-1889: Ordered by Seattle, Wash.

11-29-1889: Shipped to Engine 1, Seattle, Wash. Delivered by Vincent Chloupek (engineer of Ahrens #411 in Oakland, Cal.).

5- -1898: Changed from 2-horse to 3-horse hitch. To Engine 4.

598 3 DCN/HD/A Upright. 32"x$66^3/_4$" coil boiler (iron), heater pipes left, safety rail on back, two lanterns on sides, torch on coal bunker. $4^1/_4$" x 7" pumps, air chamber ornament. Two 4" x 10' suctions. $6^3/_4$" x 7" cyls. Wheels 16 x 56, 18 x 62. Platform springs. 62" track. Stiff pole. Gong on foot board. Whip socket on seat, tool box under seat, fresh-water tank behind seat. Brake controls for driver and stoker. Two torches on sides of frame.

8-17-1889: Ordered by Seattle, Wash.

11-29-1889: Shipped to Engine 2, Seattle, Wash. Delivered by Vincent Chloupek (engineer of Ahrens #411 in Oakland, Cal.).

2-9-1899: New boiler coils. Replaced by Metropolitan #2686 and to Engine 3.

8-25-1903: Set world's fastest-water record, of 3:00. Old record, 3:05, set by Ahrens 641.

8-24-1905: New 32" x $66^3/_4$" coil boiler.

10-11-1905: New rear springs, throttle valve.

1907: Replaced by Continental. To Engine 1.

1924: Retired.

599 3 DCN/HD/A Upright. 32"x$62^1/_2$" coil boiler (iron), heater pipes left, safety rail on back, torch on coal bunker. 4" x 7" pumps, air chamber ornament, relief valve on left. One $3^1/_2$" x 20' pre-connect suction. $6^3/_4$" x 7" cyls. Wheels 16 x 56 front, 18 x 62 rear. Platform springs. $57^1/_2$" track. Stiff pole. Gong on foot board. Whip socket on seat.

8-16-1889: Ordered by St. Louis, Mo.

1-7-1890: Shipped to Engine 4, St. Louis, Mo. Delivered by G. Fred Hawekotte.

4-12-1901: New boiler coils.

10-25-1902: New 32" x 62" Fox boiler ($850).

1909: Was in service at Engine 29. Engine 4 was using 1907 Metropolitan #3201.

600 2 DCN/HD/A Upright. $35^1/_2$"x67" coil boiler (iron), heater pipes left, safety rail on back, torch on coal bunker. $4^3/_4$" x 8" pumps, air

chamber ornament, relief valve left. One 4" x 21' pre-connect suction, 4" male and 3" female couplings. $7^3/_4$" x 8" cyls. Wheels 16 x 58, 18 x 62. Platform springs. $57^1/_2$" track. Stiff pole. Gong on foot board. Whip socket.

8-16-1889: Ordered by St. Louis, Mo.

1-7-1890: Shipped to Engine 22, St. Louis, Mo. Delivered by G. Fred Hawekotte.

9-8-1904: New 35" x 64" Fox boiler ($850).

1909: Still in service at Engine 22.

601 2 DCN/HD/A Upright. $35^1/_2$"x67" coil boiler (no casing), heater pipes right, safety rail on back, two lanterns on sides, torch on coal bunker. $4^3/_4$" x 8" pumps, relief valve right, air chamber ornament. Two $4^1/_2$" x 10', one $2^1/_2$" x 9' suctions. $7^3/_4$" x 8" cyls. Wheels 16 x 58, 18 x 62. Platform springs. $56^1/_2$" track. Stiff pole. Gong on foot board. Whip socket on suction bracket. Two torches on frame.

8-7-1889: Ordered by Chicago, Ill.

12-28-1889: Shipped to Chicago, Ill.

10-20-1903: New $35^1/_2$" x 67" coil boiler.

1906: At Engine 64. Retired 1907.

602 3 DCN/HD/A Upright. 32" x $62^1/_2$" tubular boiler (iron), heater pipes (for Silsby heater) on left, torch on coal bunker. Two $4^1/_4$" x 7" pumps, air chamber ornament, relief valve left. Two 4" x 10'6", one $2^1/_2$" x 9' suctions. $6^3/_4$" x 7" cyls. Wheels 16 x 56, 18 x 62. Platform springs. 62" track. Stiff pole. Gong on foot board. Whip socket on suction bracket. One lantern engraved "H.E. Walden, Chief".

9-20-1889: Ordered by Wichita, Kansas.

1-20-1890: Shipped to Wichita, Kansas. Delivered by Chris Ahrens. Name *Wichita #1*.

1-27-1890: Tested at Wichita. Water in 5:45. Through 600' of hose, threw two $1^1/_4$" streams over tallest building in Wichita.

4-21-1890: In service at Engine 1, Wichita.

1901: Pumped for 36 hours non-stop at Dold Packing House fire.

1928: Replaced by Ahrens-Fox motorized pumper #3332. To reserve.

c-1936: Retired and junked.

603 3 DCN/HD/A Upright. 32"x$66^3/_4$" coil boiler (no casing), heater pipes right, torch on coal bunker. Pumps $4^1/_4$" x 7", air chamber ornament, relief valve right. Two 4" x 10', one $2^1/_2$" x 9' suctions. $6^3/_4$" x 7" cyls. Wheels 16 x 56, 18 x 62. Platform springs. $56^1/_2$" track. Stiff pole. Gong on foot board. Whip socket.

10-7-1889: Ordered by Chicago, Ill.

1-14-1890: Shipped to Chicago, Ill. [Harold Walker lists as Engine 73].

Date?: Rebuilt with Metropolitan pump.

1906: At Engine 61. Still at 61 in 1917.

604 4 DCN/HD/A Upright. $28^1/_2$"x60" coil boiler

"Wichita #1" was an 1890 3rd Class Ahrens #602 of Wichita, Kansas. By the 1930's, steam fire engines were already quite rare, but no museum or collector rescued this one from the scrapheap when it was retired in 1936. Despite growing collector interest, many collectibles coming out of fire service today, still go to scrapyards. (Wichita Fire Dept. photo)

(iron), heater pipes left, two Dietz lanterns on boiler, torch and safety rail on coal bunker. Pumps 3$^{3}/_{4}$" x ??, air chamber ornament. One 3$^{1}/_{2}$" x 20' pre-connect suction. 6" x ?? cyls. Wheels 16 x 46, 18 x 58. Platform springs. 62" track. Stiff pole. Gong on foot board. Two oil cans under seat. Whip socket.

10-5-1889: Ordered by Cincinnati, Ohio.

2-21-1890: Delivered to Fire Co. #24, Cincinnati, Ohio. Name *Lewis Wisbey.*

605 4 DCN/HD/A Upright. 28$^{1}/_{2}$" x 60" tubular boiler (iron), heater pipes left, two lanterns on boiler, torch on coal bunker. 3$^{3}/_{4}$" x ?? pumps, air chamber ornament. 3$^{1}/_{2}$" x 21' pre-connect suction. 6" x ?? cyls. Wheels 16 x 48 front, 18 x 58 rear. Platform springs. 56$^{1}/_{2}$" track. Stiff pole. Gong on foot board. Two torches on frame. Whip socket on seat.

12-13-1889: Ordered by Wilmington, Ohio.

12-17-1890: Shipped to Wilmington, Ohio. Delivered by Chris Ahrens and Ed Steinway. Name *Clinton Chief #1.*

2- -1898: 28$^{1}/_{2}$" x 58$^{1}/_{2}$" Ahrens tubular boiler ($825).

12-26-1902: Rebuilt. Borrowed Ahrens #348 while repairs were made.

2-11-1903: New 28" x 58" Fox boiler.

607 2 DCN/HD/A Upright. 35$^{1}/_{2}$" x 67" tubular boiler (iron), heater pipes left (for Ahrens heater), two lanterns on sides, torch on coal bunker. 4$^{3}/_{4}$" x 8" pumps, air chamber ornament. Two 4$^{1}/_{2}$" x 10'6", one 2$^{1}/_{2}$" x 9' suctions. Two 7$^{3}/_{4}$" x 8" cyls. Wheels 16 x 48 front, 18 x 58 rear. Platform springs. 62" track. Stiff pole. Gong on foot board. Whip socket on seat. Two torches on frame.

12-16-1889: Ordered by Columbus, Ohio.

4-7-1890: Shipped to Engine 1, Columbus, Ohio. Delivered by Chris and John Ahrens.

11-2-1903: New 35$^{1}/_{2}$" x 67" coil boiler ($1100). Pump overhauled ($46.50).

608 2 DCN/HD/A Upright. 35$^{1}/_{2}$" x 67" tubular boiler (iron), heater pipes left (for Ahrens heater), two lanterns on sides, torch on coal bunker. 4$^{3}/_{4}$" x 8" pumps, air chamber ornament. Two 4$^{1}/_{2}$" x 10'6", one 2$^{1}/_{2}$" x 9' suctions. Two 7$^{3}/_{4}$" x 8" cyls. Wheels 16 x 48 front, 18 x 58 rear. Platform springs. 62" track. Stiff pole. Gong on foot board. Whip socket on seat. Two torches on frame.

12-16-1889: Ordered by Columbus, Ohio.

4-7-1890: Shipped to Columbus, Ohio. Delivered by Chris Ahrens and John P. Ahrens. In service as Engine 3.

1-19-1904: New 35$^{1}/_{2}$" x 64" Fox boiler.

610 3 DCN/HD/A Upright. 32" x 66$^{3}/_{4}$" coil boiler (iron), heater pipes left, two lanterns on boiler, torch on coal bunker. Two 4$^{1}/_{4}$" x 7" pumps, air chamber ornament. Two 4" x 10'6", one 2$^{1}/_{2}$" x 9' suctions. 6$^{3}/_{4}$" x 7" cyls. Wheels 16 x 56, 18 x 62, steel tires. Platform springs. 62" track. Gong on foot board. Whip socket on seat. Two torches on frame.

12-16-1889: Ordered by Columbus, Ohio, for Engine Co. #5.

12-22-1890: Delivered by G. Fred Hawekotte and John P. Ahrens to Columbus, Ohio.

12-26-1890: Tested & accepted by Columbus, Ohio. In service as Engine 8.

9-13-1899: New boiler coils.

10- -1902: New 32" x 62" Fox boiler.

611 3 DCN/HD/A Upright. 32" x 66$^{3}/_{4}$" coil boiler (iron), heater pipes left, two lanterns on boiler, torch on coal bunker. Two 4$^{1}/_{4}$" x 7" pumps, air chamber ornament. Two 4" x 10'6", one 2$^{1}/_{2}$" x 9' suctions. 6$^{3}/_{4}$" x 7" cyls. Wheels 16 x 56, 18 x 62, steel tires. Platform springs. 62" track. Gong on foot board. Whip socket on seat. Two torches on frame.

12-16-1889: Ordered by Columbus, Ohio, for Engine Co. #6.

12-22-1890: Delivered by G. Fred Hawekotte and John P. Ahrens to Columbus, Ohio.

12-26-1890: Tested & accepted by Columbus, Ohio. In service as Engine 2.

6-20-1903: New 32" x 62" Fox boiler ($1200).

613 3 DCN/HD/A Upright. 32" x 66$^{3}/_{4}$" coil boiler (iron), heater pipes left, two lanterns on boiler, torch on coal bunker. Two 4$^{1}/_{4}$" x 7" pumps, air chamber ornament. Two 4" x 10'6", one 2$^{1}/_{2}$" x 9' suctions. 6$^{3}/_{4}$" x 7" cyls. Wheels 16 x 56, 18 x 62, steel tires. Platform springs. 62" track. Gong on foot board. Whip socket on seat. Two torches on frame.

12-16-1889: Ordered by Columbus, Ohio.

12-22-1890: Delivered by G. Fred Hawekotte and John P. Ahrens to Columbus, Ohio.

12-26-1890: Tested & accepted by Columbus, Ohio. In service as Engine 7.

3-28-1911: Rebuilt by Ahrens-Fox. New 32" x 64" Radial Inclined Water Tube boiler.

614 3 DCN/HD/A Upright. 32" x 66$^{3}/_{4}$" coil boiler (iron), heater pipes left, two lanterns on boiler, torch on coal bunker. Two 4$^{1}/_{4}$" x 7" pumps, air chamber ornament. Two 4" x 10'6", one 2$^{1}/_{2}$" x 9' suctions. 6$^{3}/_{4}$" x 7" cyls.

Wheels 16 x 56, 18 x 62, steel tires. Platform springs. 62" track. Gong on foot board. Whip socket on seat. Two torches on frame.

12-16-1889: Ordered by Columbus, Ohio, for Engine Co. #8.

12-22-1890: Delivered by G. Fred Hawekotte and John P. Ahrens to Columbus, Ohio.

12-26-1890: Tested & accepted by Columbus, Ohio. In service as Engine 4.

1-22-1898: New boiler coils.

10-31-1904: New 32" x 62" Fox boiler.

c-1920: Motorized with Seagrave tractor.

616 2 DCN/HD/A Upright. 35 1/2" x 67" coil boiler (silver), heater pipes left, safety railing and torch on coal bunker, two lanterns on boiler. 4 3/4" x 8" pumps, air chamber ornament, relief valve left. 4 1/2" x 22'9" pre-connect suction. 6 3/4" x 8" cyls. Wheels 16 x 48, 18 x 58. Platform springs. 62" track. Gong on foot board. Whip socket. 2 torches on frame.

1-10-1890: Ordered by Quincy, Ill.

4-17-1890: Delivered by Chris and John Ahrens to Quincy, Ill. Name *Bishop #1*.

10-27-1904: New boiler coils.

617 2 DCN/HD/A Upright. 35 1/2" x 67" tubular boiler (iron), heater pipes left, torch on coal bunker. 4 3/4" x 8" pumps, air chamber ornament, oil pump on vacuum chamber, relief valve right. Two 4 1/2" x 10'6" suctions, strainer stand back of frame. 7 3/4" x 8" cyls. Wheels 16 x 58, 18 x 62. Platform springs. 62" track. Stiff pole (varnished), one extra pole. Gong on foot board. Whip socket on seat. Two torches on frame. Number plate "2" made for front of boiler, but not put on.

4-14-1890: Ordered by Milwaukee, Wis.

7-28-1890: Shipped to Milwaukee, Wis. In service at Engine 11.

4-24-1893: New 35 1/2" x 67" coil boiler. Shipped back to Milwaukee 5-25-1893.

9-1-1894: Replaced by 1877 Corbett steamer. To Engine 19.

1900: Replaced by Metropolitan #2737. To Engine 11.

6- -1900: New front platform springs ($265).

619 3 DCN/HD/A Upright. 32" x 66 3/4" coil boiler (no casing), heater pipes right, two lanterns on boiler, torch on coal bunker. 4 1/4" x 7" pumps, relief valve right, air chamber ornament, water gauge on frame. Two 4" x 10', one 2 1/2" x 9' suctions. 6 3/4" x 7" cyls.

1890 3rd Class double crane-neck Ahrens #614 served Engine 4, Columbus, Ohio. Two-wheel tractor, installed about 1920, was a Seagrave Model K (Seagrave's factory was in Columbus). Note "CFD" monogram below the seat, and numeral "4" on the hood. Tool box was in an unusual location -- forward of front fender.
(Courtesy Harold S. Walker)

Wheels 16 x 56, 18 x 62. Platform springs. $56^1/2$" track. Gong on foot board. Whip socket on seat. Two torches on frame.

12-2-1890: Shipped to Chicago, Ill. [Was at Engine 47 in 1917].

620 3 DCN/HD/A Upright. 32"x$66^3/4$" coil boiler (iron), coils chemically treated not to pit or rust, heater pipes left, torch on coal bunker, two lanterns on boiler, nameplate *Thomas G. Smith* on smokestack. $4^1/4$" x 7" pumps, air chamber ornament. $6^3/4$" x 7" cyls. 4" x 22' pre-connect suction. Wheels 16 x 56, 18 x 62. Platform springs. Stiff pole. Gong on foot board. Whip socket. Two 3-pint oil cans under seat. Two lanterns on frame.

10-20-1890: Ordered by Cincinnati, Ohio.

9-8-1891: In service at Fire Co. #5, Cincinnati, Ohio. Name *Thomas G. Smith.*

1-24-1894: After $2^1/2$ years, experimental coils (warranteed not to rust or pit) were plugged-up and pitted, had to be replaced.

4-25-1898: New boiler coils.

622 4 DCN/HD/A Upright. 30"x60" tubular boiler, heater pipes left, two lanterns on boiler, torch on coal bunker. $3^3/4$" x ?? pumps, air chamber ornament. Two $3^1/2$" x 10' suctions. 6" x ?? cyls. Wheels 16 x 54, 18 x 60. Platform springs. $56^1/2$" track. Gong on foot board. Whip socket on seat. Two torches on frame. Weight with coal, water, tools: 5660#.

4-5-1890: Ordered by Springfield, Ill.

8-29-1890: Delivered by G. Fred Hawekotte to Springfield, Ill. Two-hour test against Springfield's Button steamer. Water in 4:00. Threw a vertical stream 196', versus 175' for the Button. Threw one stream (horizontal) 216', distanced by the Button throwing 250'.

8-30-1890: Accepted by Springfield, Ill. In service as Engine 3.

1-6-1900: Two new $3^1/2$" x 11' suctions.

623 2 DCN/HD/A Upright. $35^1/2$" x 67" tubular boiler (iron), lantern on back, heater pipes right, torch on coal bunker. $4^3/4$" x 8" pumps with special cone valves (the only Ahrens built with these valves), air chamber ornament, relief valve right. Two $4^1/2$" x 10'6" suctions. $7^3/4$" x 8" cyls. Wheels 16 x 58, 18 x 62. Platform springs. $56^1/2$" track. Stiff pole, extra pole. Gong on foot board. Whip socket. Frame painted black, running-gear red.

8-16-1890: Delivered by Chris Ahrens to Engine 6, Detroit, Mich. Price $7620.

6-10-1899: New set of cone pump valves.

Date?: To Engine 7.

10- -1904: New cone pump valves. A new mold had to be made for valves, as original mold (1890) no longer existed. 8 extra valves sent so Detroit would always have a supply.

Date?: To Engine 15.

625 4 SSF/MD-HD/A Upright. 30" x 60" tubular boiler (iron & nickel), heater pipes (for Ahrens heater) left, two Dietz lanterns on boiler, torch on coal bunker. 5" x 7" pump, automatic air pump, air chamber ornament. $3^1/4$" x 20' pre-connect suction. 8" x 7" cyl. Wheels 16 x 54, 18 x 60. Whip socket on seat. Two A&W lanterns under seat. Rear brake lever. Rebuild of Ahrens #512.

1890: Built for Elwood, Indiana. They did not buy it, sold instead to Centralia, Wash.

8-15-1890: Shipped to Centralia, Wash.

626 2 DCN/HD/A Upright. $35^1/2$" x 67" tubular boiler (iron), heater pipes left, two lanterns on boiler, safety railing on back, torch on coal bunker. $4^3/4$" x 8" pumps, air chamber ornament. $4^1/2$" x 23' pre-connect suction. $7^3/4$" x 8" cyls. Wheels 16 x 58, 18 x 62. Platform springs. Stiff pole. Gong on foot board. Whip socket. Two torches on frame.

8-28-1890: Delivered by John Black and John P. Ahrens to Duluth, Minn. Name *F.B. Dougherty #1.*

10-20-1890: Relief valve sent for left side of engine, installed by Duluth.

7- -1900: New pump bottom, to carry suctions disconnected. Old pump bottom leaked.

628 2 DCN/HD/A Upright. $35^1/2$" x 67" coil boiler (iron), heater pipes on left, torch on coal bunker. $4^3/4$" x 8" pumps, air chamber ornament, relief valve left. Two $4^1/2$" x 10'6", one $2^1/2$" x 9' suctions. $7^3/4$" x 8" cyls. Wheels 16 x 58, 18 x 62. Platform springs. Poles for 3 horses. Gong on foot board. Whip socket.

5-20-1890: Ordered by Denver, Colorado.

8- -1890: Delivered by John P. Ahrens to Engine 4, Denver, Colo.

9-22-1890: 3" x 12' soft suction, female couplings on pump end and hydrant end.

Date?: New 35" x 64" Fox boiler.

629 4 DCN/HD/A Upright. 30" x 60" coil boiler (no casing), heater pipes left, safety-valve whistle on back of boiler, torch on coal bunker. $3^3/4$" x ?? pumps, air chamber ornament, nickel-plated discharge chamber. $3^1/2$" x 21' suction. 6" x ?? cyls. Wheels 16 x 54, 18 x 60. Platform springs. 59" track. Gong under foot board, portrait of William Mayer on front of foot board. Whip socket on seat.

2-16-1891: Shipped via Mail Boat to Louisville, Ky. ($15 shipping charge). Delivered by Chris Ahrens. Name *Wm. F. Mayer #14.*

631 3 DCN/HD/A Upright. 32" x 66$^{3}/_{4}$" coil boiler, heater pipes left, torch on coal bunker. 4$^{1}/_{4}$" x 7" pumps, air chamber ornament. 4" x 22' pre-connect suction. 6$^{3}/_{4}$" x 7" cylinders. Wheels 16 x 56, 18 x 62. Platform springs. Stiff pole. Gong on foot board. Whip socket on seat. Two lanterns on sides of frame.

10-20-1890: Ordered by Cincinnati, Ohio. Ahrens records do not show whether this was actually built or delivered.

632 3 DCN/HD/A Upright. 32" x 66$^{3}/_{4}$" coil boiler (iron & nickel), heater pipes on left, two FEMCO Eclipse lanterns on boiler, torch on coal bunker. 4$^{1}/_{4}$" x 7" pumps, nickel-plated air chamber with ornament atop. Two 4" x 10'6", one 2$^{1}/_{2}$" x 9' suctions. 6$^{3}/_{4}$" x 7" cyls, nickel-plated casings. Wheels 16 x 56, 18 x 62. Platform springs. 56$^{1}/_{2}$" track. Stiff pole. Gong on foot board. Whip socket on seat. Two brass 3-pint oil cans under seat. Two Eclipse lanterns on sides of frame.

10-28-1890: Shipped to Grand Rapids, Mich. Delivered by John P. Ahrens and Woehle.

11-1-1890: Accepted by Grand Rapids. In service at Engine 6.

1894: To Engine 7.

12-6-1895: New boiler coils.

634 2 DCN/HD/A Upright. 35$^{1}/_{2}$" x 67" coil boiler (no casing), heater pipes right, nickel-plated lantern on boiler, torch on coal bunker. 4$^{3}/_{4}$" x 8" pumps, air chamber ornament, relief valve right. Two 4$^{1}/_{2}$" x 10'6", one 2$^{1}/_{2}$" x 9' suctions. 7$^{3}/_{4}$" x 8" cyls. Wheels 16 x 58, 18 x 62. Platform springs. 56$^{1}/_{2}$" track. Stiff pole. Gong on foot board. Whip socket on seat. Two torches on frame.

12-18-1890: Shipped to Engine 24, Chicago, Ill. [Was at Engine 54 in 1917].

11-4-1903: 35$^{1}/_{2}$" x 67" Ahrens coil boiler.

635 2 DCN/HD/A Upright. 35$^{1}/_{2}$" x 67" coil boiler (iron), heater pipes on left, torch on coal bunker, letters *M.F.D.* back of coal bunker. 4$^{3}/_{4}$" x 8" pumps; nickel-plated air chamber with ornament atop; oil pump atop vacuum chamber; nickel-plated columns, crankshaft, heads; relief valve right. Two 4$^{1}/_{2}$" x 10'6" suctions, strainer stand back of frame. 7$^{3}/_{4}$" x 8" cyls., nickel-plated heads. Wheels 16 x 58, 18 x 62. Platform springs. 56$^{1}/_{2}$" track. Stiff pole, extra pole. Gong on foot board. Whip socket on seat. Two 3-pint oil cans under seat. Two torches on frame. Number plate *16* made for smokestack, not put on.

6-11-1891: Shipped to Milwaukee, Wis.

6-17-1891: Delivered by Chris Ahrens to

Engine 54 in Chicago, Ill., was an 1890 2nd Class double crane-neck Ahrens #634, originally Engine 24. Note that boiler was bolted to the frame, instead of attaching the boiler with the traditional plated casing. Chicago was more concerned with the cost savings this produced than with the drab appearance that resulted.

(Courtesy Harold S. Walker)

Engine 1, Milwaukee, Wis. Replaced #224.

Date?: To Engine 2.

11-1-1892: New boiler coils.

1902: To Engine 13.

637 2 DCN/HD/A Upright. 35 1/2" x 67" coil boiler (iron), heater pipes on left, two Eclipse lanterns on boiler, letters *M.F.D.* on back of coal bunker. 4 3/4" x 8" pumps, nickel-plated reflecting lamp (by White Mfg. Co.) atop air chamber, lamp atop vacuum chamber, relief valve right. Two 4" x 10'6" hard suctions, one 4" x 15' soft suction. 7 3/4" x 8" cylinders. Wheels 16 x 58, 18 x 62. Platform springs. Stiff poles for 3 horses. Gong on foot board. Two torches on sides of frame. 7/8" steel frame by Otis Steel Co., Cleveland, Ohio.

3-31-1890: Shipped to Engine 9, Minneapolis, Minn. Delivered by Chris Ahrens & Woehle.

1901: To Engine 14.

1908: Rebuilt.

1910: To Engine 7.

1922: Retired.

638 2 DCN/HD/A Upright. 35 1/2" x 67" coil boiler (iron), heater pipes on left, two Eclipse lanterns on boiler, letters *M.F.D.* on back of coal bunker. 4 3/4" x 8" pumps, nickel-plated reflecting lamp (by White Mfg. Co.) atop air chamber, lamp atop vacuum chamber, relief valve on right. Two 4" x 10'6" hard suctions, one 4" x 15' soft suction. 7 3/4" x 8" cylinders. Wheels 16 x 58, 18 x 62, extra rear wheel. Platform springs. Stiff poles for 3 horses. Gong on foot board. Two torches on frame. 7/8" frame by Otis Steel, Cleveland, Ohio.

3-31-1890: Shipped to Engine 8, Minneapolis, Minn. Delivered by Chris Ahrens & Woehle.

1915: To Engine 22.

1922: Retired.

640 3 DCN/HD/A Upright. 32" x 66 3/4" coil boiler (iron/nickel), heater pipes on right; two nickel-plated lanterns (with red lenses) on boiler, one lettered *R.W. Stewart* (Mayor), the other *Joseph Reinfried* (Chief); nameplate *R.W. Stewart* on smokestack; torch on coal bunker. 4 1/4" x 7" pumps, nickel-plated air chamber with lamp atop, ornament atop vacuum chamber, relief valve left. Two 4" x 10'6", one 2 1/2" x 9' suctions. 6 3/4" x 7" cyls., nickel-plated casings. Wheels 16 x 56, 18 x 62. Platform springs. 56 1/2" track. Stiff pole. Gong on foot board. Whip socket on seat. Tool box under seat, numeral *4* on both ends. Two lanterns on frame. Weight 7000#.

2-19-1891: Shipped to Dubuque, Iowa.

2-23-1891: Delivered by G. Fred Hawekotte to Dubuque, Iowa.

2-25-1891: Tested at Dubuque. Water in the pump from factory test froze enroute to Dubuque. Nevertheless, Ahrens had water in 4:00, two good streams from siamese at end of 500' of hose, delivered rated 700 GPM.

2-26-1891: Accepted by Dubuque. In service at Engine 4. Name *R.W. Stewart.*

9-1-1898: New boiler coils.

7-29-1901: New 3-horse hitch.

12-13-1904: New 32" x 62" Fox boiler.

641 5 SSF/MD-HD/A Upright. 25" x 56" tubular boiler (brass & silver). Two nickel-plated A&W lanterns; one had *J.H. Ortscheid* on one side, *Chief Fire Dept.* on other side. 4 1/2" x 7" pump, automatic air pump, air chamber ornament, goose-neck discharge gates. 3 1/4" x 20' pre-connect suction. 7" x 7" cylinder. Wheels 16 x 52, 18 x 60, steel tires. Rear brake lever. Rebuild of Ahrens #556.

7-29-1891: Sold to Belle Plaine, Iowa.

9-18-1891: Delivered by G. Fred Hawekotte to Belle Plaine, Iowa. Name *Rescue #1.* Set world's record for fastest water (starting from cold boiler), time of 3:05. This record broken 8-25-1903 by Ahrens #598 of Seattle, Wash., which had water in 3 minutes even.

1985: This steamer is now in the Belle Plaine Fire Dept. Museum, Belle Plaine, Iowa.

643 3 DCN/HD/A Upright. 32" x 66 3/4" coil boiler (iron), heater pipes (for Ahrens heater) on left, numeral *2* on smokestack, torch on coal bunker. 4 1/4" x 7" pumps, lamp atop vacuum chamber, relief valve left. Two 4" x 10'6", one 2 1/2" x 9' suctions. 6 3/4" x 7" cyls. Wheels 16 x 56, 18 x 62, steel tires. 62" track. Stiff pole. Gong on foot board. Whip socket on seat. Two torches on sides of frame.

4-27-1891: Delivered by John P. Ahrens to Engine 2, Dallas, Texas. Also one Ahrens heater to set on floor with this steamer.

7-22-1903: 32" x 62" Fox boiler, overhauled, replated and repainted engine.

1-25-1908: New valves from Ahrens Fire Engine Co.

12-6-1909: New 4" x 10'6" suctions from American-LaFrance.

644 3 DCN/HD/A Upright. 32" x 66 3/4" coil boiler (iron), numeral *3* on smokestack, torch on coal bunker. 4 1/4" x 7" pumps, lamp atop vacuum chamber, relief valve left. Two 4" x 10'6", one 2 1/2" x 9' suctions. 6 3/4" x 7" cyls. Wheels 16 x 56, 18 x 62, steel tires. 62" track.

The "Edward Dundon" is an 1891 2nd Class double crane-neck Ahrens #648 of Superior, Wis. Originally Engine 1, it was assigned to Engine 2 in this World War I photo. Boiler is an Ahrens Continental installed in 1907. The canvas mudguard, behind the front wheels, was typical on midwestern steamers.

(Wayne Sorenson collection)

1891 2nd Class double crane-neck #648 originally served Superior, Wis. Today, it belongs to John Kenny of St. Paul, Minn., who is tackling the monumental task of restoring it.

(John Kenny photo)

Stiff pole. Gong on foot board. Whip socket on seat. Two torches on sides of frame.

4-27-1891: Delivered by John P. Ahrens to Dallas, Texas. In service at Engine 3.

7-22-1903: 32" x 62" Fox boiler, overhauled, replated and repainted engine.

646 3 DCN/HD/A Upright. 32" x $66^3/4$" coil boiler (no casing), nickel-plated smokestack, safety-valve whistle back of boiler, safety railing and torch on coal bunker. $4^1/4$" x 7" pumps, nickel-plated air chamber with ornament atop, nickel-plated vacuum chamber. 4" x 22' pre-connect suction. $6^3/4$" x 7" cyls. Wheels 16 x 56, 18 x 62. Platform springs. 62" track. Stiff pole with brass ends. Gong under foot board, portrait of Henry S. Tyler (Mayor of Louisville) on foot board. Whip socket on seat. Two torches on frame.

6-30-1891: Shipped to Louisville, Ky. Delivered by G. Fred Ahrens.

7-2-1891: Tested & accepted by Louisville. In service at Engine 7. Name *Henry S. Tyler.*

4-19-1905: New right-side frame supplied by American-LaFrance.

647 3 DCN/HD/A Upright. 32" x 64" coil boiler (iron), heater pipes (for Ahrens heater) left, torch on coal bunker. Two $4^1/4$" x 7" pumps, nickel-plated air chamber, ornament atop. Two 4" x 10'6", one $2^1/2$" x 9' suctions. $6^3/4$" x 7" cyls. Wheels 16 x 56, 18 x 62. Platform springs. 62" track. Stiff pole. Gong on foot board. Whip socket. Two torches on frame.

5-12-1891: Shipped to Decatur, Ill.

5-19-1891: Delivered by Fred Hawekotte and John P. Ahrens to Decatur, Ill. Also Ahrens stationary heater. Accepted same date.

5-18-1908: New 32" x 64" Radial Water Tube boiler by Ahrens Fire Engine Co.

7-3-1908: New steam and exhaust pipe by Ahrens Fire Engine Co.

648 2 DCN/HD/A Upright. $35^1/2$" x 67" coil boiler (iron), heater pipes left, two Dietz lanterns on boiler, safety railing back of boiler, nameplate *Edward Dundon* on smokestack, numeral *1* on coal bunker. $4^1/2$" x 8" pumps, nickel-plated reflecting lamp (by White Mfg. Co.) atop air chamber, ornament on vacuum chamber, relief valve left. One 4" x 22' pre-connect suction. $7^1/4$" x 8" cyls. Wheels 16 x 58, 18 x 62. Platform springs. Stiff pole for 3 horses, spare pole for 2 horses. Gong on foot board. Tool box under seat, numeral *1* on both ends. Two torches on frame. $^7/8$" frame by Otis Steel Co., Cleveland, Ohio.

7-23-1891: Delivered by John P. Ahrens to Superior, Wis. Name *Edward Dundon #1.*

8-29-1894: New boiler coils.

1906: To Engine 2.

3-13-1907: 36" x 64" Radial Inclined Water Tube boiler by Ahrens Fire Engine Co.

1984: Purchased by John Kenny, St. Paul, Minn. He plans to restore it.

649 3 DCN/HD/A Upright. 32" x $66^3/4$" coil boiler (no casing), heater pipes right, two lanterns on boiler, torch on coal bunker. $4^1/4$" x 7" pumps, relief valve on right. Two 4" x 10' suctions. $6^3/4$" x 7" cyls. Wheels 16 x 56, 18 x 62. Platform springs. $56^1/2$" track. Stiff pole. Gong on foot board. Whip socket. 2 three-pint oil cans under seat. 2 torches on frame.

9-3-1891: Shipped to Chicago, Ill.

1906: At Engine 87. Retired by 1917.

651 2 DCN/HD/A Upright. $35^1/2$" x 67" coil boiler (no casing), heater pipes left, number *15* on smokestack, torch and safety railing on coal bunker. Pumps $4^1/2$" x 8", air chamber ornament, relief valve left. 4" x 21' pre-connect suction. $7^3/4$" x 8" cyls. Wheels 16 x 58, 18 x 62. Platform springs. $57^1/2$" track. Stiff pole. Gong on foot board. Whip socket on seat, two 3-pint oil cans under seat.

10-6-1891: Shipped to Engine 15, St. Louis, Mo. Delivered by G. Fred Hawekotte.

1909: Was in service at Engine 9. Had been rebuilt with a Nott boiler.

653 2 DCN/HD/A Upright. $35^1/2$" x 67" coil boiler (no casing), heater pipes on left, safety railing and torch on coal bunker. $4^1/2$" x 8" pumps, air chamber ornament, relief valve left. 4" x 21' pre-connect suction. $7^3/4$" x 8" cyls. Wheels 16 x 58, 18 x 62. Platform springs. $57^1/2$" track. Stiff pole. Gong. Whip socket. Two 3-pint oil cans under seat.

10-6-1891: Shipped to Engine 30, St. Louis, Mo. Delivered by G. Fred Hawekotte.

1909: Was a spare at Engine 9.

655 3 DCN/HD/A Upright. 32" x $66^3/4$" coil boiler (no casing), heater pipes right, two lanterns on boiler, torch on coal bunker. $4^1/4$" x 7" pumps, relief valve right. Two 4" x 10'6" suctions. $6^3/4$" x 7" cyls. Wheels 16 x 56, 18 x 62. Platform springs. $56^1/2$" track. Stiff pole. Gong on foot board. Whip socket. 2 three-pint oil cans under seat. 2 torches on frame.

10-31-1891: Shipped to Eng. 15, Chicago, Ill.

1906: Was at Engine 20.

1912: To Engine 65. Still at 65 in 1917.

656 3 DCN/HD/A Upright. 32" x $66^3/4$" coil boi-

ler (no casing), heater pipes right, two lanterns on boiler, torch on coal bunker. 4 1/4" x 7" pumps, relief valve right. Two 4" x 10'6" suctions. 6 3/4" x 7" cyls. Wheels 16 x 56, 18 x 62. Platform springs. 56 1/2" track. Stiff pole. Gong on foot board. Whip socket. 2 three-pint oil cans under seat. 2 torches on frame.

10-31-1891: Shipped to Chicago, Ill.

11-27-1906: 32" x 66 3/4" Ahrens coil boiler.

1906: At Engine 28. 1917: At Engine 114.

657 3 DCN/HD/A Upright. 32" x 66 3/4" coil boiler (no casing), heater pipes right, two lanterns on boiler, torch on coal bunker. 4 1/4" x 7" pumps, relief valve right. Two 4" x 10'6" suctions. 6 3/4" x 7" cyls. Wheels 16 x 56, 18 x 62. Platform springs. 56 1/2" track. Stiff pole. Gong on foot board. Whip socket. 2 three-pint oil cans under seat. 2 torches on frame.

11-28-1891: Shipped to Chicago, Ill.

1905: New boiler coils.

1906: Was at Engine 66. Still at 66 in 1917.

658 3 DCN/HD/A Upright. 32" x 66 3/4" coil boiler (no casing), heater pipes right, two lanterns on boiler, torch on coal bunker. 4 1/4" x 7" pumps, relief valve right. Two 4" x 10'6" suctions. 6 3/4" x 7" cyls. Wheels 16 x 56, 18 x 62. Platform springs. 56 1/2" track. Stiff pole. Gong on foot board. Whip socket. 2 three-pint oil cans under seat. 2 torches on frame.

11-28-1891: Shipped to Chicago, Ill.

1906: At Engine 55. 1917: At Engine 89.

660 3 DCN/HD/A Upright. 32" x 62 1/2" coil boiler (silver), heater pipes left, safety railing on back, nameplate *Pap Grayson* on smokestack, nickel-plated torch and numeral *3* on coal bunker. One nickel-plated A&W lantern engraved *John V. Connelly*. Pumps 4" x 7", air chamber ornament. Pre-connect suction 3 1/2" x 21'. 6 3/4" x 7" cyls. Wheels 16 x 56, 18 x 62. Platform springs. Stiff pole. Gong on foot board. Tool box under seat, numeral *3* on both ends. Whip socket. 2 nickel-plated torches on frame. Frame and running-gear painted green, wheels red. Weight 6530#.

11-2-1891: Delivered by Chris Ahrens, Charles H. Fox (former Asst. Fire Marshal, Cincinnati), and Ed Steinway (later Cincinnati Fire Marshal) to Western Fire Co. #3, Madison, Ind. Name *Pap Grayson*.

6-29-1898: Geo. Zink put in new boiler coils.

10-19-1924: Replaced by red Ahrens-Fox pumper #1231.

1926: Sold to shipyard, Madison, Ind.

1943: Scrapped.

661 6 SSF/MD-HD/A Upright. 25"x56 1/2" tubular boiler (iron), two Adams & Westlake (A&W) lanterns on boiler, torch on coal bunker. 4 1/2" x 7" pump, automatic air pump, air chamber ornament, goose-neck discharge gates. 3 1/4" x 20' pre-connect suction. 6 1/2" x 7" cylinder. Wheels 16 x 51 front, 18 x 57 rear. Rear brake lever. Two A&W lanterns under seat. Rebuild of Ahrens #514.

8-17-1891: Sold to Village of Westwood, Ohio (on west side of Cincinnati).

9-10-1891: Delivered by Chris Ahrens, G. Fred Hawekotte, and Hiram H. Nieman to Westwood, Ohio. Name *Citizens Gift #1*.

662 Blank page in Ahrens records.

664 Blank page in Ahrens records.

666 DX1 DCN/HD/A Upright. 38" x 71 1/2" coil boiler (no casing), heater pipes right, torch on coal bunker. 5 1/2" x 8" pumps, air chamber ornament, relief valve right, 3 1/2" discharge gate left. Two 5" x 10'6" suctions. 8 3/4" x 8" cyls. Wheels 16 x (64?) front, 18 x (68?) rear. Platform springs. 66" track. Stiff pole. Gong on foot board. Whip socket on seat, two 3-pint oil cans under seat.

12-5-1891: Shipped to Eng. 10, Chicago, Ill.

1906: At Engine 21. Retired by 1917.

667 6 SSF Upright. Listed as "Stock" in Ahrens records.

668 3 DCN/HD/A Upright. 32" x 62" coil boiler (iron), heater pipes left, safety railing back of boiler, torch on coal bunker. Two 4 1/4" x 7" pumps, signal lamp atop air chamber. 4" x 20' pre-connect suction. Cylinders 6 3/4" x 7". Wheels 16 x 56, 18 x 62. Platform springs. 62" track. Stiff pole. Gong on foot board. Whip socket on seat. Two torches on frame.

11- -1891: Ordered by Tower, Minn.

2-20-1892: Delivered by John Black to Tower, Minn. Name *James Tippett #2*.

1944: Retired.

1985: Tower Fire Dept. still owns this, displayed in a special outdoor pavilion.

669 3 DCN/HD/A Upright. 32" x 66 3/4" coil boiler (no casing), heater pipes left, 2 lanterns on boiler, safety railing on back, numeral *3* on smoke stack, torch on coal bunker. 4 1/4" x 7" pumps, air chamber ornament. 4" x 22' pre-connect suction. 6 3/4" x 7" cyls. Wheels 16 x 56, 18 x 62. Platform springs. 62" track. Stiff pole. Gong on foot board. Whip socket. Two torches on frame.

Tower, Minn., still owns their 1892 Ahrens 3rd Class double craneneck #668. Named the "James Tippett #2", it was retired in 1944 and is now displayed in a makeshift outdoor pavilion near the firehouse. Several years ago, vandals removed some of its brass, including the rare Globe signal lamp that once perched atop the air chamber. Hose reel at right may be an Ahrens reel delivered with the steamer.

(Courtesy Gerald Peterson, Tower Fire Dept.)

2-29-1892: Shipped to Engine 3, Indianapolis, Ind. Delivered by John P. Ahrens.

670 3 DCN/HD/A Upright. 32" x 66 3/4" coil boiler (no casing), heater pipes on left, two lanterns on boiler, safety railing back of boiler, numeral *5* on smoke stack, torch on coal bunker. 4 1/4" x 7" pumps, air chamber ornament. 4" x 22' pre-connect suction. 6 3/4" x 7" cyls. Wheels 16 x 56, 18 x 62. Platform springs. 62" track. Stiff pole. Gong on foot board. Whip socket. Two torches on frame.

3-3-1892: Shipped to Engine 5, Indianapolis, Ind. Delivered by John P. Ahrens.

1902: Rebuilt.

671 2 DCN/HD/A Upright. 35 1/2" x 67" coil boiler (no casing), heater pipes left, two lanterns on boiler, safety railing and torch on coal bunker, numeral *7* on smoke stack. 4 3/4" x 8" pumps, air chamber ornament. 4 1/2" x 23' pre-connect suction. 7 3/4" x 8" cylinders. Wheels 16 x 58, 18 x 62. First platform springs with equalizing bar. 62" track. Stiff pole. Gong on foot board. Whip socket on seat. Two torches on sides of frame.

4-21-1892: Delivered by Chris Ahrens and Kohlbecker to Engine 7, Indianapolis, Ind.

1905: Rebuilt. To Engine 8.

672 2 DCN/HD/A Upright. 35 1/2" x 67" coil boiler (no casing), safety railing and two 3-pint oil cans on back, heater pipes left, torch on coal bunker, nameplate *Robert M. Archibald #9* on smokestack. 4 3/4" x 8" pumps, air chamber ornament, relief valve left. One 4 1/2" x 23' pre-connect suction. 7 3/4" x 8" cyls. Wheels 16 x 58, 18 x 62. Platform springs. Gong on foot board. Whip socket on seat. Two lanterns on frame.

7-5-1892: In service at Fire Co. #9, Cincinnati, Ohio. Name *Robert M. Archibald*. Delivered by G. Fred Hawekotte.

673 Built as #2101 by American Fire Engine Co.

674 Built as #2102 by American Fire Engine Co.

These two Ahrens double crane-neck steamers were converted to threshing machines by J.I. Case Co. of Racine, Wis. In this 1920s photo, they still had their original seats, boilers, wheels, and rear springs, but the piston pumps had been replaced with Case threshers, and the front of the frame was reinforced to support the added weight. Note rubber tires on engine in foreground, while engine in background retained steel tires. These engines probably came from either Milwaukee or Chicago, since both had large Ahrens fleets and were not far from the Case factory. Numeral "2" on coal bunker of foreground engine suggests it might be Ahrens #635 of Milwaukee Engine 2, or #264 of Chicago Engine 2. Between the first and second world wars, dozens (perhaps hundreds) of steam fire engines were converted to threshers, dredgers, pile-drivers, power plants, etc. (Courtesy David G. Domidion)

American Fire Engine Co.

Cincinnati, Ohio
1891-1904

The American Fire Engine Company was a consolidation of Ahrens (Cincinnati, Ohio), Button (Waterford, N.Y.), Clapp & Jones (Hudson, N.Y.), and Silsby (Seneca Falls, N.Y.). The Button and C&J operations were moved into the Silsby plant, and all three makes continued to be built there. Ahrens continued to operate autonomously in Cincinnati. Serial numbers for American Fire Engine Co. started at 2101, a number which the late American-LaFrance historian T. Sloan Palmer suggested may have been arrived at by totalling the highest serial numbers of each company.

Note that there are gaps in the serial numbers in the following list. All steamers made by American Fire Engine Co. were numbered consecutively, regardless of whether they were Ahrens, Button, Clapp & Jones, or Silsby. Since this book is concerned only with the Ahrens steamers (built in Cincinnati), steamers made in Seneca Falls have deliberately been omitted from this list.

2101 3 DCN/HD/A Upright. 32" x 66$^3/_4$" coil boiler (iron), heater pipes on left, two lanterns on boiler, torch on coal bunker. 4$^1/_4$" x 7" pumps, air chamber ornament, Star gauges. Two 4" x 10'6", one 2$^1/_2$" x 9' suctions. 6$^3/_4$" x 7" cylinders. Wheels 16 x 56, 18 x 62. Platform springs. 62" track. Gong on foot board. Whip socket on seat. 2 torches on frame. Renumbered from 673.

4-11-1892: Delivered by John P. Ahrens to Evansville, Ind. Name *Indiana #4*.

2102 3 DCN/HD/A Upright. 32" x 66$^3/_4$" coil boiler (iron), heater pipes on left, two lanterns on boiler, torch on coal bunker. 4$^1/_4$" x 7" pumps, air chamber ornament, Star gauges. Two 4" x 10'6", one 2$^1/_2$" x 9' suctions. 6$^3/_4$" x 7" cylinders. Wheels 16 x 56, 18 x 62. Platform springs. 62" track. Gong on foot board. Whip socket on seat. 2 torches on frame. Renumbered from 674.

4-11-1892: Delivered by John P. Ahrens to Evansville, Ind. Name *Vanderburg #5*.

2103 3 DCN/MD-HD/A Upright. 32" x 62" coil boiler (iron), safety railing on back, heater pipes left, torch and name *Pioneer Engine #1* on coal bunker, nameplate *Anacortes #1* on smokestack. 4$^1/_4$" x 7" pumps, air chamber ornament. 4" x 22' pre-connect suction. 6$^3/_4$" x 7" cylinders. Wheels 16 x 56, 18 x 62. Platform springs. 62" track. Gong on foot board. Whip socket on seat. Water tank under seat (nickel-plated ends). Canvas mudguard. Two torches on frame.

3-16-1892: Delivered by John P. Ahrens to Pioneer Engine #1, Anacortes, Wash.

Date?: Traded-in to American-LaFrance.

9-5-1908: Shipped to Brooklyn Cooperage Company, Saint Regis Falls, N.Y.

2104 3 DCN/HD/A Upright. 32" x $66^3/_4$" coil boiler (iron), heater pipes left, two 3-pint oil cans back of boiler, torch on coal bunker, expansion smokestack. $4^1/_4$" x 7" pumps, cone valves, air chamber ornament. Two 4" x 10'6", one $2^1/_2$" x 9' suctions. $6^3/_4$" x 7" cyls. Wheels 16 x 56, 18 x 62. Platform springs. Gong on foot board. Whip socket on seat. Locking tool box under seat, end-plates engraved "Alf Handman, 4, Engineer". Seth Thomas clock on left.

5-11-1892: Delivered by Fred Hawekotte to Chattanooga, Tenn. Name *Garnett Andrews #4*. Also one Ahrens stationary heater.

2-20-1905: New boiler coils.

10-12-1909: New 32" x 62" Fox boiler.

1972: This steamer is sitting outside beside Fire Dept. headquarters in Chattanooga. Photographed there by Bill Hass.

8-10-1985: George Spencer of Chattanooga says it is now inside, but understandably is badly weathered after years outside.

2105 3 DCN/HD/A Upright. 32" x $66^3/_4$" coil boiler (iron), heater pipes left, two 3-pint oil cans back of boiler, torch and numeral *3* on coal bunker, expansion smokestack, nameplate *Henry N. Starr* front of smokestack. $4^1/_4$" x 7" pumps, air chamber ornament, relief valve right, nickel-plated valves. Two 4" x 10'6", one $2^1/_2$" x 9' suctions. $6^3/_4$" x 7" cyls. Wheels 16 x 56, 18 x 62. Platform springs. 62" track. Gong on foot board. Whip socket on seat.

5-24-1892: Delivered by Fred Hawekotte to Rockford, Ill. Name *Henry N. Starr #3*.

3-7-1902: New boiler coils.

2106 3 DCN/HD/A Upright. 32" x $66^3/_4$" coil boiler (iron), heater pipes left, two 3-pint oil cans back of boiler, torch on coal bunker. $4^1/_4$" x 7" pumps, nickel-plated air chamber with ornament atop. Two 4" x 10'6", one $2^1/_2$" x 9' suctions. $6^3/_4$" x 7" cyls. Wheels 16 x 56, 18 x 62. Platform springs. $57^1/_2$" track. Gong on foot board. Whip socket on seat. Two lanterns on frame.

6-13-1892: Delivered by Chris Ahrens to San Antonio, Texas. Name *George Clark #2*.

1910: Retired.

The "Garnett Andrews #4" is an 1892 Ahrens 3rd Class double crane neck #2104 of Chattanooga, Tenn., with a 1909 Fox boiler. In this 1972 photo, it was already showing signs of neglect from sitting outdoors, and we are told that it has deteriorated considerably in the 13 years since. There is, however, some local interest in restoring it. (Bill Hass photo)

The firefighters of San Antonio, Texas made a valiant effort to preserve its three remaining steamers in 1927. Two of the three were 3rd Class Ahrens double crane neck, Numbers 2106 and 2129 (the third was a 1908 Metropolitan). One of the two Ahrens engines was scrapped in 1944, and it was previously believed that the one scrapped was #2106. However, new information that has recently been discovered suggests that #2129 was actually the one scrapped, and #2106 the one which survives. What makes it all so confusing is that the surviving Ahrens at San Antonio Transportation Museum (shown above) no longer has its serial-number plate, and its appearance has been somewhat modified over the years.

(Ed Hass photo)

11-9-1910: Fox stationary-heater for this.

4- -1927: San Antonio retired its last steamer (Ahrens #2129), and carefully preserved Ahrens numbers 2106 and 2129, and 1908 Metropolitan #3258.

1939: This steamer donated to the Witte Museum, San Antonio, Texas.

8- -1984: Ed Hass photographed this at San Antonio Transportation Museum. No longer has serial number plate, and has been numbered as Engine 5, so identification is uncertain (based on Ahrens and San Antonio records, the museum's Ahrens is probably #2106). Suction hoses, lanterns, etc. are from 1908 Metropolitan #3258 of Texas Transportation Museum (near San Antonio Airport). Of the three steamers San Antonio intended to preserve, only this Ahrens (#2106) now survives 100% intact.

2107 3 DCN/HD/A Upright. 32" x 66$^{3}/_{4}$" coil boiler (iron), heater pipes left, two 3-pint oil cans back of boiler, torch and numeral *1* on coal bunker, nameplate *J.S. Hogg* on smokestack. 4$^{1}/_{4}$" x 7" pumps, nickel-plated air chamber with ornament atop. Two 4" x 10'6", one 2$^{1}/_{2}$" x 9' suctions. 6$^{3}/_{4}$" x 7" cyls. Wheels 16 x 56, 18 x 62. Platform springs. 57$^{1}/_{2}$" track. Gong on foot board. Whip socket on seat. Tool box under seat, numeral *1* on ends. Two lanterns on frame.

6-13-1892: Delivered by John P. Ahrens to San Antonio, Texas. Name *J.S. Hogg #1*. Replaced Ahrens #449, sent to Engine 3.

2-7-1903: New boiler coils.

8-29-1910: New boiler coils.

2108 DX1 DCN/HD/A Upright. 38" x 71" coil boiler (no casing), heater pipes left, two 3-pint oil cans and lantern back of boiler, torch on coal bunker, nickel-plated smokestack. 5$^{1}/_{2}$" x 8" pumps, two 3" discharge gates, air chamber ornament, lamp atop vacuum chamber. Two 5" x 10'6", one 2$^{1}/_{2}$" x 9' suctions. 8$^{3}/_{4}$" x 8" cyls. Wheels 16 x (64?), 18 x (68?). Platform springs. 66" track. 3-horse hitch. Gong on foot board. Whip socket on seat.

7-27-1892: Delivered by G. Fred Hawekotte to Engine 2, Nashville, Tenn.

5-22-1897: New boiler coils.

2109 3 DCN/HD/A Upright. 32" x 66$^{3}/_{4}$" coil boiler (no casing), heater pipes left, two 3-pint oil cans back of boiler, torch on coal bunker, nickel-plated smokestack. 4$^{1}/_{4}$" x 7"

1892 Ahrens 3rd Class double crane-neck #2109 served Engine 101 in Chicago, Ill. Note the rubber tires in place of traditional steel tires. Sign at right background advertising "The Fair" possibly refers to the famed Columbian Exposition, the World's Fair held in Chicago in 1893.

(Wayne Sorenson collection)

pumps; piston air pump; nickel-plated air chamber (with ornament atop), heads, columns, crankshaft, flywheel, discharge gates; relief valve right. Two 4" x 10' suctions. $6^{3}/_{4}$" x 7" cyls. (nickel-plated heads). Wheels 16 x 56, 18 x 62. Platform springs. Nickel-plated springs and axles. $56^{1}/_{2}$" track. Gong on foot board. Whip socket on seat. Two lanterns on frame.

7-19-1892: Shipped to Chicago, Ill.

1906: At Engine 97. 1917: At Engine 101.

2110 3 DCN/HD/A Upright. 32" x $66^{3}/_{4}$" coil boiler (iron), heater pipes left, lantern and committee nameplate on boiler, two 3-pint oil cans back of boiler, torch and name *Wheeling* on coal bunker, extension smokestack, numeral *4* front of smokestack. $4^{1}/_{4}$" x 7" pumps, nickel-plated air chamber with ornament atop. Two 4" x 10'6" suctions. $6^{3}/_{4}$" x 7" cyls. Wheels 16 x 56, 18 x 62. Platform springs. 61" track. New Departure gong on foot board. Whip socket on seat. Two lanterns on frame.

8-13-1892: Delivered by Chris Ahrens and Peter Ahrens to Wheeling, W. Va. Name *Wheeling #4*. Also one Ahrens heater.

12-5-1903: Colonel Braam installed new boiler coils at Wheeling.

4-1-1907: New boiler coils.

2121 2 DCN/HD/A Upright. $35^{1}/_{2}$" x 67" coil boiler (no casing), heater pipes left, safety railing and two 3-pint oil cans on back, torch and number *32* on coal bunker, *32* on smokestack. $4^{3}/_{4}$" x 8" pumps, air chamber ornament, relief valve left. 4" x 21' preconnect suction. $7^{3}/_{4}$" x 8" cyls. Wheels 16 x 58, 18 x 62. Platform springs. $57^{1}/_{2}$" track. Gong on foot board. Whip socket on seat. *32* on tool box. Two lanterns on frame.

5-30-1892: Ordered by St. Louis, Mo.

9-12-1892: Shipped to Eng 32, St. Louis, Mo.

1909: Was at Engine 26. Had Fox boiler.

2122 2 DCN/HD/A Upright. $35^{1}/_{2}$" x 67" coil boiler (nickel), heater pipes on left (for Ahrens heater), two 3-pint oil cans back of boiler, torch on coal bunker, nameplate *Schuylkill #12* and committee nameplate front of smokestack. $4^{3}/_{4}$" x 8" pumps; nickel-plated air chamber (with ornament atop), crankshaft, pump-support columns, flywheel. Two $4^{1}/_{2}$" x 10'6", one $2^{1}/_{2}$" x 9' suctions. $7^{3}/_{4}$" x 8" cyls. Wheels 16 x 58, 18 x 62. Platform springs. 62" track. Gong on foot board. Whip socket on seat. Two lanterns on frame. Wheels, axles, and springs painted dark blue, frame red.

2-11-1893: Ordered by Reading, Pa.

3-29-1893: Shipped to Schuylkill Fire Co. #12, Reading, Pa. Retired 1921.

2123 X1 DCN/HD/A Upright. 38" x 71" coil boiler (iron), heater pipes left, two 3-pint oil cans back of boiler, torch on coal bunker, number *1* on smokestack. 5 1/2" x 8" pumps, air chamber ornament, oil pump atop vacuum chamber, relief valve right, one 3 1/2" discharge gate. Two 5" x 10'6" suctions, strainer stand back of frame. 8 3/4" x 8" cyls. Wheels 16 x ?? front, 18 x ?? rear. Platform springs. Stiff pole (varnished), extra pole. Gong on foot board. Whip socket on seat.

5-20-1892: Ordered by Milwaukee, Wis.

8-31-1892: Delivered by G. Fred Hawekotte to Engine 1, Milwaukee, Wis.

5-14-1901: New pump valves.

1908: New Continental boiler by Ahrens Fire Engine Co.

2- -1910: Replaced at Engine 1.

1919: To Engine 21.

1921: To Engine 18.

1922: Replaced by 1918 Seagrave pumper.

1923: To Engine 7.

1925: Retired.

2124 1 DCN/HD/A Upright. No further details.

1893: Built for Louisville, Ky. Not sold.

1894: Rebuilt as #2258 for Columbus, Ohio.

2125 3 DCN/HD/A Upright. 32" x 66 3/4" coil boiler (no casing), heater pipes left, safety railing and two 3-pint oil cans on back, torch on coal bunker, number *11* on smokestack. 4 1/4" x 7" pumps, air chamber ornament, relief valve left. 4" x 21' pre-connect suction. 6 3/4" x 7" cyls. Wheels 16 x 56, 18 x 62. Platform springs. 57 1/2" track. Gong on foot board. Whip socket on seat. Two lanterns on frame. Canvas mudguard.

5-30-1892: Ordered by St. Louis, Mo.

10-11-1892: Shipped to Engine 11, Saint Louis, Mo.

Date?: Rebuilt with Continental boiler by Ahrens Fire Engine Co.

1909: Was at Engine 43.

2126 3 DCN/HD/A Upright. 32" x 66 3/4" coil boiler (no casing), heater pipes left, safety railing and two 3-pint oil cans on back, torch on coal bunker, number *31* on smokestack. 4 1/4" x 7" pumps, air chamber ornament, relief valve left. 4" x 21' pre-connect suction. 6 3/4" x 7" cyls. Wheels 16 x 56, 18 x 62. Platform springs. 57 1/2" track. Gong on foot board. Whip socket on seat. Two lanterns on frame. Canvas mudguard.

5-30-1892: Ordered by St. Louis, Mo.

11-19-1892: Shipped to Engine 31, Saint Louis, Mo.

Date?: Rebuilt with Fox boiler.

8- -1904: New air chamber.

1909: Was at Engine 46.

2127 3 DCN/HD/A Upright. 32" x 66 3/4" coil boiler (no casing), heater pipes left, two three-pint oil cans back of boiler, torch on coal bunker. 4 1/4" x 7" pumps, air chamber ornament, relief valve right. Two 4" x 10' suctions. 6 3/4" x 7" cyls. Wheels 16 x 56, 18 x 62. Platform springs. Gong on foot board. Whip socket. Two lanterns on frame.

6-22-1892: Ordered by Chicago, Ill.

9-5-1892: Shipped to Chicago, Ill.

Date?: Rebuilt with Metropolitan pump.

11-27-1905: 32" x 66 3/4" Ahrens coil boiler.

1906: At Engine 24. 1917: At Engine 81.

8-4-1914: New boiler.

1918: Motorized with two-wheel Christie tractor.

2128 3 DCN/HD/A Upright. 32" x 66 3/4" coil boiler (no casing), heater pipes left, two three-pint oil cans back of boiler, torch on coal bunker. 4 1/4" x 7" pumps, air chamber ornament, relief valve right. Two 4" x 10' suctions. 6 3/4" x 7" cyls. Wheels 16 x 56, 18 x 62. Platform springs. Gong on foot board. Whip socket. Two lanterns on frame.

6-22-1892: Ordered by Chicago, Ill.

10-19-1892: Shipped to Chicago, Ill. Was at Engine 51 in 1906. At Engine 85 in 1917.

2129 3 DCN/HD/A Upright. 32" x 66 3/4" coil boiler (iron), heater pipes left, two three-pint oil cans back of boiler, torch and name *Bryan Callaghan #4* on coal bunker, nameplate *Bryan Callaghan* on smokestack. 4 1/4" x 7" pumps, nickel-plated air chamber with ornament atop, piston air pump. Two

American Fire Engine Co. factory delivery photo shows the "Bryan Callaghan #4" of San Antonio, Texas. This 1892 Ahrens was a 3rd Class double crane-neck #2129. This engine was carefully preserved by San Antonio firemen after its 1927 retirement, only to be donated to the World War II scrap drive in 1944.

(Courtesy Harold S. Walker)

4" x 10'6", one $2^1/_2$" x 9' suctions. $6^3/_4$" x 7" cyls. Wheels 16 x 56, 18 x 62. Platform springs. $57^1/_2$" track. Gong on foot board. Whip socket on seat. 2 lanterns on frame.

6-28-1892: Ordered by San Antonio, Texas.

10-13-1892: Shipped to San Antonio, Texas. Name *Bryan Callaghan #4*.

4- -1927: Retired. This and two other steamers (1892 Ahrens #2106 and 1908 Metropolitan #3258) were preserved for historical purposes.

1944: Over the strong objection of the fire department, city officials of San Antonio donated the steamer *Bryan Callaghan* to the World War II scrap drive. Thus, only two steamers remain in San Antonio in 1984 (only one of the two is in good condition).

2130 3 DCN/HD/A Upright. 32" x $66^3/_4$" coil boiler (iron), heater pipes left (for Ahrens heater), committee nameplate on boiler, two 3-pint oil cans on back, torch and name *Atlantic #2* on coal bunker, expansion smokestack with numeral *2* on front. $4^1/_4$" x 8" pumps, nickel-plated air chamber with ornament atop. Two 4" x 10'6" suctions. $6^3/_4$" x 8" cyls. Wheels 16 x 56, 18 x 62. Platform springs. 62" track. 13" gong on foot board. Whip socket on seat. Numeral *2* on tool box. One lantern on frame.

8-24-1892: Ordered by Wheeling, W. Va.

11-20-1892: Delivered by Chris and Peter Ahrens to Wheeling, WV. Name *Atlantic #2*.

6- -1895: Wrecked when a train hit it. Bought #2339 to replace it. #2130 was rebuilt and sold to Columbus, Ohio.

2141 3 DCN/HD/A Upright. 32" x $66^3/_4$" coil boiler (no casing), heater pipes left, two three-pint oil cans on back, torch and name *D.S. Bates #1* on coal bunker, expansion smokestack with name *D.S. Bates #1* on front. $4^1/_4$" x 7" pumps, nickel-plated air chamber with ornament atop. Two 4" x 10'6", one $2^1/_2$" x 9' suctions. $6^3/_4$" x 7" cyls. Wheels 16 x 56, 18 x 62. Platform springs. 62" track. Gong on foot board. Whip socket on seat. Numeral *1* on tool box. Two lanterns on frame. Weight 6500#.

8-24-1892: Ordered by Piqua, Ohio.

12-13-1892: Delivered by Peter Ahrens, John P. Ahrens, and Charles H. Fox to Piqua, Ohio. Name *D.S. Bates #1*. Water in 4:00. Threw two $1^1/_8$" streams one block.

2-13-1902: Kohlbecker installed new boiler coils at Piqua.

7-14-1904: New pump valves.

2142 3 DCN/HD/A Upright. 32"x66¾" coil boiler (no casing), heater pipes right, torch on coal bunker, nickel-plated smokestack. 4¼" x 7" pumps; nickel-plated suction/discharge caps, air chamber (w/ornament atop); relief valve right. 4" x 10'6" suctions, expanding-ring couplings (nickel-plated). 6¾" x 7" cyls., nickel-plated casing. Wheels 16 x 56, 18 x 62, nickel-plated hub caps. Platform springs; nickel-plated springs, axles, fifth wheel. 56½" track. Gong on foot board. Whip socket on seat. 2 lanterns on frame.

12-6-1892: Ordered by World's Columbian Exposition, Chicago, Ill.

1-16-1893: Shipped to Columbian Exposition (World's Fair), Jackson Park, Chicago, Ill. In service as Engine 1 (Exposition celebrated the 400th anniversary of Columbus arriving in America). The exposition fire brigade (organized 7-19-1892) became the 14th Battalion of the Chicago Fire Dept. on 12-1-1892. Staffing of the World's Fair battalion was *ten* men per steamer! (By contrast, in this 1980s era of manpower shortages and tight budgets, engine companies are often forced to get by with crews of only 2 or 3.)

7-10-1893: $250,000 fire in a cold-storage warehouse at the Columbian Exposition. Several of the exposition's steamers were used, plus apparatus of Chicago Fire Dept.

1894: In service at Engine 70, Chicago, Ill. Still at Engine 70 in 1917. Later Eng. 112.

10- -1971: This steamer restored, appeared in Chicago's Fire Prevention Week parade commemorating the centennial of the Great Chicago Fire of October 8-11, 1871.

7- -1977: In storage at Chicago Fire Dept. Repair Shop. All parts that were nickel-plated for World's Fair are now painted.

2143 3 DCN/HD/A Upright. 32" x 66¾" coil boiler (no casing), heater pipes right, two three-pint oil cans on back, torch on coal bunker. 4¼" x 7" pumps; nickel-plated air chamber (with ornament atop), relief valve (on right), and discharge gates. Two 4" x 10'6" suctions, expanding-ring couplings (nickel-plated). 6¾" x 7" cyls (nickel-plated casing). Wheels 16 x 56, 18 x 62, nickel-plated hub caps. Platform springs, nickel-plated springs & axles. 56½" track. Gong on foot board. Whip socket on seat. Wheels painted cream; dark blue frame and coal bunker; red pumps, seat, and tool box.

5-18-1893: Shipped to Chicago, Ill., for display as an exhibit at the World's

This October, 1951 photo shows an 1893 Ahrens 3rd Class double crane-neck #2142 during the 80th anniversary celebration of the Great Chicago Fire of October 8-11, 1871. This is the first of seven steamers which Ahrens built for the 1893 Chicago World's Fair, and the only one which survives. When the fair ended, this steamer went in service at Engine 70 in Chicago, Ill., and today it is one of two Ahrens steamers still owned by Chicago Fire Dept.

(John McIntyre photo)

Columbian Exposition. Chief Sweenie of the World's Fair fire brigade misunderstood the purpose of this engine, and put it in service as Engine 2. As a result, Chicago was billed for, and paid for, this engine.

3-1-1894: Mr. Bringhurst of Ahrens Co., not realizing that this engine belonged to Chicago, assumed it was still the property of Ahrens. Thus, when the fair closed, Bringhurst returned this engine to the factory, it was renumbered as 2209 and sold to Ottawa, Ill. Ahrens had to give Chicago #2145, no charge, to replace it.

2144 3 DCN/HD/A Upright. 32" x 66^{3}/4" coil boiler (no casing), heater pipes right, two three-pint oil cans on back, torch on coal bunker. 4^{1}/4" x 7" pumps; nickel-plated air chamber (with ornament atop), relief valve (on right), pump caps, and discharge gates. Two 4" x 10'6", one 2^{1}/2" x 9' suctions. 6^{3}/4" x 7" cyls (nickel-plated casing). Wheels 16 x 56, 18 x 62, nickel-plated hub caps. Platform springs, nickel-plated springs & axles. 56^{1}/2" track. Gong on foot board. Whip socket on seat. Two lanterns on frame. Shipped unpainted (painted in Chicago).

5-4-1893: Ordered by Columbian Exposition fire brigade, Chicago, Ill.

7-8-1893: Shipped to Columbian Exposition fire brigade, Chicago, Ill. Engine 3.

1894: Returned to factory, rebuilt as #2225.

2145 3 DCN/HD/A Upright. 32" x 66^{3}/4" coil boiler (no casing), heater pipes right, two three-pint oil cans on back, torch on coal bunker. 4^{1}/4" x 7" pumps; nickel-plated air chamber (with ornament atop), relief valve (on right), discharge gates, and pump caps. Two 4" x 10'6", one 2^{1}/2" x 9' suctions. 6^{3}/4" x 7" cyls (nickel-plated casing). Wheels 16 x 56, 18 x 62, nickel-plated hub caps. Platform springs, nickel-plated springs & axles. 56^{1}/2" track. Gong on foot board. Whip socket on seat. Two lanterns on frame. Shipped unpainted (painted in Chicago).

7-8-1893: Loaned to Columbian Exposition, Chicago, Ill.

11- -1893: Given to World's Fair Fire Brigade no charge to replace #2143 (Engine 2), which had been returned to factory, rebuilt as #2209, and sold to Ottawa, Ill. (discrepancy in dates: 2143 didn't go to Ottawa until 3-1-1894, don't know why it was replaced four months earlier!)

1894: Rebuilt as #2224 for Chicago.

2146 3 DCN/HD/A Upright. 30" x 62^{1}/2" coil boiler (iron/silver); heater pipes left; safety railing on back; torch, name *W.B. Young*, and numeral *1* on coal bunker; nameplate *W.B. Young* on smokestack. 4" x 7" pumps, air chamber ornament, relief valve left.

The world's first combination pumper and hose car was this 1893 4th Class double crane-neck Ahrens #2147. During a demonstration in Cincinnati, Chief Sweenie of the Columbian Exposition fire brigade (Chicago World's Fair) was so impressed that he purchased this engine, and ordered two more like it. Because the first three "combination engines" were used by the Columbian Exposition, this style quickly came to be known as the "Columbian" engine. Chicago, Milwaukee, and St. Louis were among the cities that used Columbian pump and hose cars.

(Ed Hass collection)

$3^{1}/_{2}$" x 21' pre-connect suction. $6^{3}/_{4}$" x 7" cyls. Wheels 16 x 56, 18 x 62. Platform springs. $56^{1}/_{2}$" track. Gong on foot board. Whip socket on seat. Number *1* on tool box. Two lanterns on frame. Weight 5740#.

8-2-1893: Delivered by John P. Ahrens to Clarksville, Tenn. Name *W.B. Young #1.*

1894: Rebuilt as #2235. See 2235 for subsequent history.

2147 4 DCN/HD/A Columbian. 27" x 56" tubular boiler (no casing), heater pipes right, 3 lanterns on boiler, torch on coal bunker. $3^{1}/_{2}$" x 6" pumps, relief valve right. Two $3^{1}/_{2}$" x 11', one $2^{1}/_{2}$" x 9' suctions. 6" x 6" cyls. Wheels 16 x $38^{1}/_{2}$", 18 x $58^{1}/_{2}$". Platform springs. $56^{1}/_{2}$" track. Whip socket. Hose box behind seat. Two lanterns on hose box. Weight 4560#. First Columbian engine.

10-4-1892: First test at Cincinnati. Threw a 1" stream 225', two $^{3}/_{4}$" streams 178' each.

10-8-1892: Demonstrated in Cincinnati for Chief Sweenie, Columbian Exposition fire brigade. He verbally ordered this engine.

11-5-1892: Shipped to Columbian Exposition (World's Fair), Chicago, Ill. Engine 4.

12-6-1892: Contract from Columbian Exposition, for this and 2 more "Columbians".

1906: At Eng. 69, Chicago. Retired by 1917.

2148 4 DCN/HD/A Columbian. 27" x 56" tubular boiler (no casing), heater pipes right, 3 lanterns on boiler. $3^{1}/_{2}$" x 6" pumps, relief valve right, first pump with center flywheel. Two $3^{1}/_{2}$" x 11' suctions, one $2^{1}/_{2}$" x 9'. 6" x 6" cyls. Wheels 16 x $38^{1}/_{2}$", 18 x $58^{1}/_{2}$". Platform springs. $56^{1}/_{2}$" track. Whip socket on seat. Hose box behind seat (larger hosebox than #2147). 2 lanterns on hose box. Torch on right frame. Weight 4870#.

12-6-1892: Ordered by Columbian Exposition, Chicago, Ill.

3-13-1893: Shipped to Columbian Exposition, Chicago, Ill. Engine 5.

1906: At Eng. 74, Chicago. Retired by 1917.

2149 4 DCN/HD/A Columbian. 27" x 56" tubular boiler (no casing), heater pipes right, 3 lanterns on boiler. $3^{1}/_{2}$" x 6" pumps, relief valve on right, flywheel in center. Two $3^{1}/_{2}$" x 11', one $2^{1}/_{2}$" x 9' suctions. 6" x 6" cyls. Wheels 16 x $38^{1}/_{2}$", 18 x $58^{1}/_{2}$". Platform springs. $56^{1}/_{2}$" track. Whip socket. Hose box behind seat. Two lanterns on hose box. Torch on right frame. Weight 4870#.

12-6-1892: Ordered by Columbian Exposition, Chicago, Ill.

3-13-1893: Shipped to Columbian Exposition, Chicago, Ill. Engine 6.

1906: At Eng. 79, Chicago. Retired by 1917.

2150 4 DCN/HD/A Columbian. 27" x 56" tubular boiler (no casing), heater pipes right, torch on coal bunker. Pumps $3^{3}/_{4}$" x 6", relief valve right. Suctions: two $3^{1}/_{2}$" x 11', one $2^{1}/_{2}$" x 9'. 6" x 6" cyls. Wheels 16 x $38^{1}/_{2}$" front, 18 x $58^{1}/_{2}$" rear. $56^{1}/_{2}$" track. Platform springs. Whip socket on seat. Two lanterns on hose box, 3 on boiler. Nickel-plating on pump, boiler, springs, axles.

4-29-1893: Shipped to Columbian Exposition, Chicago, Ill.

2-15-1894: Rented to Evanston, Ill., for 4 months.

8-1-1895: Rebuilt as #2208 for Chicago.

2171 4 DCN/HD/A Columbian. 27" x 56" tubular boiler (iron), heater pipes right, torch on coal bunker. $3^{3}/_{4}$" x 6" pumps, relief valve right. Two $3^{1}/_{2}$" x 11', one $2^{1}/_{2}$" x 9' suctions. 6" x 6" cyls. Wheels 16 x $38^{1}/_{2}$" front, 18 x $58^{1}/_{2}$" rear. $56^{1}/_{2}$" track. Platform springs. Whip socket on seat. Two lanterns on hose box, 3 on boiler.

1893: Built for Milwaukee, Wis., not sold.

5-27-1893: Shipped to Chicago as #2222.

2172 6 SSF/MD-HD/A Upright. 25"x$56^{1}/_{2}$" tubular boiler (iron), lantern on back, name *Fountain City* on smokestack & coal bunker, number *1* on coal bunker, torch on coal bunker. $4^{1}/_{2}$" x 7" pump, air chamber ornament, automatic air pump, goose-neck discharge gates, cast-iron frame around pump instead of support columns. Suctions: one $3^{1}/_{4}$" x 20' pre-connect, one $3^{1}/_{4}$" x 6'. One $6^{1}/_{2}$" x 7" cylinder. Wheels 16 x 51 front, 18 x 57 rear. $53^{1}/_{2}$" track. Rear brake lever. Two lanterns on seat, extra lantern engraved "Chas. R. Beckmann, Mayor". Height 7'7". Length 11'6" (20'7" with pole).

2-16-1893: Shipped to Engine 1, Fountain City, Wis. Delivered by John Black.

2173 2 DCN/HD/A Upright. $35^{1}/_{2}$" x 67" coil boiler (iron), heater pipes left, two 3-pint oil cans on back of boiler, torch on coal bunker. $4^{3}/_{4}$" x 8" pumps, piston air pump, ornament atop air chamber, oil pump atop vacuum chamber, relief valve right. Two $4^{1}/_{2}$" x 10'6" suctions. $7^{3}/_{4}$" x 8" cyls. Wheels 16 x 48 front, 18 x 58 rear. Platform springs. Gong on foot board. Whip socket on seat. 3-horse hitch. 2 lanterns on frame, strainer-stand on back of frame.

3-11-1893: Ordered by Milwaukee, Wis.

5-25-1893: Shipped to Milwaukee, Wis., in

This 1911 photo shows an 1893 Ahrens 2nd Class double crane-neck #2173/2214 in front of Engine 6 in Milwaukee, Wis. on a cold winter day, sometime between 1893 and 1915. The firehouse, at Franklin Place and Brody St., probably dates from the volunteer days (as evidenced by the name "Butler No. 6" over the door).
(Courtesy Harold S. Walker)

primer (not painted).

1894: Rebuilt and painted, as #2214. See #2214 for its subsequent history.

2174 2 DCN/HD/A Upright. 35$^{1}/_{2}$"x67" coil boiler (no casing), heater pipes left, safety railing and two 3-pint oil cans on back, hook for apron on left, torch on coal bunker. 4$^{3}/_{4}$" x 8" pumps, air chamber ornament, relief valve left, piston air pump, Star gauges. 4" x 21' pre-connect suction. 7$^{3}/_{4}$" x 8" cyls. Wheels 16 x 58, 18 x 62. Platform springs. 57$^{1}/_{2}$" track. Gong on foot board. Whip socket on seat. Lanterns on frame and boiler.

7-20-1893: Ordered by St. Louis, Mo.

10-7-1893: Shipped to St. Louis, Mo. (built as #2246). Delivered by G. Fred Hawekotte. See #2246 for its subsequent history.

2175 4 DCN/HD/A Columbian. 27" x 59$^{3}/_{8}$" coil boiler (no casing), heater pipes right, nickel-plated number *17* on smokestack. Pumps 3$^{1}/_{2}$" x 6", relief valve right. Suctions: two 3$^{1}/_{2}$" x 11', one 2$^{1}/_{2}$" x 9'. 6" x 6" cyls. Wheels 16 x 38$^{1}/_{2}$ front, 18 x 58$^{1}/_{2}$ rear. 56$^{1}/_{2}$" track. Platform springs. Whip socket on seat, torch on right frame.

5-4-1893: Ordered by Chicago, Ill.

8-15-1893: Shipped to Chicago, Ill. (as #2226), in primer (not painted). See #2226 for subsequent history.

2176 4 DCN/HD/A Columbian. 27" x 59$^{3}/_{8}$" coil boiler (no casing), heater pipes right. Pumps 3$^{1}/_{2}$" x 6", relief valve right. Suctions: two 3$^{1}/_{2}$" x 11', one 2$^{1}/_{2}$" x 9'. 6" x 6" cyls. Wheels 16 x 38$^{1}/_{2}$ front, 18 x 58$^{1}/_{2}$ rear. 56$^{1}/_{2}$" track. Platform springs. Whip socket on seat, torch on right frame.

5-4-1893: Ordered by Chicago, Ill.

8-15-1893: Shipped to Chicago, Ill. (as #2227). See #2227 for subsequent history.

2177 4 DCN/HD/A Columbian. 27" x 59$^{3}/_{8}$" coil boiler (iron), first Columbian engine delivered with coil boiler, heater pipes left, nickel-plated number *17* on smokestack, torch on coal bunker. Two 3$^{1}/_{2}$" x 6" pumps, relief valve right. Two 3$^{1}/_{2}$" x 11' suctions. Two 6" x 6" cylinders (left cylinder overbored $^{1}/_{16}$"). Wheels 16 x 38$^{1}/_{2}$" front, 18 x 58$^{1}/_{2}$" rear. Platform springs. 56$^{1}/_{2}$" track. Gong on foot board. Whip socket on seat. Weight 5430#.

5-19-1893: Ordered by Milwaukee, Wis.

8-5-1893: Shipped to Milwaukee, Wis. (as

#2228) Delivered by G. Fred Hawekotte. Milwaukee Fire Dept. records say that #2177 was originally Engine 18, and #2228 was originally Engine 8, but Ahrens records show that these were the same machine and was originally Engine 17; uncertain which was original assignment. See #2228 for its history.

2178 5 SSF Upright. Tubular boiler. Records don't show if it was built or for whom.

2179 2 DCN Upright. Ahrens lists as *Stock.*

2180 2 DCN Upright. Ahrens lists as *Stock.*

2181 3 DCN/HD/W Upright. 32" x 66$^3/_4$" coil boiler (no casing), heater pipes left, safety railing & two 3-pint oil cans on back, number *16* on smokestack, hook for apron on left. 4$^1/_4$" x 7" pumps, air chamber ornament, relief valve left, piston air pump. 4" x 21' pre-connect suction. 6$^3/_4$" x 7" cyls. Wheels 16 x 56, 18 x 62. 57$^1/_2$" track. Platform springs. Gong on foot board. Lanterns on frame and boiler.

1893: Built as Stock.

7-20-1893: Ordered by St. Louis, Mo.

10-7-1893: Shipped to Saint Louis, Mo. (as #2247). Delivered by G. Fred Hawekotte. See #2247 for its history.

Last steamer listed in Ahrens Co.'s handwritten ledger, called *Book #1: Destination of Ahrens Engines* (book is now in a museum). That book recorded each Ahrens steamer built from 1869 to 1893. *Book #1* also refers to *Book #2*, which presumably covered the era 1894-1904; unfortunately, *Book #2* is not known to survive, so the rest of this list had to be pieced together from what few records **do** survive.

2182 Note penciled on #2181 indicates that there were Ahrens engines numbered 2182 through 2186, but doesn't say what they were or where they were delivered.

2203 1 DCN/HD/A Upright. 38" x 67" coil boiler. 5" x 8" pumps, piston air pump. 8" x 8" cyls. Wheels 16 x 60 front, 18 x 64 rear.

1893: In service at Reading Hose Co. #1, Reading, Pa. (Possibly rebuild of #2122?).

2208 4 DCN/HD/A Columbian. 27"x56" tube boiler (no casing), heater pipes right, torch on coal bunker. 3$^3/_4$" x 6" pumps, relief valve right. Two 3$^1/_2$" x 11', one 2$^1/_2$" x 9' suctions. 6" x 6" cyls. Wheels 16 x 38$^1/_2$", 18 x 58$^1/_2$". 56$^1/_2$" track. Platform springs. Whip socket. 2 lanterns on hose box, 3 on boiler. Nickel-plated pump, boiler, springs, axles.

8-1-1895: Shipped to Engine 63, Chicago, Ill. Rebuild of World's Fair engine #2150.

1906: At Engine 63. Retired by 1917.

2209 3 DCN/HD/A Upright. 32" x 66$^3/_4$" coil boiler (no casing), heater pipes right, two three-pint oil cans on back, torch on coal bunker. 4$^1/_4$" x 7" pumps; air chamber and ornament, relief valve (on right), and discharge gates all nickel-plated. Two 4" x 10'6" suctions, expanding-ring couplings (nickel-plated). 6$^3/_4$" x 7" cyls (nickel-plated casing). Wheels 16 x 56, 18 x 62, nickel-plated hub caps. Platform springs, nickel-plated springs & axles. 56$^1/_2$" track. Gong on foot board. Whip socket on seat. Wheels painted cream; dark blue frame and coal bunker; red pumps, seat, and tool box.

3-1-1894: Delivered to Ottawa, Ill. Rebuild of #2143 from Chicago World's Fair.

6-6-1900: New pump valves.

2214 2 DCN/HD/A Upright. 35$^1/_2$"x67" coil boiler (iron), heater pipes left, two 3-pint oil cans on back, torch on coal bunker. 4$^3/_4$" x 8" pumps, air chamber ornament, oil pump on vacuum chamber, relief valve right, piston air pump. Two 4$^1/_2$" x 10'6" suctions. 7$^3/_4$" x 8" cyls. Wheels 16 x 48, 18 x 58. Platform springs. Gong on foot board. Whip socket on seat. 3-horse hitch. 2 lanterns and strainer-stand on frame.

1894: In service at Engine 6, Milwaukee, Wis. Rebuild of #2173.

1915: Replaced by Seagrave pumper.

2222 4 DCN/HD/A Columbian. 27"x56" tube boiler (iron), heater pipes right, torch on coal bunker. 3$^3/_4$" x 6" pumps, relief valve right. Two 3$^1/_2$" x 11', one 2$^1/_2$" x 9' suctions. 6" x 6" cyls. Wheels 16 x 38$^1/_2$", 18 x 58$^1/_2$". 56$^1/_2$" track. Platform springs. Whip socket. 2 lanterns on hose box, 3 on boiler.

5-27-1893: Shipped to Chicago, Ill. Rebuild of #2171. At E. 60, 1906. Retired by 1917.

2224 3 DCN/HD/A Upright. 32" x 66$^3/_4$" coil boiler (no casing), heater pipes right, two three-pint oil cans on back, torch on coal bunker. 4$^1/_4$" x 7" pumps; air chamber and ornament, relief valve (on right), discharge gates, and pump caps all nickel-plated. Suctions: two 4" x 10'6", one 2$^1/_2$" x 9'. 6$^3/_4$" x 7" cyls (nickel-plated casing). Wheels 16 x 56, 18 x 62, nickel-plated hub caps. Platform springs, nickel-plated springs & axles. 56$^1/_2$" track. Gong on foot board. Whip socket on seat. Two lanterns on frame.

1894: Shipped to Chicago, Ill. Rebuild of World's Fair engine #2145.

1906: At Engine 88. Later Engine 108.

1917: At Engine 72.

2225 3 DCN/HD/A Upright. 32" x 66$^{3}/_{4}$" coil boiler (no casing), heater pipes right, two three-pint oil cans on back, torch on coal bunker. 4$^{1}/_{4}$" x 7" pumps; nickel-plated air chamber (with ornament atop), relief valve (on right), discharge gates, and pump caps. Suctions: two 4" x 10'6", one 2$^{1}/_{2}$" x 9'. 6$^{3}/_{4}$" x 7" cyls (nickel-plated casing). Wheels 16 x 56, 18 x 62, nickel-plated hub caps. Platform springs, nickel-plated springs & axles. 56$^{1}/_{2}$" track. Gong on foot board. Whip socket on seat. Two lanterns on frame.

1894: Shipped to Chicago, Ill. Rebuild of World's Fair engine #2144.

1906: At Engine 73.

1917: At Engine 111. Later Engine 113.

2226 4 DCN/HD/A Columbian. 27" x 59$^{3}/_{8}$" coil boiler (no casing), heater pipes right, nickel-plated number *17* on smokestack. Pumps 3$^{1}/_{2}$" x 6", relief valve right. Suctions: two 3$^{1}/_{2}$" x 11', one 2$^{1}/_{2}$" x 9'. 6" x 6" cyls. Wheels 16 x 38$^{1}/_{2}$ front, 18 x 58$^{1}/_{2}$ rear. 56$^{1}/_{2}$" track. Platform springs. Whip socket on seat, torch on right frame.

8-15-1893: Shipped to Engine 17, Chicago, Ill., in primer. Rebuild of #2175.

1906: Was at Engine 33. Retired by 1917.

2227 4 DCN/HD/A Columbian. 27" x 59$^{3}/_{8}$" coil boiler (no casing), heater pipes right. Pumps 3$^{1}/_{2}$" x 6", relief valve right. Suctions: two 3$^{1}/_{2}$" x 11', one 2$^{1}/_{2}$" x 9'. 6" x 6" cyls. Wheels 16 x 38$^{1}/_{2}$ front, 18 x 58$^{1}/_{2}$ rear. 56$^{1}/_{2}$" track. Platform springs. Whip socket on seat, torch on right frame.

8-15-1893: Shipped to Chicago, Ill. Rebuild of #2176. Harold Walker lists as Engine 38.

1906: At Engine 35. 1917: At Engine 109.

2228 4 DCN/HD/A Columbian. 27" x 59$^{3}/_{8}$" coil boiler (iron), first Columbian engine delivered with coil boiler, heater pipes left, nickel-plated number *17* on smokestack, torch on coal bunker. 3$^{1}/_{2}$" x 6" pumps, relief valve right. Two 3$^{1}/_{2}$" x 11' suctions. 6" x 6" cyls (left cyl. overbored $^{1}/_{16}$"). Wheels 16 x 38$^{1}/_{2}$", 18 x 58$^{1}/_{2}$". Platform springs. 56$^{1}/_{2}$" track. Gong on foot board. Whip socket on seat. Weight 5430#.

8-5-1893: Shipped to Milwaukee, Wis. (rbld of #2177). Delivered by Fred Hawekotte. Milwaukee F.D. records say that #2177 was originally Engine 18, and #2228 originally Engine 8; Ahrens records show #2177/2228 was originally Engine 17, then Engine 8.

9-1-1903: #372 to Eng. 8; #2228 to Eng. 25.

1906: Replaced by Ahrens #587.

3-1-1912: To E. 31. Replaced by #635, 1915.

2234 4 DCN/HD/W Columbian. 27"x56" tube boiler. 3$^{1}/_{2}$" x 6" pumps. Two 6" x 6" cylinders. Wheels 16 x 38$^{1}/_{2}$ front, 18 x 58$^{1}/_{2}$ rear.

1893: In service at Engine 72, Chicago, Ill.

1906: At Engine 72. Retired by 1917. Chicago had at least 2 more 1893 Columbians (Reg. Nos. unknown) that were serving as Engines 89 and 91 in 1906.

2235 3 DCN/HD/A Upright. 30" x 62$^{1}/_{2}$" coil boiler (iron/silver); heater pipes left; safety railing on back; torch, name *W.B. Young*, and numeral *1* on coal bunker; nameplate *W.B. Young* on smokestack. 4" x 7" pumps, air chamber ornament, relief valve left. 3$^{1}/_{2}$" x 21' pre-connect suction. 6$^{3}/_{4}$" x 7" cyls. Wheels 16 x 56, 18 x 62. Platform springs. 56$^{1}/_{2}$" track. Gong on foot board. Whip socket on seat. Number *1* on tool box. Two lanterns on frame. Weight 5740#.

1894: Rebuild of #2146, Clarksville, Tenn.

9-1-1901: Pump rebuilt.

11-12-1912: Rebuilt by American-LaFrance.

2246 2 DCN/HD/A Upright. 35$^{1}/_{2}$"x67" coil boiler (no casing), heater pipes left, safety railing and two 3-pint oil cans on back, hook for apron on left, torch on coal bunker. 4$^{3}/_{4}$" x 8" pumps, air chamber ornament, relief valve left, piston air pump. 4" x 21' pre-connect suction. 7$^{3}/_{4}$" x 8" cyls. Wheels 16 x 58, 18 x 62. Platform springs. 57$^{1}/_{2}$" track. Gong on foot board. Whip socket. Lanterns on frame and boiler.

10-7-1893: Shipped to St. Louis, Mo. (rbld of #2174). Delivered by Fred Hawekotte.

1909: Was a spare at Engine 5.

6- -1910: New 35" x 64" Fox boiler.

2247 3 DCN/HD/W Upright. 32"x66$^{3}/_{4}$" coil boiler (no casing), heater pipes left, safety railing & two 3-pint oil cans on back, number *16* on smokestack, hook for apron on left. 4$^{1}/_{4}$" x 7" pumps, air chamber ornament, relief valve left, piston air pump. 4" x 21' pre-connect suction. 6$^{3}/_{4}$" x 7" cyls. Wheels 16 x 56, 18 x 62. 57$^{1}/_{2}$" track. Platform springs. Gong on foot board. Lanterns on frame and boiler.

10-7-1893: Shipped to Engine 16, St. Louis, Mo. (rebuild of #2181). Delivered by Fred Hawekotte.

10- -1901: New pump valves & springs.

2-11-1907: New 32" x 62" Fox boiler.

1909: Was at Engine 23.

Rushville, Indiana, used this 1893 Ahrens 3rd Class double crane-neck #2250 until 1925. After 50 years in storage, it was restored in 1975 and placed on permanent display at the Firefighters Gallery of the Indianapolis Children's Museum. The pre-connect suction hose squirrel-tailed across the front was typical of Ahrens.

(Courtesy Rodger Birchfield)

Loveland, Ohio, still owns this 3rd Class double crane-neck Ahrens #2252, which they bought new in 1894. The Continental boiler was installed by Ahrens-Fox in 1915. Although painted a lighter red than the original, and lacking gold-leaf decoration, it is nevertheless remarkably well-preserved. The horse pole (lying under the steamer), squirrel-tail suction hose, and brass gong on the front, are among items that are often missing from steamers today, but somehow survived at Loveland.

(Bill Foster photo)

2248 2 DCN/HD/A Upright. 35 1/2" x 67" coil boiler. 4 1/2" x 8" pumps. 7 3/4" x 8" cyls. Wheels 16 x 58 front, 18 x 62 rear.

1893: In service at Cincinnati, Ohio. Had 28 steamers in service at the time.

7-31-1908: Ahrens Fire Engine Co. installed 34" x 64" Radial Inclined boiler on 1893 Ahrens #2248 of Fire Co. 10, Cincinnati. Blueprint on file Ahrens-Fox Co. 8-10-1985.

2249 3 DCN/HD/A Upright. 32" x 62" coil boiler. 4 1/4" x 7" pumps. 6 3/4" x 7" cylinders. Wheels 16 x 56 front, 18 x 62 rear.

1893: In service at Evanston, Ill.

2250 3 DCN/HD/A Upright. 32"x62" coil boiler. 4 1/4" x 7" pumps. 4" x 20' pre-connect suction. 6 3/4" x 7" cyls. 16x56, 18x62 wheels.

1893: In service, Rushville, Ind. ($4350).

1-8-1925: Retired and to storage.

1975: To Firefighters Gallery, Indianapolis Children's Museum, 30th & Meridian, Indianapolis, Ind. Restored during 1975-1977 and on display there ever since.

2252 3 DCN/HD/A Upright. 32" x 62" coil boiler. 4 1/4" x 7" pumps. 4" x 20' pre-connect suction. 6 3/4" x 7" cylinders. Wheels 16 x 56 front, 18 x 62 rear. Gong on foot board.

1894: In service at Loveland, Ohio.

1915: Rebuilt by Ahrens-Fox Fire Engine Co. with new Continental boiler.

1975: Loveland displays this in parades.

2253 3 DCN/HD/A Upright. 32" x 62" coil boiler. 4 1/4" x 7" pumps. 4" x 20' pre-connect suction. 6 3/4" x 7" cylinders. Wheels 16 x 56 front, 18 x 62 rear. Gong on foot board.

1894: In service at St. Louis, Mo.

Date?: Rebuilt with Fox boiler.

1909: Was at Engine 37.

2258 1 DCN/HD/A Upright. No further details.

1894: Sold to Columbus, Ohio. Rebuild of 1892 Ahrens #2124. Apparently **NOT** the same as #2258 for Manila, Philipines.

2258 X1 DCN/HD/A Upright.

1894: Delivered to Manila, Philipine Islands

2259 X1 DCN/HD/A Upright.

1894: Delivered to Manila, Philipine Islands

2260 2 DCN/HD/A Upright. 35 1/2" x 67" coil boiler. 4 1/2" x 8" pumps. 7 3/4" x 8" cyls. Wheels 16 x 58 front, 18 x 62 rear.

1894: In service at Birmingham, Alabama.

2261 2 DCN/HD/A American. Ahrens coil boiler. First American built. Differed from Upright by having flywheel in center of pump, and decreased curvature of frame.

1894: In service at Erie, Pa.

2262 2 DCN/HD/A American. Ahrens boiler.

1894: In service at St. Joseph, Mo.

2274 1 DCN/HD/A Upright. 38"x67" coil boiler. 5"x8" pumps. 8" cyls. 16x60, 18 x 64 wheels.

1893: In service, Engine 4, Dallas, Texas.

1-25-1908: New pump valves installed by Ahrens Fire Engine Co.

2277 4 DCN/HD/W Columbian. 3 1/2" x 6" pump. 6" x 6" cyl. 2-horse hitch.

1894: Shipped to Chicago, Ill.

1906: At Engine 75. 1917: At Engine 108.

2279 4 DCN/HD/W Columbian. 3 1/2" x 6" pump. 6" x 6" cyl. 2-horse hitch.

1894: Shipped to Chicago, Ill.

1906: At Engine 76. 1917: At Engine 88.

2280 2 DCN/HD/A Upright. 35 1/2" x 67" coil boiler. 4 1/2" x 8" pumps. 7 3/4" x 8" cyls. Wheels 16 x 58 front, 18 x 62 rear.

9-1-1894: In service, E. 21, Milwaukee, Wis.

1908: Replaced by new Nott. To Engine 28.

1921: Replaced by Metropolitan #2738.

2285 2 DCN/HD/A Upright. 35 1/2" x 67" coil boiler. 4 1/2" x 8" pumps. 7 3/4" x 8" cyls. Wheels 16 x 58 front, 18 x 62 rear.

1894: In service at St. Louis, Mo.

Date?: Rebuilt with Fox boiler.

1909: Was at Engine 1.

2286 4 DCN/HD/W Columbian. 3 1/2" x 6" pump. 6" x 6" cyl. 2-horse hitch.

1894: Shipped to Chicago, Ill.

1906: At Engine 77. Retired by 1917.

2287 4 DCN/HD/W Upright. 30"x60" coil boiler. 3 3/4"x6" pump. 6" cyls. 16x54, 18x60 wheels.

1894: In service at New London, Wis.

2288 4 DCN/HD/W Columbian. $3^{1}/_{2}$" x 6" pump. 6" x 6" cyl. 2-horse hitch.

1894: Shipped to Chicago, Ill.

1906: At Engine 86. Retired by 1917.

2316 2 DCN/HD/A Upright. $35^{1}/_{2}$" x 67" coil boiler. $4^{1}/_{2}$" x 8" pumps. $7^{3}/_{4}$" x 8" cyls. Wheels 16 x 58 front, 18 x 62 rear.

1894: To Springfield, Ill.

2319 2 DCN/HD/A Upright. $35^{1}/_{2}$" x 67" coil boiler. $4^{1}/_{2}$" x 8" pumps. $7^{3}/_{4}$" x 8" cyls. Wheels 16 x 58 front, 18 x 62 rear.

1894: To Rockford, Ill.

2320 4 DCN/HD/A American. Ahrens boiler.

1895: In service at St. Louis, Mo.

1909: Was at Engine 42.

2321 4 DCN/HD/A American. Ahrens boiler.

1895: In service at St. Louis, Mo.

Date?: Rebuilt with Fox boiler.

1909: Was at Engine 31.

2337 2 DCN/HD/A Upright. $35^{1}/_{2}$" x 67" coil boiler. $4^{1}/_{2}$" x 8" pumps. $7^{3}/_{4}$" x 8" cyls. Wheels 16 x 58 front, 18 x 62 rear.

1895: In service at St. Louis, Mo.

Date?: Rebuilt with Fox boiler.

1909: Was at Engine 8.

1914: Replaced by Ahrens #564.

2338 3 DCN/HD/A Upright. 32" x 62" coil boiler. $4^{1}/_{4}$" x 7" pumps. $6^{3}/_{4}$" x 7" cylinders. Wheels 16 x 56 front, 18 x 62 rear.

1895: To Kalamazoo, Mich.

12-3-1908: Rebuilt with 30" x 62" Radial Inclined Water Tube boiler by Ahrens Fire Engine Co. Blueprint for this conversion is still on file at Ahrens-Fox Co., 8-10-1985.

2339 1 DCN/HD/A Upright. 38" x 67" coil boiler. 5" x 8" pumps, relief valve on right, capacity 950 GPM. 8" x 8" cyls. Wheels 16 x 60 front, 18 x 64 rear. Weight 7500#.

8-5-1895: Delivered by Chris Ahrens to Atlantic Fire Co. #8, Wheeling, W. Va. Price $4800. Replaced wrecked Ahrens #2130. An old Ahrens loaned to Wheeling while this was built.

Date?: Continental boiler installed by Ahrens Fire Engine Co. Ed Hass saw the blueprint for this conversion at Ahrens-

Jack Selvey of Gahanna, Ohio, found this 1895 Ahrens #2339 neglected in a field near Wheeling, W. Va., in 1984, and promptly bought it. Originally Engine 8 in Wheeling, this 1st Class double crane-neck was later fitted with an Ahrens-Fox Continental boiler. Selvey is now undertaking the monumental task of restoration: note the disintegrating rubber tires, broken rims on all wheels, damaged suction hose, and faded paint. Among missing parts are: manufacturer's nameplate (on boiler), suction and intake caps (on pump), driver's footboard, seat, and horse-hitch.

(Jack Selvey photo)

This unidentified Ahrens steamer from Chicago served Engine 39. It appears to be a 3rd Class double crane-neck of about 1890-1895 vintage, with the boiler stay-bolted rather than cased.
(Courtesy Harold S. Walker)

Leipsic, Ohio, still owns and operates this 1895 Ahrens 5th Class single straight-frame #2364. The frame, seat, and coal bunker still have their original deep-red paint and gold-leaf decorations.
(Rick Knife photo)

Fox in 1980, but could not find it (to copy the date off of it), on a 1985 visit.

1918: Retired.

1967: Operated for the last time. Sat outside, neglected, for next 17 years.

1984: Sold to Jack Selvey of Gahanna, Ohio. He found it in a field near Wheeling, where it had been sitting for many years.

2360 2 DCN/HD/W American.

1895: To St. Louis, Mo.

Date?: Rebuilt with Fox boiler.

1909: Was at Engine 33.

2362 2 DCN/HD/A Upright. 35 1/2" x 67" coil boiler. 4 1/2" x 8" pumps. 7 3/4" x 8" cyls. Wheels 16 x 58 front, 18 x 62 rear.

7- -1896: In service, Eng. 14, Detroit, Mich.

Date?: To Engine 12.

Date?: To Engine 38.

1930s: Scrapped.

2364 5 SSF/MD-HD/W Upright. 25" x 56" tubular boiler. 4 1/2" x 7" pump. 7" x 7" cylinder. Wheels 16 x 52 front, 18 x 60 rear.

1895: In service at Leipsic, Ohio.

1984: Leipsic Fire Dept. still owns, operated at Greenfield Village muster.

2368 1 DCN/HD/W American.

1895: To Fort Worth, Texas.

2378 3 DCN/HD/W American.

1895: To Traverse City, Mich.

1986: Now Mich. Millers Mutual Insurance.

2381 1 DCN/HD/A Upright. 38" x 67" coil boiler. 5" x 8" pumps. Two 8" x 8" cylinders. Wheels 16 x 60 front, 18 x 64 rear.

1895: To St. Louis, Mo.

1909: Not listed on St. Louis F.D. roster.

2383 4 DCN/HD/W American.

1895: To McKinney, Texas.

2384 3 DCN/HD/A Upright. 32" x 62" coil boiler. 4 1/4" x 8" pumps. 6 3/4" x 8" cylinders. Wheels 16 x 54 front, 18 x 60 rear.

1895: In service at Engine 28, Chicago, Ill.

1906: At Engine 80. 1917: At Engine 73.

Oxford, Ohio, still owns this modified 1895 Ahrens 5th Class Upright #2386, with what appears to be a Silsby crane-neck frame and the pump of a Nott steamer (note the twin cylindrical air chambers). After decades deteriorating in a barn, it was moved in 1984 into an authentic replica of a 19th-century firehouse. Oxford plans to restore the weather-beaten Ahrens to the same beautiful condition as its quarters. The Ahrens still has the rope and reel on the front axle (for pulling by hand) and the rear brake controls for the stoker (although the brake-blocks themselves are missing).

(Courtesy Richard Keebler, Oxford Fire Dept.)

2385 4 DCN/HD/W American.

1895: To McKinney, Texas.

2386 5 DCN/MD-HD/W Upright. 25"x56" tubular boiler, heater pipes left. 3 1/2" x 6" pumps. 3 1/4" x 20' pre-connect suction. 5 1/2" x 6" cylinders. Wheels 16 x 52, 18 x 60. Platform springs. Double seat. Rear brake lever.

7-11-1895: Delivery to Oxford, Ohio. Name "Neptune". 1875 Ahrens #45 traded-in.

1916: Rebuilt, Ahrens-Fox smoke & water-tube boiler. (Retired 1928).

1983: The Ahrens is weathered from years in a barn, but will be restored.

2387 6 SSF/MD/A Upright. 25" x 56 1/2" tubular boiler. 4 1/2" x 7" pump. 6 1/2" x 7" cylinder. Wheels 16 x 51 front, 18 x 57 rear.

1895: In service at Spencerville, Ohio.

2399 1 DCN/HD/W American.

1895: To San Francisco, Cal.

2404 1 DCN/HD/W American.

1895: To San Francisco, Cal.

2405 3 DCN/HD/A Upright. 32"x62" boiler. 4 1/4" x 7" pumps. 6 3/4" cyls. 16x56, 18x62 wheels.

1895: In service at Engine 5, Louisville, Ky.

1910: Fox boiler. (retired 1917)

2406 3 DCN/HD/A Upright. 32" x 62" coil boiler. 4 1/4" x 7" pumps. 6 3/4" x 7" cylinders. Wheels 16 x 56 front, 18 x 62 rear.

1895: In service, Engine 12, Louisville, Ky.

1914: Replaced by new Metropolitan #3416.

2413 3 DCN/HD/A Upright. 32"x62" coil boiler. 4 1/4" x 7" pump. 7" x 7" cyl.

1895: To Chicago, Ill.

1906: At Engine 94. Retired by 1917.

2414 1 DCN/HD/A Upright. 38" x 67" coil boiler. Two 5" x 8" pumps. Two 8" x 8" cyls. Wheels 16 x 60 front, 18 x 64 rear.

1896: In service, Engine 19, Louisville, Ky.

1910: Rebuilt with Fox boiler.

1923: Retired.

2417 2 DCN/HD/W Columbian.

1896: To St. Louis, Mo.

1909: Not listed on St. Louis Fire Dept. apparatus roster.

Factory photo shows 1896 Columbian 2nd Class double crane-neck #2417, originally used in St. Louis, Mo. By 1909, it was no longer in service there, and its subsequent history is unknown.
(Courtesy John Peckham)

For many years, the Louisiana State Museum has been carefully preserving the remains of this 1896 Ahrens 2nd Class double crane-neck #2428, ex-New Orleans, La. This photo, taken on March 10, 1957, clearly shows how much is missing: crown around smokestack of boiler, steam chests and oilers above the pump, suction inlets and discharge gates, squirrel-tail suction hose, lanterns, seat cushion, gong, horse-hitch, and brass hubcaps from all four wheels. The innovative Ahrens platform front-springs kept the steamer stable even on the roughest terrain.

(Lester Williams photo)

2420 1 DCN/HD/A Upright. 38" x 67" coil boiler. Two 5" x 8" pumps. Two 8" x 8" cyls. Wheels 16 x 60 front, 18 x 64 rear.

1896: To Indianapolis, Ind.

2421 4 DCN/HD/W American.

1896: In service at St. Louis, Mo.

Date?: Rebuilt with Fox boiler.

1909: Was at Engine 20.

2423 2 DCN/HD/A Upright. 35 1/2" x 67" coil boiler. Two 4 1/2" x 8" pumps. Two 7 3/4" x 8" cyls. Wheels 16 x 58 front, 18 x 62 rear.

1896: In service at New Orleans, La.

2428 2 DCN/HD/A Upright. 35 1/2" x 67" coil boiler. Two 4 1/2" x 8" pumps. Two 7 3/4" x 8" cyls. Wheels 16 x 58 front, 18 x 62 rear.

1896: In service at New Orleans, La.

1940s: To State Museum (Pontalba Historical House), New Orleans, La. Photographed there by Dr. Lester Williams 9-10-1957.

1978: In storage, Historic City Mint (New Orleans), poor condition, parts missing.

2430 3 DCN/HD/A Upright. 32" x 62" coil boiler. Two 4 1/4" x 7" pumps. Two 6 3/4" x 7" cyls. Wheels 16 x 56 front, 18 x 62 rear.

1896: In service at New Orleans, La.

2438 1 DCN/HD/A Upright. 38" x 67" coil boiler. Two 5" x 8" pumps. Two 8" x 8" cyls. Wheels 16 x 60 front, 18 x 64 rear.

1896: In service, E. 3, Grand Rapids, Mich.

2445 2 DCN/HD/A Upright. 35 1/2" x 67" coil boiler. Two 4 1/2" x 8" pumps. Two 7 3/4" x 8" cyls. Wheels 16 x 58 front, 18 x 62 rear.

1-1-1898: In service, E. 22, Milwaukee, Wis.

1921: Replaced by Ahrens #396.

2446 2 DCN/HD/A Upright. 35"x67" boiler. 4 1/2" x 8" pumps. 7 3/4" cyls. 16x58, 18x62 wheels.

1896: In service, Eng. 3, Chattanooga, Tenn.

1911: American-LaFrance rebuilt at Elmira.

2447 4 DCN/HD/A Columbian. 27" x 56" boiler. 3 1/2" x 6" pump. 6" x 6" cyl.

1896: To Chicago, Ill.

1906: At Engine 81. Retired by 1917.

2458 X1 DCN/HD/W Upright.

1896: In service, E. 4, Grand Rapids, Mich.

2460 3 DCN/HD/A Upright. 32" x 62" coil boiler. Two $4^1/_4$" x 7" pumps. Two $6^3/_4$" x 7" cyls. Wheels 16 x 56 front, 18 x 62 rear.

1896: To Columbus, Ohio.

2466 3 DCN/HD/A Upright. 32"x62" boiler. $4^3/_4$" x 8" pumps. 8" cyls. 16x56, 18x62 wheels.

1896: Shipped to Chicago, Ill.

1906: At Engine 25. 1917: At Engine 115.

2467 3 DCN/HD/A Upright. 32"x62" boiler. $4^3/_4$" x 8" pumps. 8" cyls. 16x56, 18x62 wheels.

1896: Shipped to Chicago, Ill.

Date?: $4^3/_4$" x 7" Metropolitan pump.

1906: At Engine 30. 1917: At Engine 112.

2468 4 DCN/HD/W Columbian. $3^1/_2$" x 6" pump, piston air pump. 6" cyl. 2-horse hitch.

1896: Shipped to Chicago, Ill.

1906: At Engine 82. 1917: At Engine 119.

2469 4 DCN/HD/W Columbian. $3^1/_2$" x 6" pump, piston air pump. 6" cyl. 2-horse hitch.

1896: Shipped to Chicago, Ill.

1906: At Engine 83. 1917: At Engine 97.

2490 2 DCN/HD/A Upright. 35"x67" boiler. $4^1/_2$" x 8" pumps. $7^3/_4$" cyls. 16x58, 18x62 wheels.

1897: To Columbus, Ohio.

2491 2 DCN/HD/A Upright. 35"x67" boiler. $4^1/_2$" x 8" pumps. $7^3/_4$" cyls. 16x58, 18x62 wheels.

1897: To Columbus, Ohio.

2492 2 DCN/HD/W Upright. Ahrens coil boiler.

1896: In service at St. Louis, Mo.

1909: Was at Engine 32.

2493 3 DCN/HD/A Upright. 32"x62" boiler. $4^1/_4$" x 7" pumps. $6^3/_4$" cyls. 16x56, 18x62 wheels.

1897: To New Orleans, La.

2502 4 /HD/W Upright.

1897: To New Orleans, La.

2503 4 /HD/W Upright.

1897: To New Orleans, La.

1896 Ahrens 3rd Class double crane-neck #2467 was being towed out of a Chicago firehouse by an unseen truck at left in this 1920s photo. Note that nameplate on boiler says "Ahrens No. 2467", and tall smokestack is typical of Ahrens, but the central-flywheel pump is from a Metropolitan. Such hybrids were typical of modifications made by the Chicago Fire Dept. Repair Shop. Although in excellent condition when photographed, this steamer does not survive today.
(John McIntyre photo)

Another Chicago shop alteration was Engine 97's 1896 5th Class Columbian #2469. Note that the hose box was removed, and a shelf installed to carry such essential tools as playpipes, coils of rope, and a suction strainer. The patent plate on the frame, typical of Metropolitans, was rare on Upright and Columbian models.
(Dan Martin, courtesy Harold S. Walker)

2504 4 /HD/W Upright.

1897: To New Orleans, La.

2510 4 DCN/HD/W American.

1897: To Kewaunee, Wis.

2511 2 DCN/HD/A Upright. 35"x67" boiler. 4 1/2" x 8" pumps. 7 3/4" cyls. 16x58, 18x62 wheels.

1897: To Nashville, Tenn.

2516 1 DCN/HD/A Upright. 38" x 67" boiler. 5" x 8" pumps. 8" cyls. 16 x 60, 18 x 64 wheels.

4-1-1897: In service at Cincinnati, Ohio.

2517 1 DCN/HD/A Upright. 38" x 67" boiler. 5" x 8" pumps. 8" cyls. 16 x 60, 18 x 64 wheels.

4-1-1897: In service at Fire Co. 22, Cincinnati, Ohio. Name *John Goetz Jr.* Nameplate from Ahrens #550.

2519 4 S /HD/W Upright.

1897: To New Orleans, La.

2520 5 S /HD/W Upright.

1897: To New Orleans, La.

2522 5 S /HD/W Upright.

1897: To New Orleans, La.

2523 5 S /HD/W Upright.

1897: To New Orleans, La.

2524 5 S /HD/W Upright.

1897: To New Orleans, La.

2526 5 S /HD/W Upright.

1897: To New Orleans, La.

2527 3 DCN/HD/A Upright. 32"x62" boiler. 4 1/4" x 7" pumps. 6 3/4" cyls. 16x56, 18x62 wheels.

1897: To Columbus, Ohio.

2528 Stationary steam-powered fire pump.

1897: Delivered to Goodrich Transportation Co., Chicago, Ill. (a steam ship company).

2545 1 DCN/HD/A Upright. 38"x67" boiler. 5 1/2" x 8" pumps, 900 GPM. 9" cyls. 16x60, 18x64 wheels. 3-horse hitch. (Pump rebuilt 1905).

1897: In service at Engine 1, Chicago, Ill.

It is not known where the double crane-neck Ahrens in this c-1896 factory delivery photo was used; its only identifying mark is the Engine Company number "2" on the front of the smokestack. The patent plate on the frame (beside the pump) was rare for an Upright model. Note the piston air pump on the side of the boiler, the two lanterns and the water gauge at front of frame, and the suction strainer on back of frame.

(Courtesy Harold S. Walker)

1917: Still at Engine 1.

2546 1 DCN/HD/A Upright. 38"x67" boiler. $5^1/_2$" x 8" pumps. 9" cyls. 16x60, 18x64 wheels.

1897: In service at Engine 42, Chicago, Ill.

1917: Still at Engine 42.

2547 1 DCN/HD/A Upright. 38"x67" boiler. $5^1/_2$" x 8" pumps. 9" cyls. 16x60, 18x64 wheels. 3-horse hitch. (New boiler tubes 1905).

1897: In service at Engine 93, Chicago, Ill.

1917: Still at Engine 93.

2554 4 DCN/HD/A Upright. 30"x60" boiler. $3^3/_4$" x 6" pumps. 6" cyls. 16x54, 18x60 wheels.

1897: To New Orleans, La.

2555 2 DCN/HD/A American.

1897: To New Orleans, La.

2561 2 DCN/HD/A American.

1897: To New Orleans, La.

2562 3 DCN/HD/A Upright. 32"x62" boiler. $4^1/_4$" x 7" pumps. $6^3/_4$" cyls. 16x56, 18x62 wheels. Later rebuilt with Metropolitan pump.

1897: In service at St. Louis, Mo.

1909: Was at Engine 11.

2563 1 DCN/HD/A American.

1897: In service at St. Louis, Mo.

Date?: Rebuilt with Fox boiler.

1909: Was at Engine 12.

2569 2 DCN/HD/A Upright. 35"x67" boiler. $4^1/_2$" x 8" pumps. $7^3/_4$" cyls. 16x58, 18x62 wheels.

1898: To New Orleans, La.

2570 2 DCN/HD/A American.

1898: To New Orleans, La.

2577 2 DCN/HD/A American.

1898: To Cincinnati, Ohio.

2578 4 DCN/HD/A Upright. 30"x60" boiler. $3^3/_4$" x 6" pumps. 6" cyls. 16x54, 18x60 wheels.

1898: To New Orleans, La.

2584 1 DCN/HD/A Metropolitan (the first one made). Central-flywheel pump. Fox boiler.

1898: In service at St. Louis, Mo. Was at Engine 6 in 1909.

2585 1 DCN/HD/A Metropolitan. Fox boiler. Flywheel in center of pump.

1898: In service at St. Louis, Mo.

1909: Was at Engine 15.

2589 2 DCN/HD/A Metropolitan. Fox boiler. Flywheel in center of pump.

1898: To North American Trading & Transfer Co., Chicago, Ill.

2592 DX-1 fire-boat pump (3300 GPM). Two 10" x 11" pumps. Two 17" cylinders.

1898: Built for Fireboat *Illinois* of Chicago.

1-14-1899: Fireboat *Illinois* placed in service as Engine 37, Chicago, Ill.

1908: Fireboat *Illinois* retired.

2593 DX-1 fire-boat pump (3300 GPM). Two 10" x 11" pumps. Two 17" cylinders.

1898: Built for Fireboat *Illinois* of Chicago.

1-14-1899: Fireboat *Illinois* placed in service as Engine 37, Chicago, Ill.

1908: Fireboat *Illinois* retired.

2594 DX-1 fire-boat pump (3300 GPM). Two 10" x 11" pumps. Two 17" cylinders.

1898: Built for Fireboat *Illinois* of Chicago.

1-14-1899: Fireboat *Illinois* placed in service as Engine 37, Chicago, Ill.

1908: Fireboat *Illinois* retired.

2606 2 DCN/HD/A Metropolitan. Fox boiler (silver), heater pipes left. Air chamber ornament, bell on steam cyls. Driver's foot brake. Water-tank under seat.

1898: In service at Engine 1, Colorado Springs, Colorado.

1910: Rebuilt with Continental boiler by Ahrens Fire Engine Co.

1977: At Pike's Peak Ghost Town, Colorado Springs, Colo. Excellent condition.

2610 2 DCN/HD/A Metropolitan. $4^3/4$" x 8" pumps. 8" x 8" cyl. 2-horse hitch.

1898: Shipped to Chicago, Ill.

1906: At Engine 15. 1917: At Engine 36.

2611 2 DCN/HD/A Metropolitan. $4^3/4$"x8" pump. 8" cyl. 2-horse hitch. (New manifolds 1905).

1898: Shipped to Chicago, Ill.

This 1901 photo shows steamer No. 1 of Colorado Springs, Colo., when it still sported its gleaming silver-plated Fox boiler. The 1898 Metropolitan 2nd Class double crane-neck #2606 was rebuilt with an Ahrens Continental boiler (with traditional black shell) in 1910, and is now displayed at Pike's Peak Ghost Town in Colorado Springs. (Courtesy Lester Williams)

1906: At Engine 19. 1917: At Engine 44.

2612 1 DCN/HD/A Metropolitan. $5^1/_2$"x8" pump. 9" cyl. 3-horse hitch. (New manifolds 1902).

1898: Shipped to Chicago, Ill.

1906: At Engine 17. Retired by 1917.

2623 3 DCN/HD/A Metropolitan.

1899: In service at San Francisco, Cal.

1915: Motorized, 2-wheel Christie tractor.

2624 3 DCN/HD/A Metropolitan.

1899: In service at San Francisco, Cal.

2638 3 DCN/HD/A Metropolitan.

1899: In service Engine 3, Los Angeles, Cal.

1908: To Engine 5. To reserve 1916.

1919: To Engine 9. To reserve 1921.

1923: Sold.

2639 1 DCN/HD/A Metropolitan.

1899: In service Engine 4, Los Angeles, Cal.

1919: To Engine 24.

1921: To reserve. Sold 1923.

2640 2 DCN/HD/A Metropolitan.

1899: In service, Eng. 11, Los Angeles, Cal.

1905: To reserve.

1908: Returned to service at Engine 11.

1912: Motorized with two-wheel Christie tractor, serial #0013 (only the 13th tractor out of the Front Drive Motor Co. factory).

1913: To reserve.

1915: To Engine 29.

1922: To Engine 2.

1923: To Engine 15. Sold 1924.

c-1948: Steamer and tractor sold to Mr. Lee Bothwell, Los Angeles, Cal. He rented it to movie studios for use in several films.

11-18-1971: Sold to Community Fire Co. #1, Mountain View (Wayne Township), N.J.

5- -1975: Restoration completed, appears at parades and musters up and down the east coast (from Syracuse to Baltimore). Has the oldest Christie tractor known to survive.

This 1899 2nd Class Metropolitan #2640 was originally horse-drawn and assigned to Engine 11, Los Angeles, Cal. The two-wheel Christie tractor, built by Front Drive Motor Co. of Hoboken, N.J., was added in 1912. The tractor serial number is 0013, making it the oldest Christie known to survive. Sold in 1971 to Community Fire Co. #1 of Mountain View, N.J., this motorized steamer is a prize-winner at musters all over the eastern U.S. Note the three trophies on the hood, which it had just won when this photo was taken at the 1981 muster in Newark, N.J.

(Ed Hass photo)

2652 1 DCN/HD/A Metropolitan.

1899: In service at San Francisco, Cal.

1914: Motorized with American-LaFrance two-wheel tractor.

6-24-1924: Replaced. This was the last steamer in San Francisco to be retired.

2653 1 DCN/HD/A Metropolitan.

1899: In service at San Francisco, Cal.

2656 2 DCN/HD/A Metropolitan.

1899: In service at Rockford, Ill.

2658 2 DCN/HD/A Metropolitan.

8-1-1899: Delivered to Washington Fire Co. No. 2, Madison, Ind. Replaced Ahrens #12.

1916: Traded-in on Ahrens-Fox #825.

2659 X1 DCN/HD/A Metropolitan.

1899: To Guayaquil, Ecuador.

2667 2 DCN/HD/A Upright. 35"x67" boiler. 4^3/4" x 8" pumps. 7^3/4" cyls. 16x58, 18x62 wheels.

1899: Shipped to Chicago, Ill.

1904: New boiler tubes.

1906: At Engine 8. 1917: At Engine 20.

2668 2 DCN/HD/A Upright. 35"x67" boiler. 4^3/4" x 8" pumps. 7^3/4" cyls. 16x58, 18x62 wheels.

1899: Shipped to Chicago, Ill.

1904: Pump rebuilt.

1906: At Engine 2. Still at Eng. 2 in 1917.

2670 2 DCN/HD/A Upright. 35"x67" boiler. 4^3/4" x 8" pumps. 7^3/4" cyls. 16x58, 18x62 wheels.

1899: Shipped to Chicago, Ill.

1906: At Engine 12. 1917: At Engine 24.

2672 2 DCN/HD/A Upright. 35"x67" boiler. 4^3/4" x 8" pumps. 7^3/4" cyls. 16x58, 18x62 wheels.

1899: Shipped to Chicago, Ill.

1906: At Engine 18. 1917: At Engine 57.

2675 2 DCN/HD/A Metropolitan.

1899: In service at San Francisco, Cal.

2676 2 DCN/HD/A Metropolitan.

1899: In service at San Francisco, Cal.

4-18-1906: Used at Great Earthquake and Fire in San Francisco (500 dead, 28,000

San Francisco, Cal., was original home of this 1899 2nd Class Metropolitan #2676. In 1914 it was sold to San Jose, Cal. The Knox-Martin 3-wheel tractor was then attached, and it became Engine 1. San Jose retired it in 1937, and sold it to the Levin Brothers scrapyard. The Levins were so enchanted with the three-wheel tractor and steamer that, instead of cutting it up, they stored it away in a shed for the next 32 years! In 1969, San Jose firefighters formed a club (the San Jose Muster Team) to collect and preserve antique fire engines, so the Levin family donated this, the only surviving Knox-Martin. Now restored, it is a favorite all over the western U.S., as in this 1985 photo at Hillsborough, Cal.

(Ed Hass photo)

Perhaps the best-known fire engine in the Seattle, Wash., area is Seattle Fire Dept.'s Engine 14XX. This 1899 Metropolitan 1st Class #2686 is attached to a 1921 Seagrave "side-saddle" tractor (so named because the seats are side of the hood). It appears at numerous local parades and musters.

(Bill Hass collection)

buildings destroyed, $350 million loss; compare this to 1871 Great Chicago fire: 250 dead, 17,000 buildings, $88 million).

1914: Sold to San Jose, Cal. (Engine 1). Motorized, 3-wheel Knox-Martin tractor.

1937: Retired. Tractor and steamer sold to Levin Bros., San Jose scrap dealer. They liked the unique tractor, so instead of cutting it up, they stored it in a shed.

1968: Tractor and steamer donated to San Jose Fire Dept. for restoration.

1985: Owned by San Jose Muster Team, a group of San Jose firefighters. Kept in operating condition, pumps at musters all over Western U.S. (even at a muster in Connecticut in 1975!). Tractor is one of only two Knox-Martin 3-wheelers known to survive. The other Knox-Martin is ONLY the tractor (no longer on a steamer); it belonged to a lumber company in Washington state, painted dark green to match their lumber trucks, and donated to Smithsonian Institute in Washington, D.C.

2686 1 DCN/HD/A Metropolitan.

1899: In service at Engine 2, Seattle, Wash.

1914: Motorized with Christie tractor.

1921: Christie tractor replaced with a new 4-cylinder Seagrave tractor. To Engine 14.

1985: Seattle still has steamer and Seagrave tractor, kept in operable condition for parades and musters. Officially still serves as Spare Engine 14XX (for nostalgia), but fights only *imaginary* fires (at musters).

2687 2 DCN/HD/A Metropolitan.

1899: To Hammond, Ind.

2693 4 DCN/HD/A Metropolitan. Fox boiler.

1900: In service at St. Louis, Mo.

1909: Was at Engine 47.

1986: Now M. Glover, Roswell, N.M.

2694 4 DCN/HD/A Metropolitan. Fox boiler.

1900: In service at St. Louis, Mo.

1909: Was at Engine 38.

1918: Motorized with One-Wheel Auto Horse (alias "Iron Mule") tractor.

2695 4 DCN/HD/A Upright. 30"x60" boiler. $3^3/_4$" x 6" pumps. 6" cyls. 16x54, 18x60 wheels.

Aberdeen, Mississippi, bought one of the last Ahrens Upright engines built, an 1899 4th Class double crane-neck #2695. By the time this was built, the newly-introduced Metropolitan (central flywheel, Fox boiler) was rapidly cornering the steam fire engine market; the last Upright was delivered in 1902. Historians appreciate the fact that Aberdeen not only bought one of the last of this rare breed, but has so carefully preserved it to this day.

(Bill Hass collection)

1899: In service at Aberdeen, Miss.

1985: Still own.

2696 3 DCN/HD/A Metropolitan.

1900: In service at Ardmore, Oklahoma. Name *Epic*.

1971: To Oklahoma Firefighters Museum, Oklahoma City, Okla.

2705 X1 DCN/HD/A Metropolitan.

1900: In service at Duluth, Minn.

2706 X1 DCN/HD/A Metropolitan.

1900: In service at Denver, Colo.

2713 2 DCN/HD/A Metropolitan.

1900: In service at Cincinnati, Ohio.

2715 3 DCN/HD/A Metropolitan.

1900: In service at Engine 5, Oakland, Cal.

2722 5 DCN/MD-HD/A Metropolitan. Safety rail on boiler, 2 safety rails above coal bunker, *M.F.D.* on back of coal bunker. Tool box under seat. Platform springs.

1900: In service at Michigamme, Mich.

2-18-1986: Owned by Wesley Trathen, Flint, Mich. He bought it in 1960s. Still has original paint, gold-leaf, equipment.

2723 1 DCN/HD/A Metropolitan.

1900: In service at Decatur, Ill. Sold to Bryan, Ohio in 1918. At Williams County Historical Society, Montpelier, Ohio, 1986.

2737 1 DCN/HD/A Metropolitan.

1900: In service, Eng. 19, Milwaukee, Wis.

1921: To Engine 8.

1927: Retired.

1985: Now owned by Keith Franz, Slinger, Wis. Has been restored.

2738 1 DCN/HD/A Metropolitan.

1900: In service, Eng. 20, Milwaukee, Wis.

1921: To Engine 28.

1925: Retired.

2739 3 DCN/HD/A Metropolitan.

1900: In service at Atchinson, Kansas.

2741 3 DCN/HD/A Metropolitan.

1900: In service at Fort Smith, Arkansas.

2746 2 DCN/HD/A Metropolitan.

1900: In service at Dayton, Ohio.

1915: Motorized with American-LaFrance tractor #845.

2747 1 DCN/HD/A Metropolitan. Fox boiler.

1900: In service at St. Louis, Mo.

1909: Was at Engine 14.

2753 1 DCN/HD/A Metropolitan. Fox boiler.

1900: In service at Bloomington, Ill.

2754 2 DCN/HD/A Metropolitan. Fox boiler. $4^{3}/_{4}$" x 8" pump. 8" cyl. 3-horse hitch.

1900: Shipped to Chicago, Ill.

1906: At Engine 46. Still at Eng. 46 in 1917.

2755 2 DCN/HD/A Metropolitan. Fox boiler. $4^{3}/_{4}$" x 8" pump. 8" cyl. 2-horse hitch.

1900: Shipped to Chicago, Ill.

1906: At Engine 90. Still at Eng. 90 in 1917.

2756 1 DCN/HD/A Metropolitan. Fox boiler. $5^{1}/_{2}$" x 8" pump. 9" cyl. 2-horse hitch.

1900: Shipped to Chicago, Ill.

1906: At Engine 13. 1917: At Engine 53.

2762 3 DCN/HD/A (Metropolitan?).

1900: In service at Jackson, Miss.

2765 3 DCN/HD/A Metropolitan.

12-14-1900: In service, Engine 3, Nashville, Tenn. At Children's Museum, 1986.

2766 2 DCN/HD/A Metropolitan.

1900: In service at Akron, Ohio.

2767 2 DCN/HD/A Metropolitan.

1900: In service at Bakersfield, Cal.

2768 4 DCN/HD/A Upright. 30"x60" boiler. $3^{3}/_{4}$" x 6" pumps. 6" cyls. 16x54, 18x60 wheels.

1900: In service at Glenville, Ohio.

2770 3 DCN/HD/A Metropolitan.

1900: In service at Morgan City, La.

2773 2 DCN/HD/A Metropolitan.

1900: In service at Jackson, Mich.

2777 1 DCN/HD/A Metropolitan.

1900: In service, Engine 4, Spokane, Wash.

2780 1 DCN/HD/A Metropolitan.

1900: In service at Honolulu, Hawaii.

2782 X1 DCN/HD/A Metropolitan.

1901: In service at Indianapolis, Ind.

2783 X1 DCN/HD/A Metropolitan.

1901: In service at Indianapolis, Ind.

2785 1 DCN/HD/A Metropolitan.

1900: In service at Birmingham, Ala.

2786 2 DCN/HD/A Metropolitan.

1901: In service at San Francisco, Cal.

2789 1 DCN/HD/A Metropolitan.

1901: In service at El Paso, Texas.

2791 X1 DCN/HD/A Metropolitan.

1901: In service at Portland, Oregon.

1975: Owned by Dan Eagle, Spokane, Wash.

2798 4 DCN/HD/A Metropolitan.

1901: In service, Engine 7, Houston, Texas.

10-20-1954: Still own. Photoed by Dr. Lester Williams. Missing suction hoses, smokestack.

2799 1 DCN/HD/A Metropolitan. Fox boiler.

1901: In service at St. Louis, Mo.

1909: Was at Engine 18.

2804 2 DCN/HD/A Metropolitan. Fox boiler.

1901: In service at Empire Fire Co. #1, San Jose, Cal.

1907: Replaced at Engine 1 by new Ahrens Continental steamer.

2806 2 DCN/HD/A Metropolitan.

1901: To Traverse City, MI

2807 2 DCN/HD/A Metropolitan.

1901: To Matanzas, Cuba

2809 4 DCN/HD/A Metropolitan.

1901: To Babahoyo, Ecuador

2816 2 DCN/HD/A Metropolitan.

1902: In service at Cincinnati, Ohio.

2817 2 DCN/HD/A Metropolitan.

1902: In service at Cincinnati, Ohio.

2818 2 DCN/HD/A Metropolitan.

1902: In service at Cincinnati, Ohio.

2819 1 DCN/HD/A Metropolitan. Fox boiler. Two $5^1/_2$" x 8" pumps. Two 9" x 8" cyls.

1902: In service at Engine 7, Louisville, Ky.

Date?: Rebuilt.

1918: Replaced at Engine 7.

1923: Motorized with International Harvester four-wheel tractor.

Date?: Became a Spare.

2820 1 DCN/HD/A Metropolitan. Fox boiler. Two $5^1/_2$" x $8^3/_4$" pumps. Two 9" x 8" cyls.

1902: In service at Engine 8, Louisville, Ky.

1922: Retired.

2821 X1 DCN/HD/A Metropolitan with three-horse hitch.

1902: In service at Engine 3, 18th & Topeka streets, Wichita, Kansas.

Date?: To Engine 2.

c-1943: Donated to World War II scrap drive; rescued at last minute by Wichita Historical Society.

1981: Restored and now displayed at Cowtown Museum, Wichita, Kansas.

2823 2 DCN/HD/A Metropolitan.

1902: In service at Reno, Nevada.

2824 X1 DCN/HD/A Metropolitan.

1902: In service, Portland, Oregon. 2-wheel ALF tractor #1339 in 1916.

1985: Now Al Foglio, Milwaukie, Oregon.

2828 2 DCN/HD/A Metropolitan.

1902: To Manila, Philipine Islands.

2829 2 DCN/HD/A Metropolitan.

1902: To Manila, Philipine Islands.

2835 1 DCN/HD/A Metropolitan. 5"x10" pumps. 8" x 10" cyl. 3-horse hitch.

1902: In service at Engine 10, Chicago, Ill.

1917: Was at Engine 21.

2838 X1 DCN/HD/A Metropolitan.

1902: In service at Battle Creek, Mich.

2839 2 DCN/HD/A Metropolitan. 2-horse hitch.

1903: In service at Engine 2, Springfield, Ohio. Delivered by John P. Ahrens.

1916: To Engine 8. (To reserve 1926)

1-21-1941: Sold to Norman Friedman, scrap dealer (for $230). He cut it up.

2841 2 DCN/HD/A Metropolitan. 2-horse hitch.

1903: In service at Engine 4, Springfield, Ohio. Delivered by John P. Ahrens.

1913: To reserve at Engine 4.

4- -1928: Sold.

2842 3 DCN/HD/A Metropolitan.

1902: To Quincy, Ill.

2844 1 DCN/HD/A Metropolitan.

1902: To Salt Lake City, Utah.

2854 3 DCN/HD/A Metropolitan.

1902: To Seattle, Wash.

2856 3 DCN/HD/A Metropolitan.

1902: To Seattle, Wash.

2857 1 DCN/HD/A Metropolitan. 5"x10" pumps. 8" x 10" cyls. 3-horse hitch.

1902: In service at Engine 11, Chicago, Ill.

1917: Still at Engine 11.

2858 1 DCN/HD/A Metropolitan. 5"x10" pumps. 8" x 10" cyls. 3-horse hitch.

1902: In service at Engine 5, Chicago, Ill.

1917: Still at Engine 5.

2859 1 DCN/HD/A Upright. 5"x10" pumps, air chamber ornament. 8" cyls. 3-horse hitch. Gong on foot board. Tool box under seat.

1902: Shipped to Chicago, Ill.

1906: At Engine 34. 1917: At Engine 17.

Date?: To Museum of Science and Industry,

The last Ahrens "Upright" engine built was this 1902 1st Class double crane-neck #2859 of Engine 17, Chicago, Ill. It is now displayed at Harold Warp's Pioneer Village museum, Minden, Nebraska, and its appearance has changed little since this photo from the World War I era.

(Dan Martin collection, courtesy Harold S. Walker)

By the time the Museum of Arts and Sciences in Evansville, Ind., was looking for a steam fire engine to display, all of Evansville's steamers had been scrapped. So this 1902 Extra 1st Class Metropolitan #2882 was purchased from Kalamazoo, Mich. It is now parked on the museum's porch, partially exposed to the weather.

(Bill Hass photo)

Chicago, Ill.

Date?: To Harold Warp's Pioneer Village, Minden, Nebraska. Still have 1984.

2861 Metropolitan-style pump for fire-boat.

1902: To Milwaukee, Wis.

2862 Metropolitan-style pump for fire-boat.

1902: To Milwaukee, Wis.

2863 Metropolitan-style pump for fire-boat.

1902: To Milwaukee, Wis.

2866 6 SSF/MD/W Cosmopolitan (first one built)

1902: To Xenia, Ill.

2867 X1 DCN/HD/A Metropolitan.

1902: In service at Engine 2, St. Louis, Mo.

1915: Motorized, two-wheel Christie tractor.

2868 X1 DCN/HD/A Metropolitan.

1902: In service at Engine 7, St. Louis, Mo.

2869 6 SSF/MD/A Cosmopolitan.

1902: To Niles Center, Ohio.

2870 3 DCN/HD/A Metropolitan.

1902: To West Bend, Wis.

2880 1 DCN/HD/A Metropolitan.

1902: In service at Engine 13, St. Louis, Mo.

2881 1 DCN/HD/A Metropolitan.

1902: In service at Engine 24, St. Louis, Mo.

2882 X1 DCN/HD/A Metropolitan. Air chamber ornament. Tool box. Driver's foot brake.

1902: In service at Kalamazoo, Mich.

1977: Now at Museum of Arts & Sciences, Evansville, Ind.

2883 X1 DCN/HD/A Metropolitan.

1902: To Denver, Colo.

2890 4 DCN/HD/A Metropolitan.

1902: To Cavite, Philipine Islands.

2892 X1 DCN/HD/A Metropolitan.

1902: In service at Aberdeen, Wash.

1984: Still at Aberdeen.

2893 X1 DCN/HD/A Metropolitan.

Engine 7, Milwaukee, Wis., used this 1902 Metropolitan 1st Class #2907 from 1902 until 1915. Milwaukee is one of the few cities that kept horse-drawn steamers in service well into the 1920s. For example, long after this photo was taken, this steamer continued to serve, first at Engine 10 (1915-1921), then Engine 16 (1921-1927).
(Wayne Sorenson collection)

1902: To Denver, Colo.

2896 3 DCN/HD/A Metropolitan.

1902: In service, Engine 1, San Diego, Cal.

2902 X1 DCN/HD/A Metropolitan.

1902: Built for Exposition, St. Louis, Mo. Not listed on 1909 St. Louis F.D. roster.

2903 2 DCN/HD/A Metropolitan.

1902: In service at Dallas, Texas.

2904 X1 DCN/HD/A Metropolitan.

1902: In service at Dallas, Texas.

2905 1 DCN/HD/A Metro. Gauge on frame.

1902: In service, Engine 2, Milwaukee, Wis.

1917: To Engine 21.

1918: Replaced by 1908 Nott. Retired.

1920: To Engine 6.

1922: Retired

2906 1 DCN/HD/A Metro. Gauge on frame.

1902: In service, Engine 9, Milwaukee, Wis.

1926: Retired.

2907 1 DCN/HD/A Metropolitan. Gong on foot board. Air chamber ornament. Tool box.

1902: In service, Engine 7, Milwaukee, Wis.

1915: Replaced by Ahrens-Fox motorized pumper #546. To Engine 10.

1921: To Engine 16.

1927: Retired.

2908 2 DCN/HD/A Metropolitan.

1902: To Birmingham, Ala.

2909 4 DCN/HD/A Metropolitan.

1902: To Bloomington, Ill.

2910 3 DCN/HD/A Metropolitan.

1902: To Boise, Idaho.

1983: Now State Historical Society, Boise.

2911 2 DCN/HD/A Metropolitan.

1902: In service at Engine 1, Springfield, Mo. Name *Queen City*.

c-1923: Motorized, 4-wheel Int'l Harvester tractor. Rear steamer wheels converted to

Ad for Kelly-Springfield tires featured 1902 Metropolitan 2nd Class #2911 of Engine 1, Springfield, Mo. Tractor was a circa-1923 IH. The tractor has long since been junked, but the steamer was sold at an auction in Carthage, Mo., in August, 1985, to Bill Schott of Cincinnati. It still has the dual rear tires as in this photo, and its front wheels appear to be the tractor's front wheels. Schott plans to restore this steamer to original horse-drawn appearance.

(Wayne Sorenson collection)

dual solid-rubber (Kelly-Springfield) tires; featured in Kelly-Springfield ads.

Date?: Tractor removed and conventional front wheels put on the steamer (possibly the front wheels of the IH tractor?).

8- -1985: Sold at public auction in Carthage, Mo., to Bill Schott, Cincinnati, Ohio. He plans to restore it.

2912 2 DCN/HD/A Metropolitan.

1902: In service, Engine 2, Springfield, Mo.

2913 1 DCN/HD/A Metropolitan.

1902: In service at Tacoma, Wash.

2914 X1 DCN/HD/A Metropolitan.

1902: In service at Denver, Colo.

2925 3 DCN/HD/A Metropolitan.

1903: In service, Indianapolis, Ind. ($4250). Took Ahrens #164 in trade 6-13-1904.

2928 X1 DCN/HD/A Metropolitan.

1903: To Sacramento, Cal.

2931 4 DCN/HD/A Metropolitan.

1903: To Manila, Philipine Islands.

2932 4 DCN/HD/A Metropolitan.

1903: To Manila, Philipine Islands.

2933 4 DCN/HD/A Metropolitan.

1903: To Manila, Philipine Islands.

2937 5 SSF/MD/W Cosmopolitan.

1903: To Coney Island, N.Y.

1907: To Fort Brady, Mich.

This is possibly the Cosmopolitan steamer that belonged to Harrisville, Mich. in 1949.

2938 5 SSF/MD/W Cosmopolitan.

1903: To San Salvador, El Salvador.

2939 5 SSF/MD/W Cosmopolitan.

1903: To Durban, South Africa.

2942 5 SSF/MD/W Cosmopolitan.

1903: To North Lewisburg, Ohio.

This is possibly the Cosmopolitan steamer that was sold to Urbana, Ohio, by Sutphen Fire Apparatus Co., Columbus, Ohio.

2943 5 SSF/MD/W Cosmopolitan.

1903: To Fort Thomas, Ky.

2944 5 SSF/MD/W Cosmopolitan.

1903: To Millersburg, Ky.

2945 3 DCN/HD/W Metropolitan.

1903: To Wheeling, W. Va.

2946 1 DCN/HD/W Metropolitan.

1903: To Evansville, Ind.

2948 2 DCN/HD/W Metropolitan.

1903: To Portsmouth, Ohio.

The International Fire Engine Co. was formed on Aug. 4, 1900, by merger of LaFrance Fire Engine Co. (Elmira, N.Y.), Thomas Mannning Jr. & Co. (Cleveland, Ohio), Fire Extinguisher Mfg. Co. (Chicago, Ill., makers of Babcock, Champion, and Hale apparatus), Charles T. Holloway & Co. (Baltimore, Md.), Gleason & Bailey Mfg. Co. Ltd. (Seneca Falls, N.Y.), and Macomber Chemical Fire Extinguisher Co. (Worcester, Mass.). International Fire Engine Company was owned by five New York City financiers: John H. Flagler (chairman), Fred W. Shibley, August Hecksher, Alfred C. Bedford, and George H. Robinson. These five men also owned a controlling interest in American Fire Engine Co., but American and International remained two separate companies until 1904.

Due to financial mismanagement, International Fire Engine Co. went bankrupt, and a reorganization plan was drafted on Oct. 1, 1903. This plan called for merger of American and International Fire Engine companies, the resulting company was American-LaFrance Fire Engine Co.

On Jan. 6, 1904, International was placed in receivership; one of the receivers was James R. Clarke, who had been president of American Fire Engine Co. since Sept. 10, 1903.

On May 2, 1904, American Fire Engine Co. and International Fire Engine Co. were both legally dissolved, and a new company called American-LaFrance Fire Engine Co. was formed by merger of American and International. The Gleason & Bailey operation was moved into the Elmira plant, but all other factories continued operation as before.

Metropolitan steamers continued to sport the American Fire Engine Co. nameplate, rather than American-LaFrance, until Dec. 28, 1904. On that date, all factories were ordered to immediately start using the American-LaFrance nameplate on all apparatus. One can safely assume that due to the holidays, the order was not carried out until at least Jan. 2, 1905, so that all 1904 steamers would still have had the American nameplate.

It is an interesting coincidence that this was almost exactly the same time that American-LaFrance lost its Cincinnati plant. The various members of the Ahrens family, and all of their employees in Cincinnati, resigned from American-LaFrance Dec. 9, 1904, to form Cincinnati Engine

1910 factory photo shows 1904 5th Class Cosmopolitan #2959 of Claremore, Oklahoma. Cosmopolitans were rare, but this one was rarer: note the nameplate "American-LaFrance Fire Engine Co., Cincinnati, Ohio." American-LaFrance production in Cincinnati virtually ceased on Dec. 9, 1904, and the American-LaFrance nameplate first appeared on steamers on Dec. 28, 1904. Thus this was one of a very few, if not the only, steamers to bear the American-LaFrance brand name and Cincinnati manufacturing location. This steamer, now in the American-LaFrance archives, is being restored. (Courtesy John M. Peckham)

& Pump Works (later Ahrens Fire Engine Co.).

This left American-LaFrance's Cincinnati factory unstaffed, so it is unlikely that any American-LaFrance fire engines were built in Cincinnati after the end of 1904 (although ALF continued to own the old Ahrens factory until 1908). A few months before the Ahrens resignation, the Silsby family had also resigned, so that American-LaFrance may not have had much of an operation left in Seneca Falls, either. By 1905, production seems to have been almost exclusively in the former LaFrance Fire Engine Co. plant at Elmira, N.Y.

2955 2 DCN/HD/W Metropolitan.

1904: In service at Jacksonville, Ill.

1986: Now at New Salem, Ill.

2959 5 SSF/MD/W Cosmopolitan.

1904: In service at Claremore, Oklahoma. Rebuilt and repainted in 1910.

1980: Now in American-LaFrance archives (Figge International) at Elmira, N.Y.

2-20-1986: Now being restored for Figge International by Andy Swift, Jonesport, Maine (he restored 1907 Continental #131).

2960 X1 DCN/HD/W Metropolitan. Fox boiler. Rated 1000 GPM.

1904: In service at Engine 40, St. Louis, Mo.

1915: Motorized, two-wheel Christie tractor.

2961 2 DCN/HD/W Metro (700 gpm). Fox boiler.

1904: In service at Engine 5, St. Louis, Mo.

2962 2 DCN/HD/W Metro (700 gpm). Fox boiler.

1904: In service at Engine 25, St. Louis, Mo.

2963 5 DCN/HD/W Metropolitan.

1904: To Carlisle, Ky.

2964 2 DCN/HD/W "Chicago Special" Metropolitan. $4^3/4$" x 8" pump. 8" cyl. 2-horse hitch.

1904: Shipped Chicago, Ill. ($4700) 850 gpm.

1906: At Engine 32. 1917: At Engine 34.

2965 2 DCN/HD/W "Chicago Special" Metropolitan. $4^3/4$" x 8" pump. 8" cyl. 2-horse hitch.

1904: Shipped Chicago, Ill. ($4700) 850 gpm.

1906: At Engine 40. 1917: At Engine 29.

This gleaming Metropolitan, with an abundance of nickel-plating and gold-leaf decoration, was featured in a 1904 encyclopedia. What city used it, and its Registered Number, are not known. Note the fresh-water tank under the seat, the toolbox above the crane-neck of the frame, and the oil can on the frame just forward of the boiler.
(Bill Hass collection)

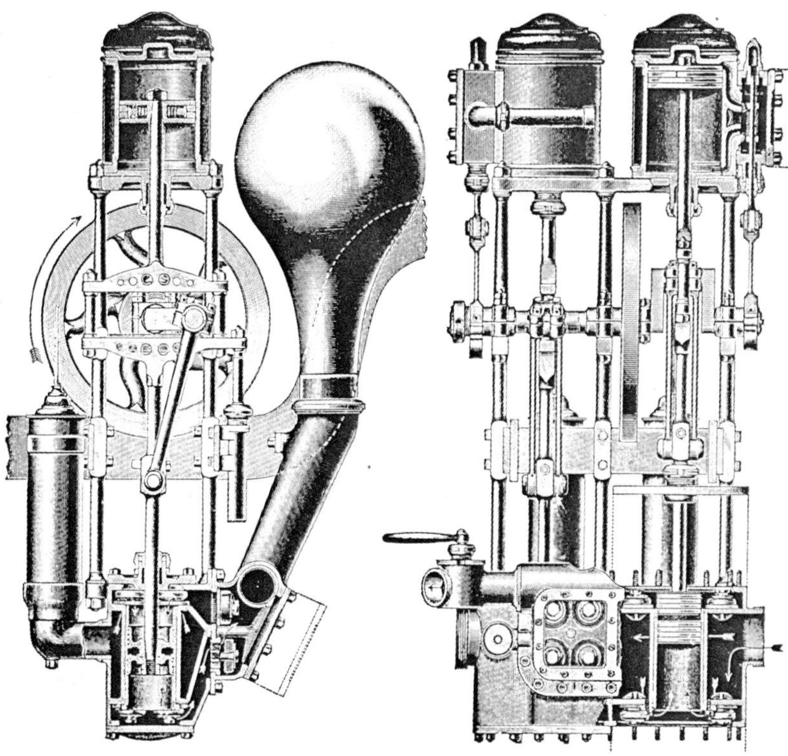

The "Metropolitan" pump was used by American Fire Engine Co. on "American", "Columbian", and "Metropolitan" steamers from 1893 to 1904, and by American-LaFrance on "Metropolitan" steamers from 1904 to 1917. Side view is at left, while front view (sans air chamber) is at right. The steam cylinders and steam chests are at top, the crankshaft and flywheel in the middle, and the actual pump (pistons, valves, suction inlets, and discharge gates) at the bottom.
(Ed Hass collection)

2966 2 DCN/HD/W "Chicago Special" Metropolitan. $4^3/_4$" x 8" pump. 8" cyl. 3-horse hitch.

1904: Shipped Chicago, Ill. ($4700) 850 gpm.

1906: At Engine 16. 1917: At Engine 27.

2968 1 DCN/HD/W Metropolitan.

1904: To Walla Walla, Wash.

2969 2 DCN/HD/W Metropolitan.

1904: In service at Reno, Nevada.

1979: Now at Hall of Flame (fire museum), Phoenix, Arizona.

2978 2 DCN/HD/W Metropolitan.

1904: In service at St. Paul, Minn.

2979 2 DCN/HD/W Metropolitan.

1904: In service at Galveston, Texas.

4-19-1975: Was in parade at Friendswood, Texas. Needs restoration. Owner unknown.

2985 3 DCN/HD/W Metropolitan.

1904: In service at Engine 5, Fresno, Cal.

11-16-1914: Motorized, Christie tractor.

2986 2 DCN/HD/W Metropolitan.

1904: To Vancouver, Wash.

2989 1 DCN/HD/W "Milwaukee Special" Metro.

1904: In service, Engine 4, Milwaukee, Wis.

1921: Retired.

2990 1 DCN/HD/W "Milwaukee Special" Metro.

1904: In service, Eng. 10, Milwaukee, Wis.

1915: Replaced by Metro #2907. To Eng. 12.

1915: To Engine 7.

1923: Replaced by Ahrens #2123.

2991 1 DCN/HD/W "Milwaukee Special" Metro.

1904: In service, Engine 3, Milwaukee, Wis.

1919: Replaced by motorized Ahrens-Fox pumper #546. To Engine 5.

1922: Retired.

1970: Restored. Displayed in Pioneer Hook & Ladder Society museum, San Diego, Cal.

2992 X1 DCN/HD/W Metropolitan.

1904: In service at Engine 2, Louisville, Ky.

Topeka, Kansas, still owns this 1904 Metropolitan 2nd Class double crane-neck #2998, seen on parade at Kansas City, Mo., in 1971. This engine was one of the few built by American-LaFrance in the old Ahrens plant on Webster St., Cincinnati. It is authentic except for non-original boiler.

(Courtesy Topeka Fire Dept.)

1918: Motorized with One-Wheel Auto-Horse tractor. To Engine 7.

1923: Retired.

2993 1 DCN/HD/W Metropolitan.

1904: In service at Engine 4, Louisville, Ky.

1-13-1908: New pump valves installed by Ahrens Fire Engine Co.

1923: Retired.

2994 2 DCN/HD/W Metropolitan.

1904: In service at Engine 41, St. Louis, Mo.

2995 4 DCN/HD/W Metropolitan.

1904: To Beaumont, Texas.

2996 4 DCN/HD/W Metropolitan.

1904: To Beaumont, Texas.

2997 2 DCN/HD/W Metropolitan.

1904: To Piqua, Ohio. Name *L.C. Cron.*

1977: This steamer is in storage at the police impound garage, Piqua, Ohio.

2998 2 DCN/HD/W Metropolitan.

1904: In service, Engine 2, Topeka, Kansas. Price $4700.

1985: Displayed at Topeka firehouse. Non-original boiler (home-made?).

2999 X1 DCN/HD/W Metropolitan.

1904: To Omaha, Nebraska.

3000 2 DCN/HD/W Metropolitan.

1904: To Marysville, Cal.

3001 2 DCN/HD/W Metropolitan.

1904: To Youngstown, Ohio ($4750)

3002 2 DCN/HD/W Metropolitan.

1904: In service, E. 3, Grand Rapids, Mich.

3003 2 DCN/HD/W Metropolitan.

1904: In service at Riverside, Cal.

3004 3 DCN/HD/W Metropolitan.

1904: To East St. Louis, Ill.

3005 3 DCN/HD/W Metropolitan.

1904: To Evansville, Ind.

3006 X1 DCN/HD/W Metropolitan.

1904: In service, Eng. 2, Eureka, Cal. $6375.

1983: Still own. Restored to operating condition and excellent appearance.

3007 1 DCN/HD/W Metropolitan.

1904: In service at Montgomery, Alabama.

1985: Now displayed in the lobby of a Montgomery municipal building.

3008 4 DCN/HD/W Metropolitan.

1904: In service at Dallas, Texas.

3009 1 DCN/HD/W Metropolitan.

1904: To Portland, Oregon.

3010 1 DCN/HD/W Metropolitan.

1904: In service at Cincinnati, Ohio.

3011 1 DCN/HD/W Metropolitan.

1904: In service at Cincinnati, Ohio.

Note: Cincinnati also purchased a 3rd Class "Cincinnati Special" Metropolitan #3114 in 1905. This was apparently built in Elmira, N.Y., which made it the only steamer ever used by the Cincinnati Fire Dept. that was not built in Cincinnati.

Ahrens Fire Engine Co.

Alfred & Cook Streets, Cincinnati
Feb. 1905 to Aug. 1910

No fewer than 174 blueprints of various features on Continental steamers still survive at Ahrens-Fox Company. Many are labeled with the serial number, city, and date. This information has been included in the following Continental listing, as it provides dates during which each steamer was under construction (it took about 6 months to build a Continental steamer). Obviously, delivery dates were some months after the blueprint dates. In some cases, a blueprint date is several years after the steamer was built, indicating that the blueprint was for a repair part.

101 3 DCN/HD/W Continental. 32"x64" inclined boiler. $4^3/_8$" x 8" pumps, 600 GPM, 4" inlet. $7^1/_4$" x 8" cyls. Length 12'11" (23'4" w/pole). Height 9'3". $56^1/_2$" track. Weight 6800#.

Blueprint dates: 7-3-1905 boiler stay-bolt pattern; 9-23-1905 suction inlet; 11-15-1905 boiler; 8-17-1910 copper exhaust pipe.

1905: Delivered by John P. Ahrens to East St. Louis, Ill. Steam in 7:00. Threw two $1^{1}/_{2}$" streams 200' each.

102 3 DCN/HD/W Continental. 32"x64" inclined boiler. $4^{3}/_{8}$" x 8" pumps, 600 GPM. $7^{1}/_{4}$" x 8" cyls. Length 12'11" (23'4" with pole). Height 9'3". $56^{1}/_{2}$" track. Weight 6800#.

Blueprints: 7-3-1905 boiler stay-bolt pattern.

11-1905: In service, Eng. 2, Indianapolis, Ind.

Date?: To Engine 7.

103 2 DCN/HD/W Continental. 36" radial boiler. $4^{3}/_{4}$" x 8" pumps, 700 GPM, $4^{1}/_{2}$" suction inlet. 8" x 8" cyls. Length 13'6" (23'11" with pole). Height 9'3". 59" track. Weight 7800#.

Blueprint dates: 10-11-1905 $4^{1}/_{2}$" suction inlet; 11-18-1905 boiler; 5-21-1907 rear spring bracket; 8-17-1910 copper exhaust pipe for boiler.

1905: In service at Cincinnati, Ohio ($8000).

104 1 DCN/HD/W Continental. 38" radial boiler. $5^{1}/_{2}$" x 8" pumps, 900 GPM. 9" x 8" cylinders. Length 14'1" (24'6" with pole). Height 9'6". 62" track. Weight 8800#.

Blueprint dates: 10-9-1905 boiler stay-bolt pattern; 8-17-1910 copper exhaust pipe.

1905: To Charleston, West Virginia.

105 1 DCN/HD/W Continental. 38" x 64" radial inclined boiler. $5^{1}/_{2}$" x 8" pumps, 900 GPM. Two 9" x 8" cylinders. Length 14'1" (24'6" with pole). Height 9'6". 60" track.

Blueprints: 10-12-1905, $2^{7}/_{8}$" axles; 7-24-1907 blanket holder on back of seat.

1905: In service at Engine 9, Louisville, Ky.

Date?: To Engine 14.

1923: Motorized with One-Wheel Auto Horse tractor.

106 3 DCN/HD/W Continental. 32" x 64" radial inclined boiler. $4^{3}/_{8}$" x 8" pumps, 600 GPM. $7^{1}/_{4}$" x 8" cyls. Length 12'11" (23'4" w/pole). Height 9'3". $56^{1}/_{2}$" track. Weight 6800#.

10-16-1905: Ordered by Galion, Ohio, to replace 1873 Ahrens #25.

Blueprint dates: 11-13-1905 boiler stay-bolt pattern.

1906: Delivered to Galion, Ohio. Name *H.C. Sponhauer*. Pumped 700 GPM.

1965: Still at Galion, per Harold Walker.

107 1 DCN/HD/W Continental. Radial boiler. Two $5^{1}/_{2}$" x 8" pumps, rated 900 GPM. Two 9" x 8" cylinders. Length 14'1" (24'6" with pole). Height 9'6". 62" track. Weight 8800#.

1906: In service at Engine 1, Toledo, Ohio. Pumped 1090 GPM. Had 210# nozzle pressure on one line with a $2^{1}/_{4}$" nozzle, 180# nozzle pressure with 4 lines siamesed into a $1^{3}/_{4}$" nozzle, 165# nozzle pressure with 4 lines siamesed into a 2" nozzle. Threw one stream to height of 230'.

1912: Motorized, two-wheel Christie tractor.

108 3 DCN/HD/W Continental. Radial boiler. $4^{3}/_{8}$" x 8" pumps, 600 GPM. Two $7^{1}/_{4}$" x 8" cylinders. Length 12'11" (23'4" with pole). Height 9'3". $56^{1}/_{2}$" track. Weight 6800#.

1906: Delivered by John P. Ahrens to Terre Haute, Ind. Name *G.F.E. Kanfleiter*. Through 100' of 4" hose, threw a 1" stream 208' high. Through 1500' of hose, threw one stream 100' vertically, 185' horizontally.

109 3 DCN/HD/W Continental. Radial boiler. Two $4^{3}/_{8}$" x 8" pumps, 600 GPM. Two $7^{1}/_{4}$" x 8" cylinders. Length 12'11" (23'4" with pole). Height 9'3". $56^{1}/_{2}$" track. Weight 6800#.

1906: Delivered by John P. Ahrens to Terre Haute, Ind. Name *Edwin J. Bidaman*. Threw a 1" stream 208' high, through 100' of 4" hose. Through 1500' of hose, threw one stream 100' vertically, 185' horizontally.

110 X1 DCN/HD/W Continental. 40" radial boiler, safety-rails on coal bunker. $5^{3}/_{4}$" x 8" pumps, 1100 GPM, ornament on air chamber. One 6" x 20' pre-connect suction. $9^{1}/_{2}$" x 8" cyls. Length 14'7" (25' with pole). Height 9'11". 66" track. Gong and foot-brake on foot board. Tool box under seat. Weight 9800#.

1906: In service at Superior, Wis. Name *John Schafer Jr. #1*. Pumped 1275 gpm on test.

Blueprints: 8-19-1910 copper exhaust pipe.

1981: Local private owner returned this steamer to Superior F.D. for restoration.

111 2 DCN/HD/W Continental. Radial boiler. Two $4^{3}/_{4}$" x 8" pumps, rated 700 GPM. Two 8" x 8" cylinders. Length 13'6" (23'11" with pole). Height 9'3". 59" track. Weight 7800#.

1906: In service at Engine 16, Louisville, Ky.

1922: Briefly at Engine 8, then retired.

112 2 DCN/HD/W Continental. 34" radial boiler. $4^{3}/_{4}$" x 8" pumps, 700 GPM, bell atop pump, ornament atop air chamber. Two $4^{1}/_{2}$" x 10' suctions. Two 8" x 8" cylinders. Length 13'6" (23'11" with pole). Height 9'3". 59" track. Foot-brake on foot board. Water tank under seat. Hose reel behind seat. Weight 7800#.

The "John Schafer Jr. #1" is an Extra 1st size Continental #110, shown shortly after its 1906 delivery to Superior, Wis. Note lavish use of gold-leaf, and the long squirrel-tail suction hose.
(Courtesy Superior Fire Dept.)

Superior's 1906 Continental #110 as it looked when returned to Superior Fire Dept. in 1981. Gold-leaf decoration is gone, and two short suction hoses have replaced the single long suction.
(Courtesy Superior Fire Dept.)

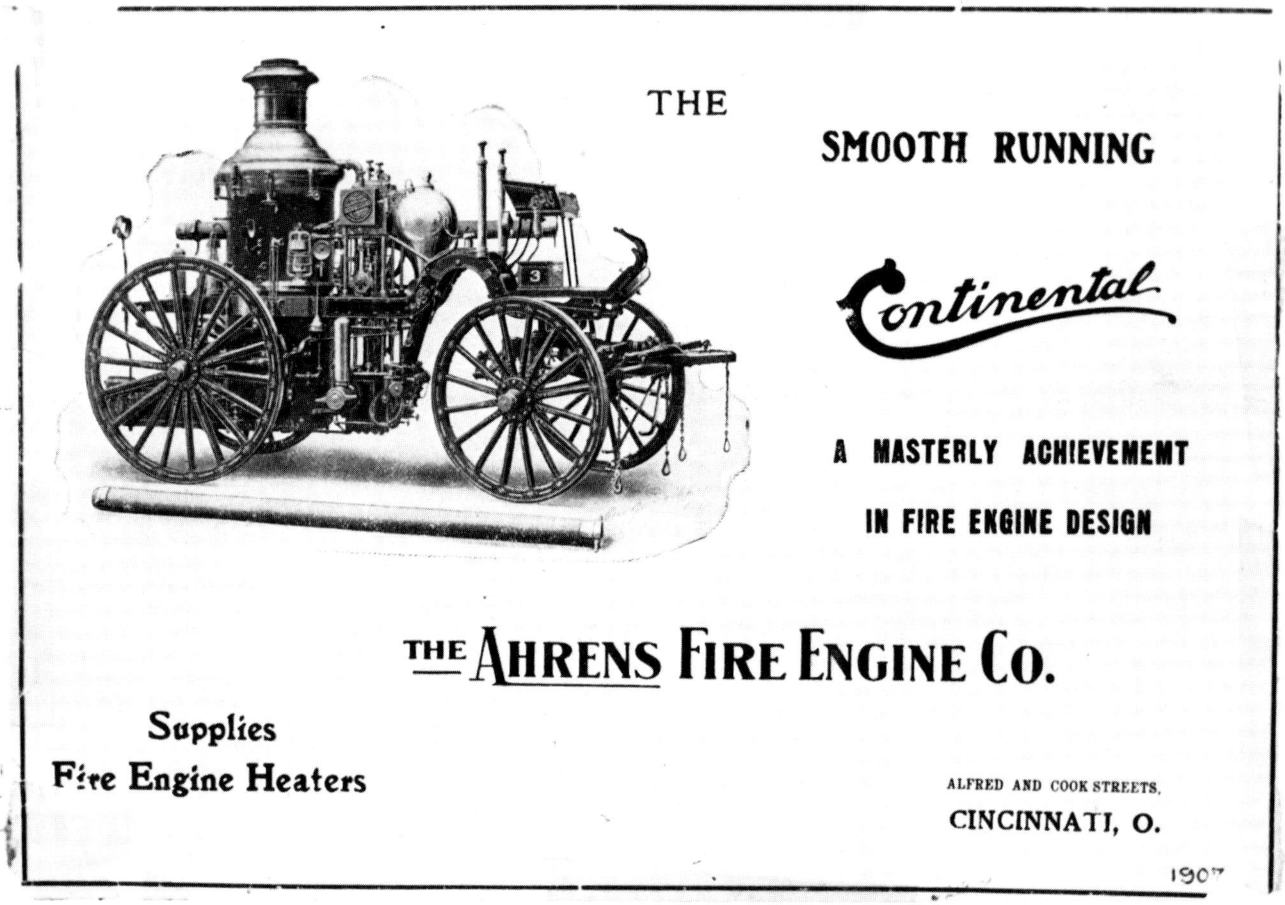

This 1907 Ahrens Fire Engine Co. ad featured 1906 2nd size Continental #115 of Engine 3, Dallas, Texas. This engine set a world's record for discharge from a 700-GPM pumper, delivering a steady 930 GPM for 6 hours at the fire chiefs convention in Dallas in Oct., 1906; another Continental broke the record two months later. (Ed Hass collection)

1906: In service at Engine 4, Seattle, Wash. Ahrens sales agent was H.W. Mix of Mix Fire Apparatus Co., Seattle.

Blueprints: 8-17-1910 copper exhaust pipe.

113 3 DCN/HD/W Continental. Radial boiler. $4^{3}/_{8}$" x 8" pumps, 600 GPM. Two $7^{1}/_{4}$" x 8" cylinders. Length 12'11" (23'4" with pole). Height 9'3". Track $56^{1}/_{2}$". Weight 6800#.

7-23-1906: Delivered to North Montgomery Fire Co. #5, Montgomery, Alabama ($4250).

1916: Was at Company 6.

114 1 DCN/HD/W Continental. Radial boiler. $5^{1}/_{2}$" x 8" pumps, 900 GPM. Two 9" x 8" cylinders. Length 14'1" (24'6" with pole). Height 9'6". 62" track. Weight 8800#.

1906: In service, Engine 1, Denver, Colorado.

1918: Motorized with two-wheel American-LaFrance tractor.

115 2 DCN/HD/W Continental. Radial boiler. Two $4^{3}/_{4}$" x 8" pumps, rated 700 GPM. Two 8" x 8" cylinders. Length 13'6" (23'11" with pole). Height 9'3". 59" track. Weight 7800#.

10- -1906: Delivered by John P. Ahrens to Dallas, Texas. On test at International Assn. of Fire Engineers conference in Dallas, pumping through two 100' lines with $2^{1}/_{2}$" nozzles, pumped 930 GPM. This set a new world's record for a 2nd Class steamer, only to be beaten two months later by Continental #118 at San Diego, Cal.

11- -1906: In service, Eng. 3, Dallas, Texas.

116 2 DCN/HD/W Continental. Radial boiler. Two $4^{3}/_{4}$" x 8" pumps, rated 700 GPM. Two 8" x 8" cylinders. Length 13'6" (23'11" with pole). Height 9'3". 59" track. Weight 7800#.

1906: In service at Engine 18, Buffalo, N.Y.

Blueprint dates: 1-15-1907, $2^{1}/_{2}$" and 4" Gaylord snatch couplings.

117 1 DCN/HD/W Continental. Radial boiler. Two $5^{1}/_{2}$" x 8" pumps, rated 900 GPM. Two 9" x 8" cylinders. Length 14'1" (24'6" with pole). Height 9'6". 62" track. Weight 8800#.

1906: In service, E. 10, Grand Rapids, Mich.

118 2 DCN/HD/W Continental. Radial boiler. Two $4^{3}/_{4}$" x 8" pumps, rated 700 GPM. Two 8" x 8" cylinders. Length 13'6" (23'11" with pole). Height 9'3". 59" track. Weight 7800#.

12-8-1906: Delivered by John P. Ahrens to San Diego, Cal. Name *George R. Harrison.*

1906 2nd Class Continental #118 is seen setting a new world's record for a 700-gallon-per-minute pumper. The "George R. Harrison" was throwing 975 GPM over city hall in San Diego, Cal., when this photo was taken on Dec. 8, 1906. The previous record of 930 GPM was set by Continental #115 two months earlier. Another photo of same test bears date 12-11-06. (Cincinnati Historical Society)

The first of three Ahrens-Foxes to serve Everett, Wash., was this 1907 3rd size Continental #123, now in the Everett Firefighters Museum. The second (a 1927 piston pumper) was junked, but the fire museum displays an identical 1926 Ahrens-Fox ex-Wilmerding, Pa. Everett's third Ahrens-Fox, a 1937 quad, is also in the museum. (Everett Fire Dept. photo)

Pumped 975 GPM, new world's record for a 2nd Class steamer (beating the record of 930 GPM set two months earlier by Continental #115 at Dallas, Texas). Water in 5:30. Through two hose lines siamesed into a $1^1/_8$" nozzle, threw one stream 200' vertically. Threw two $1^1/_8$" vertical streams 160' each.

119 X1 DCN/HD/W Continental. Radial boiler. $5^3/_4$" x 8" pumps, 1100 GPM, bell atop pump. Two 6" x 10' suctions. $9^1/_2$" x 8" cyls. Length 14'7" (25' with pole). Height 9'11". 66" track. Tool box under seat. Weight 9800#.

1907: In service, Engine 2, Kansas City, Mo.

120 2 DCN/HD/W Continental. Radial boiler. $4^3/_4$" x 8" pumps, 700 GPM. 8" x 8" cylinders. Length 13'6" (23'11" with pole). Height 9'3". 59" track. Canvas mud shield. Weight 7800#.

Blueprints: 12-21-1906 mud shield (apron).

1907: In service at Engine 59, Chicago, Ill.

121 2 DCN/HD/W Continental. Radial boiler. $4^3/_4$" x 8" pumps, 700 GPM. 8" x 8" cylinders. Length 13'6" (23'11" with pole). Height 9'3". 59" track. Canvas mud shield. Weight 7800#.

Blueprints: 12-21-1906 mud shield (apron).

1907: In service at Engine 64, Chicago, Ill.

122 Unknown.

123 3 DCN/HD/W Continental. Radial boiler. $4^3/_8$" x 8" pumps, 600 GPM, ornament atop air chamber. Two 4" x 10' suctions. $7^1/_4$" x 8" cylinders. Length 12'11" (23'4" with pole). Height 9'3". $56^1/_2$" track. Water tank under seat, hose reel behind seat. Weight 6800#.

1907: In service at Everett, Wash.

1981: Restored by the Everett Firefighters Assn., on display in their fire museum.

124 2 DCN/HD/W Continental. Radial boiler. $4^3/_4$" x 8" pumps, $4^1/_2$" suction inlet, 700 GPM. 8" x 8" cyls. Length 13'6" (23'11" with pole). Height 9'3". 59" track. Weight 7800#.

Blueprint dates: 3-28-1907 $4^1/_2$" Morse gate valve; 3-29-1907 $4^1/_2$" suction inlet.

1907: In service at Engine 15, Louisville, Ky.

125 3 DCN/HD/W Continental. Radial boiler. Two $4^3/_8$" x 8" pumps, 600 GPM. Two $7^1/_4$" x 8" cylinders. Length 12'11" (23'4" with pole). Height 9'3". $56^1/_2$" track. Weight 6800#.

1907: In service at Canal Dover, Ohio. This town later shortened its name to Dover.

1976: Tim Elder reports that this steamer is now displayed at Ohio Historical Society, Columbus. Its playpipe is on runningboard of Tim's 1925 Ahrens-Fox, ex-Dover, Ohio.

This 1907 Ahrens Continental 2nd Size #131 of Valdez, Alaska, was in service until 1935, then was kept in storage for 40 years. It now belongs to Valdez Heritage Center, and was restored by Valdez firefighter Andy Swift to the pristine condition seen here. Andy now lives in Jonesport, Maine, where he runs a steamer restoration shop. (Courtesy Valdez Heritage Center)

126 DX1 Continental fireboat pump. Two 6" x 8" pumps, rated 1300 GPM.

1907: Delivered to Goodrich Transportation Co., Chicago, Ill. (for Steamship *Alabama*).

127 Unknown.

128 4 DCN/HD/W Continental. 30" x 62" radial inclined boiler. 4" x 8" pumps, 500 GPM. $6^{1}/_{2}$" x 8" cyls. Length 12'6" (22'11" with pole). Height 8'10". 60" track. Weight 5800#.

Blueprints: 12-27-1906 $2^{7}/_{16}$" front & rear axles; 1-21-1907 brake bracket; 2-4-1907 boiler stay-bolt pattern; 2-4-1907 cylinder and pump brace; 2-5-1907 single suction holders for seat & crane neck; 2-6-1907 single suction holders for boiler and rear; 2-8-1907 double suction holders for seat and crane neck; 2-11-1907 double suction holders for boiler and rear; 3-14-1907 double suction holders for boiler and rear; 3-15-1907 double suction holders for seat and crane neck; 5-25-1907 boiler top connection; 5-27-1907 boiler bottom connection; 6-28-1907 boiler; 8-23-1910 copper exhaust pipe for boiler.

1907: In service, Eng. 21, Los Angeles, Cal.

1916: To reserve.

1918: To Engine 8.

1920: To reserve.

1923: Sold.

129 2 DCN/HD/W Continental. 34" x 64" nest tube boiler. $4^{3}/_{4}$" x 8" pumps, 700 GPM. Two 8" x 8" cylinders. Length 13'6" (23'11" with pole). Height 9'3". 59" track. Weight 7800#.

1907: In service at Engine 3, Toledo, Ohio.

1912: Motorized, Christie two-wheel tractor.

130 2 DCN/HD/W Continental. 34" x 64" nest tube boiler. $4^{3}/_{4}$" x 8" pumps, 700 GPM. Two 8" x 8" cylinders. Length 13'6" (23'11" with pole). Height 9'3". 59" track. Weight 7800#.

1907: In service at Engine 7, Toledo, Ohio.

1912: Motorized, Christie two-wheel tractor.

131 2 DCN/HD/W Continental. Radial boiler. $4^{3}/_{4}$" x 8" pumps, 700 GPM, ornament atop air chamber. $4^{1}/_{2}$" x 10' suctions. 8" x 8" cyls. Length 13'6" (23'11" w/pole). Height 9'3". 59" track. Water tank under seat. Weight 7800#.

10-1-1907: Sold and delivered by H.W. Mix (Mix Fire Apparatus Co., Seattle, Wash.) to Valdez, Alaska. In service as Engine 2.

1976: To Valdez Heritage Center.

1984: Restored by firefighter and steam enthusiast Andy Swift of Valdez (now of Jonesport, Maine). He says some of its parts

Chicago, Ill., still owns their 1907 Continental 2nd size #135. Originally Engine 39, it later served Engine 56, and is now stored at Chicago Fire Dept. Repair Shop. It is one of only three ex-Chicago Ahrens steamers known to survive (the other two are 1892 3rd size #2142 and 1902 1st size #2859).

(John Doyle photo)

Engine 120, Chicago, Ill., ran this 1907 Continental 2nd size #136 with 1914 Christie tractor. Note the gong on the frame between the tractor and steamer, and the horn next to the driver's seat. (Wayne Sorenson collection)

have the Reg. No. *124* stamped in them.

132 Unknown.

133 2 DCN/HD/W Continental. Radial boiler. $4^{3}/_{4}$" x 8" pumps, 700 GPM, ornament atop air chamber. Two $4^{1}/_{2}$" x 10' suctions. 8" x 8" cyls. Length 13'6" (23'11" w/pole). Height 9'3". 59" track. Gong and foot-brake on foot board. Tool box under seat. Weight 7800#.

Blueprint dates: 5-22-1907 rear springs.

1907: In service at Engine 38, Chicago, Ill.

1915: Motorized, Christie two-wheel tractor.

134 2 DCN/HD/W Continental. Radial boiler. $4^{3}/_{4}$" x 8" pumps, 700 GPM, ornament atop air chamber. Two $4^{1}/_{2}$" x 10' suctions. 8" x 8" cyls. Length 13'6" (23'11" w/pole). Height 9'3". 59" track. Gong and foot-brake on foot board. Tool box under seat. Weight 7800#.

1907: In service at Engine 10, Chicago, Ill.

Date?: To Engine 90.

135 2 DCN/HD/W Continental. Radial boiler. $4^{3}/_{4}$" x 8" pumps, 700 GPM, ornament atop air chamber. Two $4^{1}/_{2}$" x 10' suctions. 8" x 8" cyls. Length 13'6" (23'11" w/pole). Height 9'3". 59" track. Gong and foot-brake on foot board. Tool box under seat. Weight 7800#.

1907: In service at Engine 39, Chicago, Ill.

Date?: To Engine 56.

1977: Restored, kept at Chicago Fire Dept. Repair Shop (still horse-drawn).

136 2 DCN/HD/W Continental. Radial boiler. $4^{3}/_{4}$" x 8" pumps, 700 GPM, ornament atop air chamber. Two $4^{1}/_{2}$" x 10' suctions. 8" x 8" cyls. Length 13'6" (23'11" w/pole). Height 9'3". 59" track. Gong and foot-brake on foot board. Tool box under seat. Weight 7800#.

1907: In service at Engine 120, Chicago, Ill.

1914: Motorized with Christie two-wheel tractor. Gong moved to left side of frame.

137 3 DCN/HD/W Continental. Radial boiler. $4^{3}/_{8}$" x 8" pumps, 600 GPM. Two $7^{1}/_{4}$" x 8" cylinders. Length 12'11" (23'4" with pole). Height 9'3". Track $56^{1}/_{2}$". Weight 6800#.

1907: Delivered to Everett, Wash. Not accepted.

1907: Sold by H.W. Mix (Mix Fire Apparatus Co., Seattle) to Engine 3, Seattle, Wash.

138 Unknown.

Firemen and officials of Van Wert, Ohio pose proudly with their new 3rd size Continental #139 in October, 1907. Restored in 1984, the steamer is now on a 99-year loan to Central Mutual Insurance Co.'s fire museum in Van Wert. A 1926 Ahrens-Fox piston pumper (ex-Northville, Mich.) is also in the museum.
(Deal Studio photo, courtesy Bill Purmort)

Van Wert, Ohio's 1907 Ahrens 3rd size Continental #139 has been restored to this painstakingly-accurate condition by expert fire apparatus restorer Ken Soderbeck of Jackson, Mich. Note that paint and plating, and all of the gold-leaf patterns, exactly duplicate those in the original 1907 photo. It is also fully-operable.
(Rick Knife photo)

139 3 DCN/HD/W Continental. Radial boiler. $4^{3}/_{8}$" x 8" pumps, 600 GPM, air chamber ornament. Two 4" x 10' suctions. $7^{1}/_{4}$" x 8" cyls. Length 12'11" (23'4" w/pole). Height 9'3". $56^{1}/_{2}$" track. Tool box under seat. Weight 6800#.

10-1-1907: In service Eng. 1, Van Wert, Ohio.

1-9-1981: Letter from F.W. Purmort Jr., of Van Wert, Van Wert has leased this steamer for 99 years to Central Mutual Insurance Co.'s fire museum in Van Wert. Purmort is Chairman of the Board of Central Mutual.

7- -1984: Restoration of this steamer was nearly complete when displayed at antique fire apparatus muster at Greenfield Village, Dearborn, Mich. Restored for Central Mutual by Ken Soderbeck, Jackson, Mich.

140 Unknown.

141 1 DCN/HD/W Continental. Radial boiler. Two $5^{1}/_{2}$" x 8" pumps, rated 900 GPM. Two 9" x 8" cylinders. Length 14'1" (24'6" with pole). Height 9'6". 62" track. Weight 8800#.

1908: In service at Omaha, Nebraska.

142 1 DCN/HD/W Continental. Radial boiler. Two $5^{1}/_{2}$" x 8" pumps, rated 900 GPM. Two 9" x 8" cylinders. Length 14'1" (24'6" with pole). Height 9'6". 62" track. Weight 8800#.

1908: In service, Engine 1, Saint Paul, Minn.

1914: Motorized with two-wheel American-LaFrance Type 31 tractor. First Continental steamer motorized by American-LaFrance. As of 1985, ALF still has the blueprint for its conversion in their archives.

c-1938: Retired and sold.

143 1 DCN/HD/W Continental. Radial boiler. Two $5^{1}/_{2}$" x 8" pumps, rated 900 GPM. Two 9" x 8" cylinders. Length 14'1" (24'6" with pole). Height 9'6". 62" track. Weight 8800#.

1908: In service at Engine 16, St. Paul, Minn.

144 1 DCN/HD/W Continental. Radial boiler. Two $5^{1}/_{2}$" x 8" pumps, rated 900 GPM. Two 9" x 8" cylinders. Length 14'1" (24'6" with pole). Height 9'6". 62" track. Weight 8800#.

Blueprint dates: 3-10-1908, $2^{1}/_{2}$" Prunty relief valve.

1908: In service at Houston, Texas.

145 Unknown.

146 Unknown.

147 Unknown.

148 1 DCN/HD/W Continental. Radial boiler. Two $5^{1}/_{2}$" x 8" pumps, rated 900 GPM. Two

1908 Continental 2nd size #155 served Engine 13, then Engine 29 in Detroit, Mich. A museum in Pennsylvania acquired it in 1959, and returned it to Detroit Fire Dept. in 1970. Box 42 Associates, a local fire buffs club, restored it. The boiler was installed by Detroit Fire Dept. in the 1920s; while the home-made boiler's exaggerated height and slanted smokestack mar an otherwise-excellent restoration, an authentic boiler has not been found.

(Ed Hass photo)

9" x 8" cylinders. Length 14'1" (24'6" with pole). Height 9'6". 62" track. Weight 8800#.

Blueprint: 5-5-1908 $2^1/_2$" Prunty relief valve.

1908: In service at Cincinnati, Ohio.

149 Unknown.

150 Unknown.

151 Unknown.

152 Unknown.

153 Unknown.

154 Unknown.

155 2 DCN/HD/W Continental. $36^1/_2$" x 68" nest tube boiler. $4^3/_4$" x 8" pumps, 700 GPM. Two 8" x 8" cyls. Length 13'6" (23'11" with pole). Height 9'3". $56^1/_2$" track. Weight 7800#.

Blueprint dates: 6-3-1908 boiler stay-bolt pattern; 7-23-1908 cross bar and evener for front springs; 7-28-1908 yoke (for 3-horse pole); 8-1-1908 horse poles and single trees; 8-3-1908 drag irons for three-horse hitch; 10-15-1908 hinge box and lock; 7-23-1909 exhaust pipe for boiler.

1908: In service at Engine 13, Detroit, Mich.

Date?: To Engine 29.

8- -1959: Sold by P.W. Henault of Detroit to a museum in Pennsylvania.

1970: Returned to Detroit, restored and preserved by Box 42 Fire Buffs of Detroit.

156 2 DCN/HD/W Continental. $34^1/_2$" x 64" nest tube boiler. $4^3/_4$" x 8" pumps, 700 GPM, Mayer relief valve. Two 8" x 8" cyls. Length 13'6" (23'11" with pole). Height 9'3". 106" wheelbase. Amoskeag-style axles. 60" track. Weight 7800#.

Blueprints: 9-18-1907 boiler stay-bolt pattern; 8-24-1908 $1^5/_8$" single throttle valve; 9-8-1908 reinforcing plate; 9-12-1908 rear spring brackets; 9-15-1908 3" Mayer relief valve; 9-24-1908 Amoskeag-style axles; 9-25-1908 Amoskeag-style axle; 11-16-1908 nest tube boiler; 7-7-1909 inside pump strainer.

1908: In service, Eng. 34, San Francisco, Cal.

157 2 DCN/HD/W Continental. 34" x 64" tube boiler. Two $4^3/_4$" x 8" pumps, 700 GPM. Two 8" x 8" cylinders. Length 13'6" (23'11" with pole). Height 9'3". 59" track. Weight 7800#.

Blueprint dates: 9-17-1908 boiler.

1908: In service at Toledo, Ohio.

158 3 DCN/HD/W Continental. Radial boiler. $4^3/_8$" x 8" pumps. 4"x20' pre-connect suction. $7^1/_4$" x 8" cyls. Length 12'11" (23'4" w/pole).

Fate of the 1908 Continental 3rd size #160 from Merced, Cal., is shrouded in mystery. Merced Fire Dept. reports that it was sold for $60 to a movie studio in 1927, yet at least two fire-engine enthusiasts recall seeing this steamer in Merced around 1945-1946. The style of uniform the firemen are wearing in this photo is 1930s-vintage, which also suggests that the Ahrens was sold later than 1927. Whether it survives, and its current whereabouts, remain unknown.
(Courtesy Chief Arthur Buckingham, Merced Fire Dept.)

Engine 2 in Denver, Colorado, used this 1909 3rd size Continental #165. A unique feature was the "Continental" script gold-leafed on the front end of the frame. Note the numeral "2" on front of smokestack, and the bell atop the pump instead of the traditional gong. Garden hose from boiler was for thawing frozen hydrants.
(Courtesy Joe Yeaman)

Height 9'3". 56 1/2" track. Weight 6800#.

1908: In service at Saint Paul, Minn.

159 Unknown.

160 3 DCN/HD/W Continental. Radial boiler. 4 3/8" x 8" pumps. 4"x20' pre-connect suction. 7 1/4" x 8" cyls. Length 12'11" (23'4" w/pole). Height 9'3". 56 1/2" track. Foot brake on foot board. Tool box under seat. Weight 6800#.

1908: In service at Merced, Cal.

1980: Letter from Merced Fire Dept. states that Continental was sold 1927 to Mr. Cohen of a movie studio near Los Angeles (possibly Harry Cohn, the head of Columbia Studios).

6- -1985: Tom Flanagan of Chowchilla, Cal., owner of a 1923 Ahrens-Fox pumper, recalls seeing Merced's Continental in a parade in Merced when he and his wife were dating in 1946. His wife also remembered the steamer. The author's father, Bill Hass, also recalls seeing a steamer at Merced when he was stationed in the area during WW II. Bill Hass points out that Merced's photo shows a style of firemen's uniform from the 1930s. So Continental #160 apparently was not sold in 1927 as Merced wrote, but its ultimate fate remains unknown.

161 Unknown.

162 1 DCN/HD/W Continental Special. 38 1/2" nest tube boiler. 5 1/2" x 8" pumps, 900 GPM, ornament atop air chamber, 2 1/2" Prunty relief valve. 5" x 20' pre-connect suction. 9" x 8" cyls. Length 14'1" (24'6" with pole). Height 9'6". Warner wheels. 3" Timken roller axle. 62" track. 22"-wide single seat, with clothing box underneath. Weight 8800#.

Blueprints: 2-18-1909 1 1/2" Lunken three-way stop cocks; 2-22-1909 left suction inlet; 2-24-1909, 2 1/2" Prunty relief valve; 3-8-1909 rear spring shackle; 3-18-1909 clothing box; 5-1-1909 boiler fuel pan; 6-19-1909 rear spring saddle and braces; 6-23-1909 grate straps for dump grates; 6-23-1909 rear spring swivel shackles; 7-7-1909 inside pump strainer; 7-14-1909 safety rail and bracket; 7-31-1909 rear spring chock; 8-5-1909 tongs, poker, clips; 8-27-1909 wrench for axle cap.

1909: In service at Engine 4, Baltimore, Md.

1914: Motorized, two-wheel Christie tractor.

163 3 DCN/HD/W Continental. 36 1/2" x 68" nest tube boiler. 4 3/8" x 8" pumps, 600 GPM. One 2 1/2" pre-connect suction. Two 7 1/4" x 8" cylinders. Length 12'11" (23'4" with pole). Height 9'3". Track 56 1/2". Weight 6800#.

Blueprints: 1-13-1909 $2^{1}/_{2}$" squirrel-tail suction; 1-22-1909 Brown roller-bearing boxes; 3-8-1909 boiler fittings and trimmings.

1909: In service at New Orleans, La.

164 3 DCN/HD/W Continental. $36^{1}/_{2}$" x 68" nest tube boiler. $4^{3}/_{8}$" x 8" pumps, 600 GPM. One $2^{1}/_{2}$" pre-connect suction. Two $7^{1}/_{4}$" x 8" cylinders. Length 12'11" (23'4" with pole). Height 9'3". Track $56^{1}/_{2}$". Weight 6800#.

Blueprints: 1-13-1909 $2^{1}/_{2}$" squirrel-tail suction; 1-22-1909 Brown roller-bearing boxes; 3-8-1909 boiler fittings and trimmings.

1909: In service at New Orleans, La.

165 3 DCN/HD/W Continental. Radial boiler. $4^{3}/_{8}$" x 8" pumps, air chamber ornament. $7^{1}/_{4}$" x 8" cyls. Length 12'11" (23'4" w/pole). Height 9'3". $56^{1}/_{2}$" track. Foot-brake. Hose reel under seat. Weight 6800#.

1909: In service at Engine 5, Denver, Colo.

166 DCN/HD/W Continental.

1909: In service at Columbus, Ohio.

167 Unknown.

168 Unknown.

169 3 DCN/HD/W Continental. $32^{1}/_{2}$" x 64" nest tube boiler. $4^{3}/_{8}$" x 8" pumps, 600 GPM. $7^{1}/_{4}$" x 8" cyls. Length 12'11" (23'4" with pole). Height 9'3". Track $56^{1}/_{2}$". Weight 6800#.

Blueprints: 9-26-1907 boiler stay-bolt pattern (last use of this old pattern); 9-15-1909 sheet-iron smokestack for boiler.

2-   -1910: In service at Engine 41, San Francisco, Cal.

170 1 DCN/HD/W Continental. Radial boiler. $5^{1}/_{2}$" x 8" pumps, 900 GPM, air chamber ornament. Two 5" x 10' suctions. 9" x 8" cyls. Length 14'1" (24'6" w/pole). Height 9'6". 62" track. Gong and foot-brake. Tool box under seat. Drop poles (3 horses). Weight 8800#.

1909: Delivered by John P. Ahrens to Columbia Fire Co. #4, Allentown, Pa. Water in 6:00. Pumped 1300 GPM. Threw one $1^{7}/_{8}$" stream 220' horizontally. Threw one 2" stream 200' vertically. Pumped a 2" stream over the Allentown Bank skyscraper.

1978: Restored, operated at antique fire engine muster in Allentown.

171 2 DCN/HD/W Continental. Radial boiler. Two $4^{3}/_{4}$" x 8" pumps, rated 700 GPM. Two 8" x 8" cylinders. Length 13'6" (23'11" with pole). Height 9'3". 59" track. Weight 7800#.

3-7-1910: Delivered to Seattle, Wash. Sale made by H.W. Mix of Mix Fire Apparatus

1909 Continental **1st size #170 still belongs to Allentown, Pa. Here,** it is preparing to draft from Lake Muhlenberg at Allentown's 1984 antique fire engine muster. The original paint and gold-leaf are fading and peeling after 75 years, and the nickel-plating no longer gleams, but to a true fire engine afficianado this fully-original condition is far more valuable than even the most painstakingly-authentic restoration.

(Ed Hass collection)

Atlanta, Ga., accepts delivery of 1st Size Continental #172 from John P. Ahrens (the tall civilian in striped shirt at far left) in June, 1910. Although rated 800 gallons per minute by the Fire Underwriters, the "William A. Hancock #4" had just delivered a walloping 1160 GPM on its delivery test, moments before this photo was taken. (Ed Hass collection)

Co., Seattle.

172 1 DCN/HD/W Continental. Radial boiler. $5^1/_2$" x 8" pumps, 900 GPM, air chamber ornament. 9" x 8" cyls. Length 14'1" (24'6" w/pole). Height 9'6". 62" track. Foot brake. Tool box under seat. Weight 8800#.

6- -1910: Delivered by John P. Ahrens to Atlanta, Ga. Name *William A. Hancock #4*. ($6325). Water in 6:50. Pumped 1160 GPM.

173 3 DCN/HD/W Continental. Radial boiler. $4^3/_8$" x 8" pumps, rated 600 GPM. Two $7^1/_4$" x 8" cyls. Length 12'11" (23'4" with pole). Height 9'3". Track $56^1/_2$". Weight 6800#.

1910: In service, Engine 6, Los Angeles, Cal.

1916: To Engine 20.

1924: Sold.

174 DCN/HD/W Continental. 10" bell atop air chamber.

Blueprint: 1-28-1910 hanging detail, 10" bell.

1910: In service at Key West, Florida.

175 Unknown.

176 3 DCN/HD/W Continental. Radial boiler. $4^3/_8$" x 8" pumps, rated 600 GPM. Two $7^1/_4$" x 8" cyls. Length 12'11" (23'4" with pole). Height 9'3". $56^1/_2$" track. Weight 6800#.

Blueprint: 3-25-1910 squirrel-tail suction.

6- -1910: Delivered to New Orleans, La.

177 3 DCN/HD/W Continental. Radial boiler. Two $4^3/_8$" x 8" pumps, 600 GPM. Two $7^1/_4$" x 8" cylinders. Length 12'11" (23'4" with pole). Height 9'3". $56^1/_2$" track. Weight 6800#.

Blueprint: 3-25-1910 squirrel-tail suction.

6- -1910: Delivered to New Orleans, La.

178 DCN/HD/W Continental.

1910: Delivered to Spokane, Wash. Sale by H.W. Mix, Mix Fire Apparatus Co., Seattle.

179 1 DCN/HD/W Continental. Radial boiler, safety rails on coal bunker. Two $5^1/_2$" x 8" pumps, 900 GPM, ornament and eagle on air chamber. Two 5" x 10' suctions. 9" x 8" cyls. Length 14'1" (24'6" w/pole). Height 9'6". 62" track. Gong and foot brake. Weight 8800#.

7-4-1910: Delivered to Memphis, Tenn. Name *E.H. Crump #5*. Price $6000.

1976: Restored, displayed at the Memphis

1911 Ahrens-Fox 3rd size Continental #184 of Friendship Fire Co. #1, Woodbury, N.J., on parade in the 1940s. Note that it was being towed by a modern pumper, rather than horses. This steamer now belongs to the Supplee brothers of Orleand, Pa., and will soon be restored for display in a museum.

(Wayne Sorenson collection)

Fire & Police Museum. It is operated at various public events in Memphis.

180 2 DCN/HD/W Continental. Radial boiler. Two $4^3/_4$" x 8" pumps, rated 700 GPM. Two 8" x 8" cylinders. Length 13'6" (23'11" with pole). Height 9'3". 59" track. Weight 7800#.

7-4-1910: Delivered, Memphis, Tenn. ($5500).

Ahrens-Fox Fire Engine Co.

Aug. 1910 to 1916

181 Unknown.

182 Unknown.

183 Unknown.

184 3 DCN/HD/W Continental. Radial boiler. $4^3/_8$" x 8" pumps, 600 GPM, air chamber ornament. Two 4" x 10' suctions. $7^1/_4$" x 8" cyls. Length 12'11" (23'4" w/pole). Height 9'3". $56^1/_2$" track. Gong and foot brake. Tool box under seat. Weight 6800#.

1911: Delivered to Friendship Fire Co. #1, Woodbury, N.J.

1973: Owned by Howard and Len Supplee, Oreland, Pa.

185 2 DCN/HD/W Continental. Radial boiler. Two $4^3/_4$" x 8" pumps, rated 700 GPM. Two 8" x 8" cylinders. Length 13'6" (23'11" with pole). Height 9'3". 59" track. Weight 7800#.

1911: In service at Engine 32, Chicago, Ill.

Date?: To Engine 75.

186 Unknown.

187 DX1 DCN/HD/W Continental. 42x70" radial inclined boiler. 6" x 8" pumps, 1300 GPM, air chamber ornament, two flywheels. Two 6" x 10' suctions. Two 10" x 8" cylinders. Length 14'9" (25'2" with pole). Height 9'11". 68" track. Gong and foot brake on foot board. Tool box under seat. Weight 10800#.

Blueprint dates: 7-15-1911 3" and $3^1/_4$" axles; 9-14-1911 boiler.

10- -1911: Demonstrated at International Assn. of Fire Engineers convention in Milwaukee, Wis. There it was towed by the first Ahrens-Fox motorized pumper.

11- -1911: Delivered to Beaumont, Texas. In

Model "A" Pumping Engine and Hose Carrier Towing Double Extra Continental Steamer.

The biggest Continental steamer ever built, this Double Extra 1st size (1300 GPM) was displayed at the 1911 fire chiefs convention in Milwaukee, Wis., prior to delivery to Beaumont, Texas. Here #187 is being towed by Ahrens-Fox's first motor fire engine, which was later rebuilt with a larger pump and sold to Rockford, Ill.
(Courtesy Harold S. Walker)

Engine 122 of Chicago, Ill. Steamer was a 1912 Ahrens-Fox 2nd size Continental #190; tractor was a 1914 Christie. Note the cycle fenders over rear wheels and the horn beside the driver's seat. The hand-crank on side of frame (below the hood) was actually at the front of motor, since Christie motors were transverse to the frame.
(Wayne Sorenson collection)

1912 Ahrens-Fox 2nd size Continental #191 served Engine 43 in Chicago, Ill. Tractor was a 1916 Christie. Note unusual location of Engine Co. number, on tractor frame below crane-neck of steamer.

(Wayne Sorenson collection)

Engine 106, Chicago, Ill., used this 1912 Ahrens-Fox 2nd size Continental #192. It was never motorized. Note the intricate gold-leaf patterns on the crane-neck frame, and the abundance of nickel-plating (especially on the boiler, pump, flywheel, rear of frame, and hubs of the wheels).

(Wayne Sorenson collection)

service as Engine 1.

188 2 DCN/HD/W Continental. Radial boiler. Two 4^3/4" x 8" pumps, rated 700 GPM. Two 8" x 8" cylinders. Length 13'6" (23'11" with pole). Height 9'3". 59" track. Weight 7800#.

10-30-1911: In service, E. 1, Washington, D.C.

1918: Motorized with two-wheel Christie tractor. To Engine 14.

1930: Junked.

189 2 DCN/HD/W Continental. Radial boiler. 4^3/4" x 8" pumps. 8" x 8" cyls. Length 13'6" (23'11" w/pole). Height 9'3". 59" track. Weight 7800#.

1911: Delivered to Indianapolis, Ind.

190 2 DCN/HD/W Continental. Radial boiler. 4^3/4" x 8" pumps, 700 GPM, air chamber ornament, two flywheels. 8" x 8" cylinders. Length 13'6" (23'11" with pole). Height 9'3". 59" track. Gong and foot brake on foot board. Tool box under seat. Weight 7800#.

1912: In service at Engine 122, Chicago, Ill.

1914: Motorized, Christie two-wheel tractor.

191 2 DCN/HD/W Continental. Radial boiler. 4^3/4" x 8" pumps, 700 GPM, air chamber ornament, two flywheels. 8" x 8" cylinders. Length 13'6" (23'11" with pole). Height 9'3". 59" track. Gong and foot brake on foot board. Tool box under seat. Weight 7800#.

1912: In service at Engine 43, Chicago, Ill.

1916: Motorized, Christie two-wheel tractor.

192 2 DCN/HD/W Continental. Radial boiler. 4^3/4" x 8" pumps, 700 GPM, air chamber ornament, two flywheels. 8" x 8" cylinders. Length 13'6" (23'11" with pole). Height 9'3". 59" track. Gong and foot brake on foot board. Tool box under seat. Weight 7800#.

1912: In service at Engine 106, Chicago, Ill.

193 2 DCN/HD/W Continental. Radial boiler. 4^3/4" x 8" pumps, 700 GPM, air chamber ornament, two flywheels. 8" x 8" cylinders. Length 13'6" (23'11" with pole). Height 9'3". 59" track. Gong and foot brake on foot board. Tool box under seat. Weight 7800#.

1912: In service at Engine 123, Chicago, Ill.

1914: Motorized, Christie two-wheel tractor.

194 Unknown.

195 Unknown.

196 Unknown.

197 1 DCN/TD/W Continental. Radial boiler.

This Ahrens-Fox Continental steamer of Chicago, Ill. was identified as 1912 2nd size #193 of Engine 123. Tractor is a 1914 Christie. It is not clear why the Engine Co. number gold-leafed on the seat was 124 instead of 123. Note the cycle fenders over the rear wheels, and the gong under the crane-neck of the frame.

(Wayne Sorenson collection)

This 1913 1st Class Continental #197 was delivered to Engine 23, Rochester, N.Y., with the Christie tractor already attached. Nevertheless, it had the crane-neck frame of a horse-drawn steamer. The next Continental built with a Christie (#199) would have a single straight frame from front of tractor to rear of steamer.
(Cincinnati Historical Society)

This 1913 Ahrens-Fox 2nd size Continental #198 originally went in service at Engine 3, Paterson, N.J., on Sept. 16, 1913 with a 1913 Christie tractor. In this 1931 photo, it was assigned to Engine 10, and had a 1917 electric tractor built by Commercial Truck Co. of Philadelphia, Pa. Steamer and tractor were junked July 16, 1931.
(Courtesy Thomas Dayspring)

1913 Continental 2nd size #199 of Engine 2, Somerville, Mass. This was the only Continental steamer originally built with a Christie tractor: note the continuous frame from tractor to boiler, rather than a crane-neck frame on the steamer bolted to the frame of the tractor. Somerville sold this unique steamer to Taunton, Mass.; it now sits outdoors at Edaville Railroad Museum, South Carver, Mass. The Christie motor is gone, but tractor and steamer are otherwise intact. (Bob Ward collection)

$5^{1}/_{2}$" x 8" pumps, air chamber ornament, 900 GPM. Two 5" x 10' suctions. 9" x 8" cyls. Length 14'1" (without tractor). Height 9'6". 62" track. Bell on steam cylinders (no gong). Propelled by two-wheel Christie tractor.

1913: In service at Engine 23, Rochester, N.Y. Name *P.V. Crittendon.*

198 2 DCN/HD/W Continental. Radial boiler. $4^{3}/_{4}$" x 8" pumps, 700 GPM, air chamber ornament, bell atop steam chest. Two $4^{1}/_{2}$" x 10' suctions. 8" x 8" cylinders. Length 13'6" (23'11" with pole). Height 9'3". 59" track.

7-11-1913: Ordered by Paterson, N.J. ($9000).

9-16-1913: In service at Engine 3, Paterson, N.J. Although delivered horse-drawn, a two-wheel Christie tractor was installed before it went in service.

1915: The Christie tractor was wrecked, and this steamer reverted to horse-drawn.

1917: Two-wheel electric tractor, made by Commercial Truck Co., Philadelphia, Pa.

10-28-1930: Replaced by 1927 Ahrens-Fox pumper #3310. Retired.

1-7-1931: Returned to service as Engine 10.

7-16-1931: Replaced by Ahrens-Fox pumper #3401. Retired and scrapped.

199 2 DSF/TD/W Continental. Radial boiler. $4^{3}/_{4}$" x 8" pumps, 700 GPM, air chamber ornament, two flywheels. Two $4^{1}/_{2}$" x 10' suctions. 8" x 8" cylinders. Length 13'6" (not including tractor). Height 9'3". 59" track. Propelled by two-wheel Christie tractor.

1913: In service, Engine 2, Somerville, Mass.

Date?: Sold to Taunton, Mass.

1965: This steamer is now sitting outdoors at Edaville Railroad Museum, South Carver, Mass. Looks complete except the motor has been removed from the Christie tractor.

200 4 DCN/HD/W Continental. Radial boiler. Two 4" x 8" pumps, 500 GPM. Two $6^{1}/_{2}$" x 8" cylinders. Length 12'6" (22'11" with pole). Height 8'10". $56^{1}/_{2}$" track. Weight 5800#.

1914: In service, Engine 39, Philadelphia, Pa.

1915: Motorized, two-wheel Christie tractor.

201 4 DCN/HD/W Continental. Radial boiler. Two 4" x 8" pumps, 500 GPM. Two $6^{1}/_{2}$" x 8" cylinders. Length 12'6" (22'11" with pole). Height 8'10". $56^{1}/_{2}$" track. Weight 5800#.

One of the most unusual steamers ever built was this 1915 Ahrens-Fox 1st size Continental Special #204, mounted on a 1915 Peerless chassis and used in Hamtramck, Mich. This was its original appearance from the factory, not a conversion. This engine was painted red; it only appears white because the photo has aged.

(Bob Schierle collection)

The last steam fire engine built in Cincinnati is this 1916 Ahrens-Fox 2nd size Continental Special #205, mounted on a 1916 White Trucks chassis. Only one company (American-LaFrance) built steamers later than this one, their last being in 1917. This Continental served American Fire Co. #1 of Lansford, Pa., from 1916 until 1968, and was the last steamer to be retired from active service in a fire department (it fought its last fire in 1967). To this day, Lansford -- in the heart of Pennsylvania coal country -- keeps a full load of coal aboard #205, just in case it is needed at a fire.

(Bill Hass photo)

In addition to building "Continental" steam and gasoline fire engines, Ahrens-Fox did a brisk business rebuilding steamers of various makes. In Jan., 1916, Ahrens-Fox installed a Continental boiler and band brakes on this 1905 Nott #633 of Engine 8, Paterson, N.J. Paterson also installed a 1916 Christie tractor at that time. Retired in Dec., 1927, the Nott/Ahrens-Fox hybrid was sold to a Pennsylvania coal mine, which scrapped the Christie tractor. Today, the steamer belongs to Wyckoff, N.J., and as this 1977 photo shows, is horse-drawn once more.

(Courtesy Wyckoff Fire Dept.)

1914: In service, Engine 54, Philadelphia, Pa.

1915: Motorized, two-wheel Christie tractor.

202 4 DCN/HD/W Continental. Radial boiler. Two 4" x 8" pumps, 500 GPM. Two 6 1/2" x 8" cylinders. Length 12'6" (22'11" with pole). Height 8'10". 56 1/2" track. Weight 5800#.

1914: In service, Engine 35, Philadelphia, Pa.

1916: Motorized, two-wheel Boyd tractor.

203 4 DCN/HD/W Continental. Radial boiler. Two 4" x 8" pumps, 500 GPM. Two 6 1/2" x 8" cylinders. Length 12'6" (22'11" with pole). Height 8'10". 56 1/2" track. Weight 5800#.

1914: In service, Engine 19, Philadelphia, Pa.

1916: Motorized, two-wheel Boyd tractor.

204 1 DSF/TD/W Continental Special. Radial boiler, safety rails on coal bunker. 5 1/2" x 8" pumps, 900 GPM, air chamber ornament. Two 5" x 10' suctions. 9" x 8" cyls. Mounted on 4-wheel Peerless chain-drive automobile chassis (right-hand drive) with wooden artillery wheels (12 spokes front, 14 spokes rear) and solid rubber tires.

1915: In service at Hamtramck, Mich.

205 2 DSF/TD/W Continental Special. Radial boiler, safety rails on coal bunkers. 4 3/4" x 8" pumps, 700 GPM, air chamber ornament. Two 8" x 8" cyls. Mounted on four-wheel White Trucks chain-drive chassis (left-hand drive) with steel spoked wheels (8 spokes front, 10 spokes rear) and solid rubber tires.

1916: In service at American Fire Co. #1, Lansford, Pa.

1968: Retired from active service. Became a spare, used primarily for testing hose.

1973: Lansford still owns, no longer in service but its coal bunkers are kept full of coal just in case it is needed at a fire.

1980: Appraised value of steamer $180,000.

Ahrens-Fox Continentals

Serial Numbers Unknown

1905 to 1916

AAA 2 DCN/HD/W Continental. Radial boiler. 4 3/4" x 8" pumps, 700 GPM, air chamber ornament. Two 4 1/2" x 10' suctions. 8" x 8" cyls. Length 13'6" (23'11" with pole). Height

This 1906 2nd size Continental steamer originally served in Tacoma, Wash., and now sits in a park in McCleary, Wash. Reg. No. is not known; it is listed in this book as number "AAA". (Wayne Sorenson photo)

9'3". 59" track. Foot brake on foot board. Water tank under seat, hose reel behind seat. Poles for three horses. Weight 7800#.

1906: In service at Engine 2, Tacoma, Wash. (Possibly number 122??)

1980: Now in a park at McCleary, Wash. Missing horse poles, air chamber, suction hoses, end-caps from the water tank; otherwise very complete and good condition.

BBB 3 DCN/HD/W Continental. Radial boiler. $4^3/8$" x 8" pumps, 600 GPM. $7^1/4$" x 8" cyls. Length 12'11" (23'4" with pole). Height 9'3". $56^1/2$" track. Weight 6800#.

10- -1906: In service at Columbus, Ohio. Water in 6:10. (Possibly number 127??)

CCC 1 DCN/HD/W Continental. Radial boiler. $5^1/2$" x 8" pumps, 900 GPM, air chamber ornament. 9" x 8" cyls. Length 14'1" (24'6" with pole). Height 9'6". 62" track. Water tank under seat. Weight 9800#.

(Possibly number 132??)

1907: Delivered by John P. Ahrens to Norfolk & Western Railroad Co., Roanoke, Va. ($5959). They gave to Roanoke (Engine 2). Water in 6:45. 1100 GPM through two 100' hose lines with $1^1/4$" nozzles. Threw four streams 150' vertically. Through four lines siamesed into 2" nozzle, threw a stream 230' vertically, 300' horizontally. Through three 600' hose lines siamesed into a $1^5/8$" nozzle, threw a stream 15' over church steeple.

1965: This steamer is reportedly at a privately-owned zoo in or near Roanoke.

DDD 1 DCN/HD/W Continental. Radial boiler. $5^1/2$" x 8" pumps, 900 GPM, air chamber ornament, bell atop pump. 9" x 8" cyls. 62" track. Height 9'6". Length 14'1" (24'6" with pole). Foot brake on foot board. Water tank under seat. Poles for 3 horses. Weight 8800#.

1907: In service at Engine 1, San Jose, Cal. (Possibly number 138??)

1914: Replaced by 1899 Metropolitan #2676 purchased second-hand from San Francisco.

6- -1985: Steam gauge from this now on San Jose's Metropolitan #2676. Gauge lettered "Ahrens Fire Engine Co., Cincinnati, Ohio."

EEE 3 DCN/HD/W Continental. Radial boiler. $4^3/8$" x 8" pumps, 600 GPM. $7^1/4$" x 8" cyls. Length 12'11" (23'4" with pole). Height 9'3". $56^1/2$" track. Weight 6800#.

1907: In service at East St. Louis, Ill. (possibly number 140??).

1930 Ahrens-Fox delivery photo shows ex-Covington, Ky. 1908 Ahrens Continental steamer shortly before it was sold to Bluefield, West Virginia. Tractor is a circa-1921 American-LaFrance. Registered No. of this steamer is not known; it is listed in this book as number "GGG".

(Wayne Sorenson collection)

FFF 2 DCN/HD/W Continental. Radial boiler. $4^{3}/4$" x 8" pumps, 700 GPM. 8" x 8" cyls. 59" track. Length 13'6". Height 9'3". 7800#.

1907: In service at Engine 1, Berkeley, Cal. (possibly number 145??).

1913: 3-wheel Knox-Martin tractor added.

GGG 2 DCN/HD/W Continental. Radial boiler. $4^{3}/4$" x 8" pumps, 700 GPM, air chamber ornament, bell atop pump. Two $4^{1}/2$" x 10' suctions. 8" x 8" cyls. Length 13'6" (23'11" with pole). Height 9'3". 59" track. 7800#.

1908: In service at Engine 1, Covington, Ky. (possibly number 146??)

c-1921: 2-wheel American-LaFrance tractor.

3-25-1930: Traded-in on Ahrens-Fox #3385. Resold to Bluefield, West Virginia.

HHH 2 DCN/HD/W Continental. Radial boiler. $4^{3}/4$" x 8" pumps, 700 GPM. 8" x 8" cyls. 59" track. Length 13'6". Height 9'3". 7800#.

1908: In service at Engine 4, Dayton, Ohio (possibly number 147??).

III 2 DCN/HD/W Continental. Radial boiler. $4^{3}/4$" x 8" pumps, 700 GPM. 8" x 8" cyls. 59" track. Length 13'6". Height 9'3". 7800#.

1908: In service at Engine 6, Dayton, Ohio (not certain this was built).

JJJ 2 DCN/HD/W Continental. Radial boiler. $4^{3}/4$" x 8" pumps, 700 GPM. 8" x 8" cyls. 59" track. Length 13'6". Height 9'3". 7800#.

1908: In service at Engine 13, Dayton, Ohio (not certain this was built).

KKK 2 DCN/HD/W Continental. Radial boiler. $4^{3}/4$" x 8" pumps, 700 GPM. 8" x 8" cyls. 59" track. Length 13'6". Height 9'3". 7800#.

1908: In service at Engine 5, Berkeley, Cal. (possibly number 149??)

LLL 2 DCN/HD/W Continental. Radial boiler. $4^{3}/4$" x 8" pumps, 700 GPM. 8" x 8" cyls. 59" track. Length 13'6". Height 9'3". 7800#.

1908: In service at Engine 3, Louisville, Ky. (possibly number 150??)

1924: Four-wheel International (IH) tractor.

1930s: Junked.

MMM 4 DCN/HD/W Continental. Radial boiler, 4" x 8" pumps, 500 GPM. $6^{1}/2$" x 8" cyls. $56^{1}/2$" track. Length 12'6" (22'11" with pole). Height 8'10". Weight 5800#.

1908: In service at Engine 30, Cleveland, Ohio (possibly number 151??)

NNN 3 DCN/HD/W Continental. Radial boiler. 4 3/8" x 8" pumps, 600 GPM. 7 1/4" x 8" cyls. Length 12'11" (23'4" with pole). Height 9'3". 56 1/2" track. Weight 6800#.

1908: In service at Engine 1, Charleston, West Virginia (possibly number 152??)

OOO DCN/HD/W Continental.

Date?: In service at Wheeling, West Virginia (possibly number 153??).

PPP DCN/HD/W Continental.

Date?: In service at Wheeling, West Virginia (possibly number 154??).

QQQ 2 DCN/HD/W Continental. Radial boiler. 4 3/4" x 8" pumps, 700 GPM. 8" x 8" cyls. 59" track. Length 13'6". Height 9'3". 7800#.

1909: In service at Janesville, Wis. (possibly number 159??).

RRR 1 DCN/HD/W Continental. Radial boiler. 5 1/2" x 8" pumps, 900 GPM. 9" x 8" cyls. 62" track. Length 14'1". Height 9'6". 8800#.

1909: In service at Engine 2, Baltimore, Md. (possibly number 161??).

SSS DCN/HD/W Continental.

Date?: In service at Chicago Heights, Ill. (possibly number 167??)

TTT 1 DCN/HD/W Continental. Radial boiler. 5 1/2" x 8" pumps, 900 GPM. 9" x 8" cylinders. Length 14'1". Height 9'6". 62" track. 8800#.

1909: In service at Engine 23, Baltimore, Md. (possibly number 168??)

1918: Motorized with two-wheel Christie tractor. To Engine 1.

UUU 1 DCN/HD/W Continental. Radial boiler. 5 1/2" x 8" pumps, 900 GPM. 9" x 8" cylinders. Length 14'1". Height 9'6". 62" track. 8800#.

1909: In service at Engine 38, Baltimore, Md. (possibly number 175??)

1916: Motorized with two-wheel Christie tractor.

VVV 2 DCN/HD/W Continental. Radial boiler. 4 3/4" x 8" pumps, 700 GPM. 8" x 8" cylinders. Length 13'6". Height 9'3". 59" track. 7800#.

2- -1910: In service, Eng. 1, Milwaukee, Wis.

WWW 1 DCN/HD/W Continental. Radial boiler. Two 5 1/2" x 8" pumps, rated 900 GPM. Two 9" x 8" cylinders. Length 14'1" (24'6" with pole). Height 9'6". 62" track. Weight 8800#.

1910: In service at Engine 13, Denver, Colo..

XXX DCN/HD/W Continental.

1910: In service at Clarksburg, West Virginia (possibly number 181??).

YYY 2 DCN/HD/W Continental. Radial boiler. 4 3/4" x 8" pumps, air chamber ornament, 700 GPM. 8" x 8" cyls. Length 13'6". Height 9'3". 59" track. Tool box under seat. 7800#.

11- -1910: In service at Engine 7, St. Paul, Minn. Name *Herbert P. Keller*. Price $5500. Pumped 840 GPM. Threw a vertical stream 200'. (Possibly #182??).

ZZZ 2 DCN/HD/W Continental. Radial boiler. 4 3/4" x 8" pumps, 700 GPM, air chamber ornament. 8" x 8" cyls. Length 13'6". Height 9'3". 59" track. Tool box under seat. 7800#.

11- -1910: In service, Eng. 14, St. Paul, Minn. ($5500). Pumped 840 GPM. Threw vertical stream 200'. (Possibly number 183??).

aaa 2 DCN/HD/W Continental. Radial boiler. 4 3/4" x 8" pumps, 700 GPM. 8" x 8" cyls. 59" track. Length 13'6". Height 9'3". 7800#.

1911: In service at Engine 3, St. Paul, Minn. (possibly number 186??)

bbb DCN/HD/W Continental.

Date?: In service at Springfield, Mo. Harold Walker lists as 1910 (possibly #194 of 1912).

ccc DCN/HD/W Continental.

Date?: In service at Springfield, Mo. Harold Walker lists as 1910 (possibly #195 of 1912).

ddd 1 DCN/TD/W Continental. Radial boiler. 5 1/2" x 8" pumps, air chamber ornament, bell atop pump, 900 GPM. Two 5" x 10' suctions. 9" x 8" cyls. Length 14'1" (without tractor). Height 9'6". 62" track. Propelled by two-wheel electric tractor made by Couple Gear Freight Wheel Co., Grand Rapids, Mich.

10- -1912: Demonstrated at International Association of Fire Engineers convention, Denver, Colorado. (Possibly number 196??).

11- -1912: In service at Denver, Colorado.

Unique 1912 Continental steamer of Denver, Colo. featured an electric tractor built by a Grand Rapids, Mich., manufacturer with the unlikely name of Couple Gear Freight Wheel Co. Note the Columbian-style hosebox between the tractor and steamer. Reg. No. is not known; it is listed in this book as number "ddd".
(Dan Martin collection)

Denver's 1912 Continental steamer "ddd" with Couple Gear tractor was demonstrated at the 1912 fire chiefs convention in Denver, Colo., before it went in service. Here it poses in front of city hall during that fire chiefs conference. Tall man in overalls standing at left is John P. Ahrens; seated beside the driver is Charles H. Fox.
(Courtesy Lester Williams)

AHRENS MANUFACTURING CO., CINCINNATI, O.

Although the focus of this book is on Ahrens steam fire engines, Ahrens also built other types of fire department apparatus. Top left is a "riding" hose reel (equipped with a driver's seat), of the type usually pulled by horses. Top right is a reel with a bell arch, designed to be pulled by hand by volunteer firemen in small villages; the motion of the cart caused the arch to sway, making the bell ring.

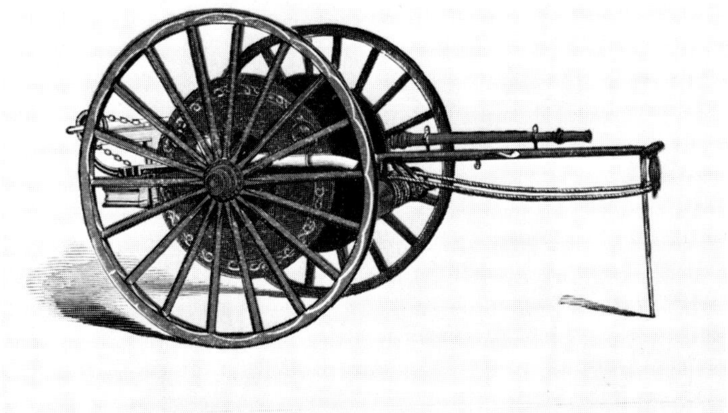

AHRENS MANUFACTURING CO., CINCINNATI, O.

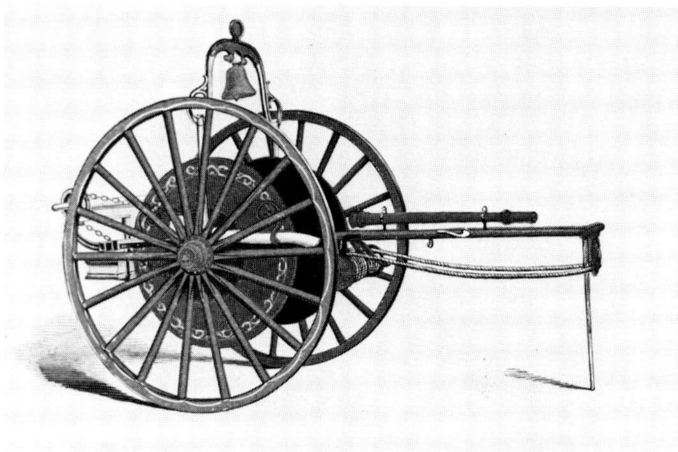

AHRENS MANUFACTURING CO., CINCINNATI, O.

Bottom left is a very utilitarian hand-drawn reel, with no arch and no seat, for communities on a tight budget; note the folding brace on the front, to keep the reel stable when not in motion. Bottom right is an Ahrens stationary heater, vintage 1907, which kept a steamer's boiler hot in the firehouse so there would be no delay at a fire.

(Ed Hass collection)

Numerical Index of Engines

This index, and the alphabetical index, include only the original name, company, and city of each Cincinnati-built steam fire engine. For a complete history of each engine indexed, see the text of this book. An asterisk (*) indicates an engine that survives today.

Latta

Reg. #	Class	Style	Name	Co.	City & State	Date
1	2nd	SSF			Prototype	3-2-1852
2	1st	SSF	Uncle Joe Ross		Cincinnati, OH	2-1-1853
3	1st	SSF	Citizens Gift	3	Cincinnati, OH	7-27-1854
4		SSF	Hope	2	Harrisburg, PA	1854
5	1st	SSF	Miles Greenwood		Boston, MA	3- -1855
6	1st	SSF	Young America		New Orleans, LA	3- -1855
7	1st	SSF	A.B. Latta	4	Cincinnati, OH	11- -1854
8	2nd	SSF	Marion	8	Cincinnati, OH	1855
9	2nd	SSF	Deluge	10	Cincinnati, OH	1855
10	2nd	SSF	Miles Greenwood	12	Cincinnati, OH	1855
11	2nd	SSF	Washington	1	Cincinnati, OH	1856
12	3rd	SSF		2	Evansville, IN	1856
13	4th	SSF			Columbus, OH	1856
14	3rd	SSF	George Kyler	1	St. Louis, MO	1856
15	3rd	SSF	David Moore	3	St. Louis, MO	1857
16	3rd	SSF	John M. Winn	4	St. Louis, MO	1857
17	1st	SSF	Eclipse	1	Louisville, KY	3-1-1857
18	3rd	SSF	Underwriter	6	St. Louis, MO	9-14-1857
19	3rd	SSF	Deluge	7	St. Louis, MO	1858
20	2nd	SSF	Western	14	Cincinnati, OH	1858
21	2nd	SSF	Joseph A. Gilliss	2	Louisville, KY	1858
22	3rd	SSF		3	Louisville, KY	1858
23	2nd	SSF		4	Louisville, KY	1859
24	2nd	SSF		5	Louisville, KY	1859
25	4th	SSF			Toledo, OH	1860
26	4th	SSF		2	Indianapolis, IN	1860
27	2nd	SSF	West End	2	Cincinnati, OH	1861
28	2nd	SSF	Jefferson	5	Cincinnati, OH	1861
29	2nd	SSF	Eastern	6	Cincinnati, OH	1861
30	2nd	SSF	Phil Sheridan	7	Cincinnati, OH	1861
31	2nd	SCN			Memphis, TN	1861

Shawk

Reg. #	Class	Style	Name	Co.	City & State	Date
1	4th	SRT	Young America	1	Philadelphia, PA	5-23-1855
2	3rd	SRT	Old Union	2	St. Louis, MO	10- -1855
3	3rd	SRT	Missouri	5	St. Louis, MO	2- -1856

Lane & Bodley

Reg. #	Class	Style	Name	Co.	City & State	Date
1		SSF		1	Dayton, OH	1863
2		SSF		2	Dayton, OH	1863
3		SSF		3	Dayton, OH	1863
4		SSF		1	Nashville, TN	1863
5		SSF			Cleveland, OH	1863
6		SSF			Battle Creek, MI	1863
7		SSF			Hamilton, OH	1864
8	2nd	SSF	John F. Torrence		Cincinnati, OH	1864
9	3rd	SCN		6	Louisville, KY	1865
10		SSF			Grand Rapids, MI	1865
11	2nd	SCN	Hampton Woodruff	8	St. Louis, MO	1865
12		SSF			Akron, OH	1866
13		SSF			Leavenworth, KS	1866
14	2nd	SCN			Memphis, TN	1866
15		SSF			Galveston, TX	1866
16		SSF			Quincy, IL	1866
17		SSF			Omaha, NE	1866
18		SSF			Racine, WI	1866
19		SSF			Oshkosh, WI	1866
20		SSF			St. Paul, MN	1866
21	3rd	SSF	Theodore Chambers		Cincinnati, OH	1867
22		SSF			Newark, OH	1867
23		SSF		9	St. Louis, MO	1867
24		SSF		10	St. Louis, MO	1867

Ahrens

Reg. #	Class	Style	Name	Co.	City & State	Date
1	2nd	SSF	A.B. Latta	10	Cincinnati, OH	1-1-1869
2	2nd	SSF		3	Hamilton, OH	4-23-1869
3	1st	SSF	Brighton	13	Cincinnati, OH	7-1-1869
4	1st	SSF	Marion	8	Cincinnati, OH	1-1-1870
5	2nd	SSF	Fair Play	1	Madison, IN	4-9-1870
6	2nd	SSF	Walnut Hills	16	Cincinnati, OH	8-1-1870
7	2nd	SSF	Queen City	6	Cincinnati, OH	9-1-1870
8	2nd	SSF		2	St. Louis, MO	10-10-1870
9	2nd	SSF	James T. Drummond		Alton, IL	1-21-1871
10	2nd	SSF	Storrs	17	Cincinnati, OH	1-1-1871
11	2nd	SSF	Jefferson	5	Cincinnati, OH	8- -1871
12	2nd	SSF	Washington	2	Madison, IN	5-17-1871
13	2nd	SSF	Reliable	18	Cincinnati, OH	6-15-1871
14	2nd	SSF	Lincoln	19	Cincinnati, OH	11- -1871
15	2nd	SSF		8	St. Louis, MO	10- -1871
16	2nd	SSF		11	St. Louis, MO	10- -1871
17	2nd	SSF	Miles Greenwood	12	Cincinnati, OH	12-30-1871
18	2nd	SSF	Eaton	1	Eaton, OH	2-21-1872
19	2nd	SSF	Mound	9	St. Louis, MO	4-27-1872
20	2nd	SSF	George W. Levi	7	Louisville, KY	6-14-1872
21	2nd	SSF	New Franklin	13	St. Louis, MO	8-7-1872
22	2nd	SSF	James Leach	8	Louisville, KY	9-13-1872
23	2nd	SSF		10	St. Louis, MO	10-26-1872
24	2nd	SSF	J.A. Koach	9	Louisville, KY	12-2-1872
25	2nd	SSF			Galion, OH	1-17-1873
26	2nd	SSF	Ben Bransford	1	Owensboro, KY	3-15-1873
27	2nd	SSF	Union	9	Cincinnati, OH	4-18-1873
28	2nd	SSF		15	St. Louis, MO	6-24-1873
29	2nd	SSF		16	St. Louis, MO	8-9-1873
30	2nd	SSF		17	St. Louis, MO	10-1-1873
31	2nd	SSF		25	Chicago, IL	11-10-1873
32	2nd	SSF	Neptune	1	Hamilton, OH	12-4-1873 *
33	2nd	SSF	John M. Letterle	10	Louisville, KY	1-1-1874
34	2nd	SSF	Union	9	Cincinnati, OH	3-16-1874
35	2nd	SSF	Al Bebee	3	Quincy, IL	4-27-1874
36	2nd	SSF	C.W. Johnston	5	Cincinnati, OH	5-30-1874
37	2nd	SSF	Suaaks	5	Indianapolis, IN	7-8-1874
38	2nd	SSF		2	Nashville, TN	12-12-1874
39	2nd	SSF	Bash Wilson	1	Greensburg, IN	10-20-1874
40	2nd	SSF		3	Memphis, TN	12-10-1874
41	1st	SSF	Thomas D. Kingan		Indianapolis, IN	11-12-1874
42	2nd	SSF		3	Shelbyville, IN	1-22-1875
43	2nd	SSF	Wooden Shoe	7	Cincinnati, OH	5-17-1875
44	2nd	SSF	Future City	18	St. Louis, MO	4-5-1875
45	3rd	SSF	Pap Ringwood		Oxford, OH	5-13-1875
146	3rd	SSF	Harry Wilcox	1	Chattanooga, TN	6-14-1875
147	3rd	SSF		1	St. Louis, MO	6-25-1875
148	1st	SSF		6	St. Louis, MO	7-30-1875
149	1st	SSF		4	St. Louis, MO	8-20-1875
150	1st	SSF		7	St. Louis, MO	10-1-1875
151	3rd	SSF			Hillsboro, OH	10-25-1875
152	1st	SSF		3	St. Louis, MO	11-8-1875
153	3rd	SSF			Waverly, OH	11-16-1875
154	1st	SSF		9	St. Louis, MO	12-1-1875
155	1st	SSF		7	Indianapolis, IN	1-7-1876
156	1st	SSF		5	St. Louis, MO	2-8-1876
157	3rd	SSF	Kansas	3	Leavenworth, KS	12-17-1875
158	3rd	SSF	Citizens Gift	3	Nashville, TN	12-30-1875
159	3rd	SSF	Cynthiana	1	Cynthiana, KY	2-17-1876
160	2nd	SSF		2	Indianapolis, IN	3-8-1876
161	2nd	SSF	Pioneer	1	Watertown, WI	6-7-1876
162	3rd	SSF		10	St. Louis, MO	5-4-1876
163	3rd	SSF	General Meredith		Cambridge City, IN	5-19-1876
164	3rd	SSF		4	Indianapolis, IN	11-2-1876
165	1st	SSF	Charles D. Jacobs	1	Louisville, KY	3-9-1877
166	3rd	SSF	Thomas Gaff		Aurora, IN	9-11-1876
167	3rd	SSF	Van Wert	1	Van Wert, OH	5-31-1877
168	3rd	SSF			Cleveland, OH	3-1-1877
169	2nd	SSF	Ashbrook		Covington, KY	12-23-1876
170	4th	SSF			Georgetown, KY	6-22-1877
171	2nd	SSF	Robert Howe		Covington, KY	4-25-1877
172	3rd	SSF	J.L. Thompson	18	Cincinnati, OH	3-29-1877
173	3rd	SSF	William Stockell	4	Nashville, TN	4- -1877
174	3rd	SSF	Mechanics	6	Galveston, TX	7-4-1877
176	4th	SSF			Uniontown, PA	6-9-1877
178	4th	SSF			Franklin, OH	7-6-1877
180	4th	SSF			Marysville, OH	5-5-1879

Reg. #	Class	Style	Name	Co.	City & State	Date
182	3rd	SSF			Parkersburg, WV	5-29-1878
184	3rd	SSF			Gallipolis, OH	12-8-1877
186	3rd	SSF	Corryville	19	Cincinnati, OH	11-5-1877
188	4th	SSF	Rescue		Paris, KY	1-8-1878
190	4th	SSF			Columbia, TN	6-5-1878
192	4th	SSF	Protector	2	New Orleans, LA	3-30-1878
194	4th	SSF		1	Topeka, KS	9- -1877
196	2nd	SSF		19	St. Louis, MO	12-28-1878
198	2nd	DCN	Franklin	4	Cincinnati, OH	5-1-1878
200	3rd	SSF		8	Cleveland, OH	10-1-1878
202	4th	SSF	Elisha Hyatt	1	Washington, IN	8-17-1878 *
204	4th	SSF	Lady Morgan	1	Lake Providence, LA	3-1-1879
206	4th	SSF	Young America	11	New Orleans, LA	3-17-1879
208	3rd	SSF		20	St. Louis, MO	11-26-1878
210	3rd	SSF		4	Omaha, NE	3-26-1879
212	3rd	SSF	Rescue		Council Bluffs, IA	6-4-1880
214	3rd	DSF	Jackson	18	New Orleans, LA	5- -1879
216	2nd	SSF			Newark, OH	7-7-1879
218	4th	SSF			Built as #244	
220	3rd	DSF	George W. Frantz	6	Louisville, KY	8-20-1879
222	3rd	DSF	Palmetto	10	New Orleans, LA	11-30-1879
224	2nd	DCN		1	Milwaukee, WI	10-24-1879
226	3rd	DSF	Dexter	1	Montgomery, AL	12-11-1879
228	3rd	DSF	L.C. Weir	3	Cincinnati, OH	12-15-1879
230	3rd	DSF		7	New Orleans, LA	4-11-1880
232	3rd	DSF	G.W.A. Clough	1	Chillicothe, OH	2-7-1880
234	3rd	DSF			Vincennes, IN	3-5-1880
236	3rd	DSF		21	St. Louis, MO	8-9-1880
238	4th	DSF	Columbia	5	New Orleans, LA	4-25-1880
240	4th	DSF		21	New Orleans, LA	4-25-1880
242	4th	SSF	U.S. Army		Camp Carlin, MO	10-8-1880
244	4th	SSF	Lena	1	Lena, IL	3-28-1881
246	2nd	DCN		4	Milwaukee, WI	6-24-1880
248	2nd	DCN		7	Chicago, IL	7-24-1880
250	2nd	DCN	Major Hughes	5	Louisville, KY	9-6-1880
252	2nd	DCN		13-1	New York, NY	9-4-1880
254	2nd	DCN		5	Chicago, IL	9-25-1880
256	3rd	SCN			Hot Springs, AR	10- -1880
258	4th	SSF	Creole	9	New Orleans, LA	11-14-1880
260	3rd	DCN		30	Chicago, IL	11-10-1880
262	3rd	DCN		31	Chicago, IL	11-10-1880
264	3rd	DCN		2	Chicago, IL	11-15-1880
266	3rd	SCN			Wausau, WI	11-13-1880
268	1st	DSF		13	Chicago, IL	2-19-1881
270	3rd	DSF	J.L. Mason	3	Louisville, KY	12-24-1880
272	3rd	SSF	Harold G. Hayes	1	Delphos, OH	1-28-1881
274	2nd	DCN	Brooklyn		Oshkosh, WI	3-10-1881
276	2nd	DSF	B.F. Bache	2	Louisville, KY	4-2-1881
278	2nd	DSF	Frank Roberts	4	Louisville, KY	4-2-1881
280	1st	DCN	Citizens Gift	3	Cincinnati, OH	5-24-1881
282	4th	SSF	Rushville	1	Rushville, IN	4-19-1881
284	4th	SSF			Paducah, KY	6-24-1881
286	4th	SSF	Industrial School		Waukesha, WI	5-25-1882
288	4th	DSF		13	New Orleans, LA	6-20-1881
290	3rd	DCN			Battle Creek, MI	8-3-1881
292	3rd	SSF			Frankfort, KY	11-5-1881
294	3rd	DSF		3	New Orleans, LA	8-20-1881
296	1st	DSF		32	Chicago, IL	9-3-1881
298	2nd	DCN		8	Chicago, IL	11-23-1881
300	2nd	DCN		31	Chicago, IL	11-23-1881
302	3rd	SSF			Greencastle, IN	4-6-1882
304	5th	SSF			West Belleville, IL	10-8-1881
306	1st	DSF		11	Chicago, IL	10-12-1881
308	2nd	DCN		14	Chicago, IL	12-22-1881
310	4th	DSF	Florida	2	Pensacola, FL	12-19-1881
312	2nd	DCN		5	Town of Lake, IL	1-19-1882
314	1st	DSF	Marion	8	Cincinnati, OH	4-1-1882
316	1st	DSF	Relief	2	Cincinnati, OH	3-1-1882
318	2nd	DCN		1	Memphis, TN	7-12-1882
320	2nd	DCN		2	Memphis, TN	8-12-1882
322	2nd	DSF	Washington	1	Cincinnati, OH	5-1-1883
324	3rd	DSF	Volunteer	1	New Orleans, LA	10-12-1882
326	5th	SSF	Philadelphia	14	New Orleans, LA	10-12-1882
328	3rd	SSF			Huntington, WV	10- -1882
330	1st	DCN		12	St. Louis, MO	10-13-1882
332	3rd	DCN	Washington	3	Vicksburg, MS	10-10-1882
334	3rd	DCN		15	Chicago, IL	11-8-1882
336	3rd	DCN		56	Chicago, IL	11-8-1882
338	5th	SSF	St. Bernard	1	New Orleans, LA	4-12-1882

Reg. #	Class	Style	Name	Co.	City & State	Date	
340	3rd	SSF		2	Lawrenceburg, IN	1-24-1883	
342	3rd	DCN		26	Chicago, IL	12-2-1882	
344	3rd	DSF	Vincent Bradas	11	Louisville, KY	12-22-1882	
346	1st	DSF	Samuel Cofflin		Covington, KY	12-28-1882	
348	3rd	SSF		1	Lawrenceburg, IN	1-24-1883	
350	3rd	DCN			Sheboygan, WI	3-14-1883	
352	3rd	DCN		3	Minneapolis, MN	4-12-1883	
354	4th	SSF		45	New York, NY	7-9-1883	
356	4th	SSF		42	New York, NY	7-9-1883	
358	3rd	DCN		13-2	New York, NY	6-26-1883	
360	3rd	DCN		20	New York, NY	6-26-1883	
362	3rd	DCN			Portsmouth, OH	6-30-1883	
364	3rd	DCN			Peoria, IL	8-23-1883	
366	3rd	DCN		15	Chicago, IL	12-19-1883	
367	3rd	DCN		32	Chicago, IL	12-19-1883	
370	4th	DSF	Jefferson	23	New Orleans, LA	6-6-1883	
372	1st	DCN		4	Milwaukee, WI	9-14-1883	
374	2nd	DCN		3	Chicago, IL	11-12-1883	
376	4th	SSF	Washington	1	Delphos, OH	10-3-1883	
378	3rd	SSF	Henry Young	1	Sidney, OH	9-27-1883	*
380	2nd	DCN	John M. Lea	5	Nashville, TN	11- -1883	
382	2nd	DCN	James Geddes	6	Nashville, TN	11- -1883	
384	1st	DSF	Jacob Thome	12	Louisville, KY	8-30-1883	
386	3rd	DCN			Ottawa, KS	12-4-1883	
388	4th	SSF	Brackenridge		Harrison, OH	11-27-1883	
390	4th	SSF			Celina, OH	1884	
391	1st	DCN		2	St. Paul, MN	2-29-1884	
392	3rd	DCN	Old Tige	1	Dallas, TX	2-8-1884	*
393	3rd	DCN	Penetentiary		Jefferson, MO	2-5-1884	
394	3rd	DSF	Aurora		Aurora, IN	2-20-1885	*
395	2nd	SSF		5	Memphis, TN	4-18-1884	
396	1st	DCN		9	Milwaukee, WI	5-4-1884	
397	4th	SSF	C.B. Blackburn	13	Louisville, KY	3-4-1884	
398	3rd	DSF		11	St. Louis, MO	6-10-1884	
399	3rd	DSF		22	St. Louis, MO	12-22-1884	
400	3rd	DSF	John M. Letterle	10	Louisville, KY	5-29-1884	
401	1st	DCN		3	Milwaukee, WI	6- -1884	
402	3rd	DSF		23	St. Louis, MO	12-22-1884	
403	2nd				Unknown	1884	
404	2nd				Unknown	1884	
405	4th	DSF			Pekin, IL	9-12-1884	
406	4th	SSF	C.L. Pullen	6	Memphis, TN	12- -1884	
407	3rd	DCN		31,74	Chicago, IL	9-6-1884	
408	3rd	DCN		54,67	Chicago, IL	9-6-1884	
409	3rd	DCN		57,96	Chicago, IL	10-15-1884	
410	3rd	DSF	Chris Kiechler	16	Cincinnati, OH	9-5-1884	
411	3rd	DCN	Alameda	5	Oakland, CA	2-4-1885	
412	3rd	DCN	Stephan Freeman		Racine, WI	1- -1885	*
413	3rd	DSF	Louisiana Hose		New Orleans, LA	11-22-1884	
414	4th	SSF	James Speed	4	Memphis, TN	12- -1884	
415	5th	SSF	Saint Paris		St. Paris, OH	1-7-1885	*
416	5th	SSF	Auburn		Auburn, IN	7-6-1885	
417	4th	SSF			Catlettsburg, KY	11-22-1884	
418	4th	SSF	Thomas J. Carlisle	2	Chattanooga, TN	3-17-1885	
419	3rd	DSF		24	St. Louis, MO	5-30-1885	
420	3rd	DSF		25	St. Louis, MO	5-30-1885	
422	4th	SSF	Phoenix	2	Vicksburg, MS	8-1-1885	
423	4th	SSF			Lebanon, OH	2-22-1887	
425	4th	SSF	Arcanum		Arcanum, OH	4-2-1885	*
427	3rd	DCN		29,116	Chicago, IL	6-22-1885	
429	3rd	DCN		23,77	Chicago, IL	8-14-1885	
431	4th	DSF		6	New Orleans, LA	12-30-1885	
432	4th	DCN			Pendleton, OR	12-8-1885	
433	5th	SSF	Shields	2	Seymour, IN	12-31-1885	*
435	5th	SSF			Sullivan, IN	1-7-1886	
437	2nd	DCN		6	Milwaukee, WI	8-13-1885	
439	3rd	DCN		94	Chicago, IL	9-7-1885	
440	3rd	DCN		62,99	Chicago, IL	9-28-1885	
441	3rd	DCN		91	Chicago, IL	9-28-1885	
443	3rd	DSF	Chris Kinsinger	8	Cincinnati, OH	11-5-1885	
445	3rd	DCN	E.F. Pryor	3	Dayton, OH	11-12-1885	
446			(Rblt. 1873 Silsby)		Adrian, MI	1885	
448	1st	DCN		8	St. Paul, MN	12-8-1885	
449	3rd	DCN	Sam Houston	1	San Antonio, TX	6-14-1886	
451	3rd	DCN		9	Milwaukee, WI	9-21-1886	
453	5th	SSF	Governor Johnson		Bardstown, KY	12-18-1885	
455	5th	SSF	H.J. Topky		Ashtabula, OH	3-6-1886	
457	5th	SSF			Waynesville, OH	3-10-1886	
458	3rd	SSF	(Rbld of #158)	1	Lawrenceburg, IN	4-17-1901	

Reg. #	Class	Style	Name	Co.	City & State	Date	
459	4th	DCN			Birmingham, AL	2-16-1886	*
461	4th	DCN			Rebuilt as #520		
463	4th	DCN			Unknown		
465	3rd	DCN	B.F. Manderbach	1	Akron, OH	10-16-1886	
467	2nd	DCN		26	St. Louis, MO	3-29-1886	
470	4th	SSF			Miamisburg, OH	5-4-1886	
475	5th	SSF			Germantown, OH	5-3-1886	
477	5th	SSF			North Vernon, IN	11-18-1886	
479	5th	SSF	Merchants Cotton		Memphis, TN	3-1-1887	
481	5th	SSF	Archibald Liggett		Ripley, OH	12-5-1887	*
483	1st	DCN		3	St. Paul, MN	8-2-1886	
485	1st	DCN			Unknown		
487	6th	SSF			Deshler, OH	3-26-1887	*
489	1st	DCN			Unknown		
491	3rd	DCN	Tim Monaghan	49,91	Chicago, IL	12-7-1886	
493	3rd	DCN	Walter S. Moore	4	Los Angeles, CA	4-14-1887	
499	2nd	DCN		11	Minneapolis, MN	10-21-1886	
500	2nd	DCN		3	Minneapolis, MN	10-21-1886	
501	2nd	DCN		40	Chicago, IL	12-7-1886	
502	3rd	DCN		27	St. Louis, MO	4- -1887	
503	3rd	DSF	James M. Doherty	13	Cincinnati, OH	6-2-1887	
505	2nd	DCN		7	Milwaukee, WI	5-3-1887	
506	4th	DCN	Little Steamer	1	Phoenix, AZ	5-10-1887	
508	2nd	DCN		21	St. Louis, MO	5-10-1887	
509	3rd	DCN		10	St. Paul, MN	8-30-1887	
511	3rd	DCN		15	St. Paul, MN	8-30-1887	
512	4th	SSF	(Rblt as #625)		Morrow, OH	5-18-1888	
514	6th	SSF	(Rblt as #661)		Demonstrator		
515	6th	SSF			Unknown		
517	6th	SSF			Unknown		
518	2nd	DCN		18	Detroit, MI	10-12-1887	
520	4th	DCN			Birmingham, AL	6- -1887	
521	2nd	DCN		27,105	Chicago, IL	11-8-1887	
523	3rd	DCN		22,76	Chicago, IL	11-8-1887	
524	3rd	DCN		78	Chicago, IL	11-8-1887	
526	4th	DCN	Alert	1	Fresno, CA	11-30-1887	
527	4th	DCN			Lake View, IL	1-14-1888	
529	3rd	DCN		9	St. Paul, MN	12-29-1887	
530	3rd	DCN		6	Indianapolis, IN	2-8-1888	
532	4th	DCN		26	St. Louis, MO	2-13-1888	
533					Unknown		
534	3rd	DSF	Abe Furst	3	Cincinnati, OH	3-20-1888	
535	3rd	DCN	Five Brothers	9	Louisville, KY	2-23-1888	
537	4th	DCN		7	Los Angeles, CA	8-8-1888	
538	4th	DCN		15	Detroit, MI	8- -1888	
540	5th	SSF			Versailles, KY	5-10-1888	
541	5th	SSF			North Amherst, OH	7-24-1888	*
543	3rd	DCN		5	Los Angeles, CA	6-16-1888	*
544	3rd	DCN		6	Los Angeles, CA	8-8-1888	
546	3rd	DCN		12	Milwaukee, WI	6-26-1888	
547	3rd	DCN		13	Milwaukee, WI	6-26-1888	
549	3rd	DCN	Daniel Worley		Canton, OH	12-3-1888	*
550	3rd	DCN	John Goetz Jr.	22	Cincinnati, OH	2-25-1889	
552	4th	DCN	Edwin Stevens	23	Cincinnati, OH	10- -1888	
553	4th	DCN	Chas. Fleishman	26	Cincinnati, OH	2-19-1889	
555	5th	SSF	Hope		Morrow, OH	6-9-1889	*
556	5th	SSF			Demonstrator	1888	
558	4th	DCN	W.L. Dugger	3	Chattanooga, TN	11-10-1888	
559					Unknown		
561	3rd	DCN	Hamilton	2	San Diego, CA	9-20-1888	
562	3rd	DCN		28	St. Louis, MO	11-14-1888	
564	2nd	DCN		27	St. Louis, MO	12- -1888	*
565	2nd	DCN		29	St. Louis, MO	12- -1888	
567	3rd	DCN		44,75	Chicago, IL	12-19-1888	
568	3rd	DCN		53,69	Chicago, IL	12-19-1888	
570	3rd	SSF	Harry Wilcox	1	Chattanooga, TN	11-10-1888	
571	3rd	DCN	J.R. Megrue	29	Cincinnati, OH	2-25-1889	
573	3rd	DCN	Abe Steinau	21	Cincinnati, OH	5-4-1889	
574	3rd	DCN		2	Columbus, OH	12-6-1889	
576	3rd	DCN		85	Chicago, IL	12-2-1889	
577	4th	DCN	Tacoma	4	Tacoma, WA	9-21-1889	*
579	4th	DCN			Unknown		
580	4th	DCN	J.W. Barnes	2	Mt. Sterling, KY	5-17-1889	
582	4th	DCN	T.S. Hillard	4	Wilkes-Barre, PA	8-21-1889	
583	3rd	DCN	A.T. Stewart	1	Pueblo, CO	8-7-1889	
585	3rd	DCN			Unknown		
587	2nd	DCN		14	Milwaukee, WI	6-25-1889	
589	2nd	DCN	Chas. H. Ware	4	Dayton, OH	9-18-1889	
591	2nd	DCN		5	Milwaukee, WI	10-12-1889	

Reg. #	Class	Style	Name	Co.	City & State	Date	
593	3rd	DCN		2	Toledo, OH	12-11-1889	
594	3rd	DCN		5	Toledo, OH	12-11-1889	
596	3rd	DCN		4	Columbus, OH	8-12-1889	
597	3rd	DCN		1	Seattle, WA	11-29-1889	
598	3rd	DCN		2	Seattle, WA	11-29-1889	
599	3rd	DCN		4	St. Louis, MO	1-7-1890	
600	2nd	DCN		22	St. Louis, MO	1-7-1890	
601	2nd	DCN		64	Chicago, IL	12-28-1889	
602	3rd	DCN	Wichita	1	Wichita, KS	1-20-1890	
603	3rd	DCN		73	Chicago, IL	1-14-1890	
604	4th	DCN	Lewis Wisbey	24	Cincinnati, OH	2-21-1890	
605	4th	DCN	Clinton Chief	1	Wilmington, OH	12-17-1890	
607	2nd	DCN		1	Columbus, OH	4-7-1890	
608	2nd	DCN		3	Columbus, OH	4-7-1890	
610	3rd	DCN		8	Columbus, OH	12-26-1890	
611	3rd	DCN		2	Columbus, OH	12-26-1890	
613	3rd	DCN		7	Columbus, OH	12-26-1890	
614	3rd	DCN		4	Columbus, OH	12-26-1890	
616	2nd	DCN	Bishop	1	Quincy, IL	4-17-1890	
617	2nd	DCN		11	Milwaukee, WI	7-28-1890	
619	3rd	DCN		47	Chicago, IL	12-2-1890	
620	3rd	DCN	Thos. G. Smith	5	Cincinnati, OH	9-8-1891	
622	4th	DCN		3	Springfield, IL	8-30-1890	
623	2nd	DCN		6	Detroit, MI	8-16-1890	
625	4th	SSF			Centralia, WA	8-15-1890	
626	2nd	DCN	F.B. Dougherty	1	Duluth, MN	8-28-1890	
628	2nd	DCN		4	Denver, CO	8- -1890	
629	4th	DCN	Wm. F. Mayer	14	Louisville, KY	2-16-1891	
631	3rd	DCN	(Not built?)		Cincinnati, OH		
632	3rd	DCN		6	Grand Rapids, MI	10-28-1890	
634	2nd	DCN		24	Chicago, IL	12-18-1890	
635	2nd	DCN		1	Milwaukee, WI	6-17-1891	
637	2nd	DCN		9	Minneapolis, MN	3-31-1890	
638	2nd	DCN		8	Minneapolis, MN	3-31-1890	
640	3rd	DCN	R.W. Stewart	4	Dubuque, IA	2-23-1891	
641	5th	SSF	Rescue	1	Belle Plaine, IA	9-18-1891	*
643	3rd	DCN		2	Dallas, TX	4-27-1891	
644	3rd	DCN		3	Dallas, TX	4-27-1891	
646	3rd	DCN	Henry S. Tyler	7	Louisville, KY	7-2-1891	
647	3rd	DCN			Decatur, IL	5-19-1891	
648	2nd	DCN	Edward Dundon	1	Superior, WI	7-23-1891	*
649	2nd	DCN		87	Chicago, IL	9-3-1891	
651	2nd	DCN		15	St. Louis, MO	10-6-1891	
653	2nd	DCN			St. Louis, MO	10-6-1891	
655	3rd	DCN		15	Chicago, IL	10-31-1891	
656	3rd	DCN		28,114	Chicago, IL	10-31-1891	
657	3rd	DCN		66	Chicago, IL	11-28-1891	
658	3rd	DCN		55,89	Chicago, IL	11-28-1891	
660	3rd	DCN	Pap Grayson	3	Madison, IN	11-2-1891	
661	6th	SSF	Citizens Gift	1	Westwood, OH	9-10-1891	
662					Unknown		
664					Unknown		
666	DX 1st	DCN		10	Chicago, IL	12-5-1891	
667	6th				"Stock"	1892	
668	3rd	DCN	James Tippett	2	Tower, MN	2-20-1892	*
669	3rd	DCN		3	Indianapolis, IN	2-29-1892	
670	3rd	DCN		5	Indianapolis, IN	2-29-1892	
671	2nd	DCN		7,8	Indianapolis, IN	3-3-1892	
672	2nd	DCN	R.M. Archibald	9	Cincinnati, OH	7-5-1892	
673	Built as 2101						
674	Built as 2102						

American (Ahrens Division)

Reg. #	Class	Style	Name	Co.	City & State	Date	
2101	3rd	DCN	Indiana	4	Evansville, IN	4-11-1892	
2102	3rd	DCN	Vanderburg	5	Evansville, IN	4-11-1892	
2103	3rd	DCN	Pioneer	1	Anacortes, WA	3-16-1892	
2104	3rd	DCN	Garnett Andrews	4	Chattanooga, TN	5-11-1892	*
2105	3rd	DCN	Henry N. Starr	3	Rockford, IL	5-20-1892	
2106	3rd	DCN	Geo. Clark	2	San Antonio, TX	6-13-1892	*
2107	3rd	DCN	J.S. Hogg	1	San Antonio, TX	6-13-1892	
2108	DX 1st	DCN		2	Nashville, TN	7-27-1892	
2109	3rd	DCN		97,101	Chicago, IL	7-19-1892	
2110	3rd	DCN	Wheeling	4	Wheeling, WV	8-10-1892	
2121	2nd	DCN		32,26	St. Louis, MO	9-12-1892	
2122	2nd	DCN	Schuylkill	12	Reading, PA	3-29-1893	
2123	X 1st	DCN		1	Milwaukee, WI	8-31-1892	

Reg. #	Class	Style	Name	Co.	City & State	Date
2124	1st	DCN			Louisville, Ky.	1892
2125	3rd	DCN		11,43	St. Louis, MO	10-11-1892
2126	3rd	DCN		31	St. Louis, MO	11-19-1892
2127	3rd	DCN		24,81	Chicago, IL	9-5-1892
2128	3rd	DCN		51,85	Chicago, IL	10-19-1892
2129	3rd	DCN	Bryan Callaghan	4	San Antonio, TX	10-13-1892
2130	3rd	DCN	Atlantic	2	Wheeling, WV	11-20-1892
2141	2nd	DCN	D.S. Bates	1	Piqua, OH	12-13-1892
2142	3rd	DCN	Columbian Fair	1,70	Chicago, IL	1-16-1893 *
2143	3rd	DCN	Columbian Fair	2	Chicago, IL	5-18-1893
2144	3rd	DCN	Columbian Fair	3	Chicago, IL	7-8-1893
2145	3rd	DCN	Columbian Fair	2	Chicago, IL	7-8-1893
2146	3rd	DCN	W.B. Young	1	Clarksville, TN	8-2-1893
2147	4th	DCN	Columbian Fair	4,69	Chicago, IL	11-5-1892
2148	4th	DCN	Columbian Fair	5,74	Chicago, IL	3-13-1893
2149	4th	DCN	Columbian Fair	6,79	Chicago, IL	3-13-1893
2150	4th	DCN	Columbian Fair		Chicago, IL	4-29-1893
2171	4th	DCN	(Not sold)		Milwaukee, WI	1893
2172	6th	SSF	Fountain City	1	Fountain City, WI	2-16-1893
2173	2nd	DCN		6	Milwaukee, WI	5-25-1893
2174	2nd	DCN			St. Louis, MO	10-7-1893
2175	4th	DCN			Chicago, IL	8-15-1893
2176	4th	DCN			Chicago, IL	8-15-1893
2177	4th	DCN			Milwaukee, WI	8-5-1893
2178	5th	SSF			Unknown	1893
2179	2nd	DCN			Stock	1893
2180	2nd	DCN			Stock	1893
2181	3rd	DCN			St. Louis, MO	10-7-1893
2182					Unknown	1893
2183					Unknown	1893
2184					Unknown	1893
2185					Unknown	1893
2186					Unknown	1893
2203	1st	DCN	Reading Hose	1	Reading, PA	1893
2208	4th	DCN	Rbld of #2150	89	Chicago, IL	8-1-1895
2209	3rd	DCN	Rbld of #2143		Ottawa, IL	3-1-1894
2214	2nd	DCN	Rbld of #2173	6	Milwaukee, WI	5-25-1893
2222	4th	DCN	Rbld of #2171	60	Chicago, IL	5-27-1893
2224	3rd	DCN	Rbld of #2145	88	Chicago, IL	11- -1893
2225	4th	DCN	Rbld of #2144	73,111	Chicago, IL	11- -1893
2226	4th	DCN	Rbld of #2175	17	Chicago, IL	8-15-1893
2227	4th	DCN	Rbld of #2176	38	Chicago, IL	8-15-1893
2228	4th	DCN	Rbld of #2177	18	Milwaukee, WI	8-5-1893
2234	4th	DCN		72	Chicago, IL	1893
2235	3rd	DCN	Rbld of #2146		Clarksville, TN	1894
2246	2nd	DCN	Rbld of #2174		St. Louis, MO	10-7-1893
2247	3rd	DCN	Rbld of #2181	16,23	St. Louis, MO	10-7-1893
2248	2nd	DCN		10	Cincinnati, OH	1893
2249	3rd	DCN			Evanston, IL	1893
2250	3rd	DCN			Rushville, IN	1893 *
2252	3rd	DCN			Loveland, OH	1894 *
2253	3rd	DCN		37	St. Louis, MO	1894
2258	1st	DCN	Rbld of #2124		Columbus, OH	1894
2258	X 1st	DCN			Manila, Philipines	1894
2259	X 1st	DCN			Manila, Philipines	1894
2260	2nd	DCN			Birmingham, AL	1894
2261	2nd	DCN			Erie, PA	1894
2262	2nd	DCN			St. Joseph, MO	1894
2274	1st	DCN			Dallas, TX	1893
2277	4th			75,108	Chicago, IL	1894
2279	4th			76,88	Chicago, IL	1894
2280	2nd	DCN		21	Milwaukee, WI	9-1-1894
2285	2nd	DCN		1	St. Louis, MO	1894
2286	4th	DCN		77	Chicago, IL	1894
2287	4th	DCN			New London, WI	1894
2288	4th	DCN		86	Chicago, IL	1894
2316	2nd	DCN			Springfield, IL	1894
2319	2nd	DCN			Rockford, IL	1894
2320	4th	DCN		42	St. Louis, MO	1895
2321	4th	DCN		31	St. Louis, MO	1895
2337	2nd	DCN		8	St. Louis, MO	1895
2338	3rd	DCN			Kalamazoo, MI	1895
2339	1st	DCN	Atlantic	8	Wheeling, WV	8-5-1895 *
2360	2nd	DCN		33	St. Louis, MO	1895
2362	2nd	DCN		14	Detroit, MI	7- -1896
2364	5th	DCN			Leipsic, OH	1895 *
2368	1st	DCN			Fort Worth, TX	1895
2378	3rd	DCN			Traverse City, MI	1895 *

Reg. #	Class	Style	Name	Co.	City & State	Date	
2381	1st	DCN			St. Louis, MO	1895	
2383	4th	DCN			McKinney, TX	1895	
2384	4th	DCN		28	Chicago, IL	1895	
2385	4th	DCN			McKinney, TX	1895	
2386	5th	SSF	Neptune		Oxford, OH	7-11-1895	*
2387	6th	SSF			Spencerville, OH	1895	
2399	1st	DCN			San Francisco, CA	1895	
2404	1st	DCN			San Francisco, CA	1895	
2405	3rd	DCN		5	Louisville, KY	1895	
2406	3rd	DCN		12	Louisville, KY	1895	
2413	3rd	DCN		94	Chicago, IL	1895	
2414	1st	DCN		19	Louisville, KY	1896	
2417	2nd	DCN			St. Louis, MO	1896	
2420	1st	DCN			Indianapolis, IN	1896	
2421	4th	DCN		20	St. Louis, MO	1896	
2423	2nd	DCN			New Orleans, LA	1896	
2428	2nd	DCN			New Orleans, LA	1896	*
2430	3rd	DCN			New Orleans, LA	1896	
2438	1st	DCN		3	Grand Rapids, MI	1896	
2445	2nd	DCN		22	Milwaukee, WI	1-1-1898	
2446	2nd	DCN		3	Chattanooga, TN	1896	
2447	4th	DCN		81	Chicago, IL	1896	
2458	X 1st	DCN		4	Grand Rapids, MI	1896	
2460	3rd	DCN			Columbus, OH	1896	
2466	3rd	DCN		25,115	Chicago, IL	1896	
2467	3rd	DCN		30,112	Chicago, IL	1896	
2468	4th	DCN		82,119	Chicago, IL	1896	
2469	4th	DCN		83,97	Chicago, IL	1896	
2490	2nd	DCN			Columbus, OH	1897	
2491	2nd	DCN			Columbus, OH	1897	
2492	2nd	DCN		32	St. Louis, MO	1896	
2493	3rd	DCN			New Orleans, LA	1897	
2502	4th				New Orleans, LA	1897	
2503	4th				New Orleans, LA	1897	
2504	4th				New Orleans, LA	1897	
2510	4th	DCN			Kewaunee, WI	1897	
2511	2nd	DCN			Nashville, TN	1897	
2516	1st	DCN			Cincinnati, OH	4-1-1897	
2517	1st	DCN	John Goetz Jr.	22	Cincinnati, OH	4-1-1897	
2519	4th				New Orleans, LA	1897	
2520	5th				New Orleans, LA	1897	
2522	5th				New Orleans, LA	1897	
2523	5th				New Orleans, LA	1897	
2524	5th				New Orleans, LA	1897	
2526	5th				New Orleans, LA	1897	
2527	3rd	DCN			Columbus, OH	1897	
2528		Pump	Goodrich Transp.		Chicago, IL	1897	
2545	1st	DCN		1	Chicago, IL	1897	
2546	1st	DCN		42	Chicago, IL	1897	
2547	1st	DCN		93	Chicago, IL	1897	
2554	4th	DCN			New Orleans, LA	1897	
2555	2nd	DCN			New Orleans, LA	1897	
2561	2nd	DCN			New Orleans, LA	1897	
2562	3rd	DCN		11	St. Louis, MO	1897	
2563	1st	DCN		12	St. Louis, MO	1897	
2569	2nd	DCN			New Orleans, LA	1898	
2570	2nd	DCN			New Orleans, LA	1898	
2577	2nd	DCN			New Orleans, LA	1898	
2578	4th	DCN			New Orleans, LA	1898	
2584	1st	DCN		6	St. Louis, MO	1898	
2585	1st	DCN		15	St. Louis, MO	1898	
2589	2nd	DCN	N. American Co.		Chicago, IL	1898	
2592	DX 1st	Pump	Illinois	37	Chicago, IL	1-14-1899	
2593	DX 1st	Pump	Illinois	37	Chicago, IL	1-14-1899	
2594	DX 1st	Pump	Illinois	37	Chicago, IL	1-14-1899	
2606	2nd	DCN		1	Colorado Springs, CO	1898	*
2610	2nd	DCN		15	Chicago, IL	1898	
2611	2nd	DCN		19	Chicago, IL	1898	
2612	1st	DCN		17	Chicago, IL	1898	
2623	3rd	DCN			San Francisco, CA	1899	
2624	3rd	DCN			San Francisco, CA	1899	
2638	3rd	DCN		3	Los Angeles, CA	1899	
2639	1st	DCN		4	Los Angeles, CA	1899	
2640	2nd	DCN		11	Los Angeles, CA	1899	*
2652	1st	DCN			San Francisco, CA	1899	
2653	1st	DCN			San Francisco, CA	1899	
2656	2nd	DCN			Rockford, IL	1899	
2658	2nd	DCN	Washington	2	Madison, IN	8-1-1899	
2659	X 1st	DCN			Guayaquil, Ecuador	1899	

Reg. #	Class	Style	Name	Co.	City & State	Date	
2667	2nd	DCN		8	Chicago, IL	1899	
2668	2nd	DCN		2	Chicago, IL	1899	
2670	2nd	DCN		12	Chicago, IL	1899	
2672	2nd	DCN		18	Chicago, IL	1899	
2675	2nd	DCN			San Francisco, CA	1899	
2676	2nd	DCN			San Francisco, CA	1899	*
2686	1st	DCN		2	Seattle, WA	1899	*
2687	2nd	DCN			Hammond, IN	1899	
2693	4th	DCN		47	St. Louis, MO	1900	*
2694	4th	DCN		38	St. Louis, MO	1900	
2695	4th	DCN			Aberdeen, MS	1899	*
2696	3rd	DCN	Epic		Ardmore, OK	1900	*
2705	X 1st	DCN			Duluth, MN	1900	
2706	X 1st	DCN			Denver, CO	1900	
2713	2nd	DCN			Cincinnati, OH	1900	
2715	3rd	DCN		5	Oakland, CA	1900	
2722	5th	DCN			Michigamme, MI	1900	*
2723	1st	DCN			Decatur, IL	1900	*
2737	1st	DCN		19	Milwaukee, WI	1900	*
2738	1st	DCN		20	Milwaukee, WI	1900	
2739	3rd	DCN			Atchinson, KS	1900	
2741	3rd	DCN			Fort Smith, AR	1900	
2746	2nd	DCN			Dayton, OH	1900	
2747	1st	DCN		14	St. Louis, MO	1900	
2753	1st	DCN			Bloomington, IL	1900	
2754	2nd	DCN		46	Chicago, IL	1900	
2755	2nd	DCN		90	Chicago, IL	1900	
2756	1st	DCN		13	Chicago, IL	1900	
2762	3rd	DCN			Jackson, MS	1900	
2765	3rd	DCN		3	Nashville, TN	12-14-1900	*
2766	2nd	DCN			Akron, OH	1900	
2767	2nd	DCN			Bakersfield, CA	1900	
2768	4th	DCN			Glenville, OH	1900	
2770	3rd	DCN			Morgan City, LA	1900	
2773	2nd	DCN			Jackson, MI	1900	
2777	1st	DCN		4	Spokane, WA	1900	
2780	1st	DCN			Honolulu, HI	1900	
2782	X 1st	DCN			Indianapolis, IN	1901	
2783	X 1st	DCN			Indianapolis, IN	1901	
2785	1st	DCN			Birmingham, AL	1900	
2786	2nd	DCN			San Francisco, CA	1901	
2789	1st	DCN			El Paso, TX	1901	
2791	X 1st	DCN			Portland, OR	1901	*
2798	4th	DCN		7	Houston, TX	1901	
2799	1st	DCN		18	St. Louis, MO	1901	
2804	2nd	DCN	Empire	1	San Jose, CA	1901	
2806	2nd	DCN			Traverse City, MI	1901	
2807	2nd	DCN			Matanzas, Cuba	1901	
2809	4th	DCN			Babahoyo, Ecuador	1901	
2816	2nd	DCN			Cincinnati, OH	1902	
2817	2nd	DCN			Cincinnati, OH	1902	
2818	2nd	DCN			Cincinnati, OH	1902	
2819	1st	DCN		7	Louisville, KY	1902	
2820	1st	DCN		8	Louisville, KY	1902	
2821	X 1st	DCN		3	Wichita, KS	1902	*
2823	2nd	DCN			Reno, NV	1902	
2824	X 1st	DCN			Portland, OR	1902	*
2828	2nd	DCN			Manila, Philipines	1902	
2829	2nd	DCN			Manila, Philipines	1902	
2835	1st	DCN		10,21	Chicago, IL	1902	
2838	X 1st	DCN			Battle Creek, MI	1902	
2839	2nd	DCN		2	Springfield, OH	1903	
2841	2nd	DCN		4	Springfield, OH	1903	
2842	3rd	DCN			Quincy, IL	1902	
2844	1st	DCN			Salt Lake City, UT	1902	
2854	3rd	DCN			Seattle, WA	1902	
2856	3rd	DCN			Seattle, WA	1902	
2857	1st	DCN		11	Chicago, IL	1902	
2858	1st	DCN		5	Chicago, IL	1902	
2859	1st	DCN		34,17	Chicago, IL	1902	*
2861		Pump	Fireboat		Milwaukee, WI	1902	
2862		Pump	Fireboat		Milwaukee, WI	1902	
2863		Pump	Fireboat		Milwaukee, WI	1902	
2866	6th	SSF			Xenia, IL	1902	
2867	X 1st	DCN		2	St. Louis, MO	1902	
2868	X 1st	DCN		7	St. Louis, MO	1902	
2869	6th	SSF			Niles Center, OH	1902	
2870	3rd	DCN			West Bend, WI	1902	
2880	1st	DCN		13	St. Louis, MO	1902	

Reg. #	Class	Style	Name	Co.	City & State	Date	
2881	1st	DCN		24	St. Louis, MO	1902	
2882	X 1st	DCN			Kalamazoo, MI	1902	*
2883	X 1st	DCN			Denver, CO	1902	
2890	4th	DCN			Cavite, Philipines	1902	
2892	X 1st	DCN			Aberdeen, WA	1902	*
2893	X 1st	DCN			Denver, CO	1902	
2896	3rd	DCN		1	San Diego, CA	1902	
2902	X 1st	DCN	Exposition		St. Louis, MO	1902	
2903	2nd	DCN			Dallas, TX	1902	
2904	X 1st	DCN			Dallas, TX	1902	
2905	1st	DCN		2	Milwaukee, WI	1902	
2906	1st	DCN		9	Milwaukee, WI	1902	
2907	1st	DCN		7	Milwaukee, WI	1902	
2908	2nd	DCN			Birmingham, AL	1902	
2909	4th	DCN			Bloomington, IL	1902	
2910	3rd	DCN			Boise, ID	1902	*
2911	2nd	DCN	Queen City	1	Springfield, MO	1902	*
2912	2nd	DCN		2	Springfield, MO	1902	
2913	1st	DCN			Tacoma, WA	1902	
2914	X 1st	DCN			Denver, CO	1902	
2925	3rd	DCN			Indianapolis, IN	1903	
2928	X 1st	DCN			Sacramento, CA	1903	
2931	4th	DCN			Manila, Philipines	1903	
2932	4th	DCN			Manila, Philipines	1903	
2933	4th	DCN			Manila, Philipines	1903	
2937	5th	SSF			Coney Island, NY	1903	
2938	5th	SSF			San Salvador, El Salvador	1903	
2939	5th	SSF			Durban, South Africa	1903	
2942	5th	SSF			North Lewisburg, OH	1903	
2943	5th	SSF			Fort Thomas, KY	1903	
2944	5th	SSF			Millersburg, KY	1903	
2945	3rd	DCN			Wheeling, WV	1903	
2946	1st	DCN			Evansville, IN	1903	
2948	2nd	DCN			Portsmouth, OH	1903	

American-LaFrance (Cincinnati Plant)

Reg. #	Class	Style	Name	Co.	City & State	Date	
2955	2nd	DCN			Jacksonville, IL	1904	*
2959	5th	SSF			Claremore, OK	1904	*
2960	X 1st	DCN		40	St. Louis, MO	1904	
2961	2nd	DCN		5	St. Louis, MO	1904	
2962	2nd	DCN		25	St. Louis, MO	1904	
2963	5th	DCN			Carlisle, KY	1904	
2964	2nd	DCN		32,34	Chicago, IL	1904	
2965	2nd	DCN		40,29	Chicago, IL	1904	
2966	2nd	DCN		16,27	Chicago, IL	1904	
2968	1st	DCN			Walla Walla, WA	1904	
2969	2nd	DCN			Reno, NV	1904	*
2978	2nd	DCN			St. Paul, MN	1904	
2979	2nd	DCN			Galveston, TX	1904	*
2985	3rd	DCN		5	Fresno, CA	1904	
2986	2nd	DCN			Vancouver, WA	1904	
2989	1st	DCN		4	Milwaukee, WI	1904	
2990	1st	DCN		10	Milwaukee, WI	1904	
2991	1st	DCN		3	Milwaukee, WI	1904	*
2992	X 1st	DCN		2	Louisville, KY	1904	
2993	1st	DCN		4	Louisville, KY	1904	
2994	2nd	DCN		41	St. Louis, MO	1904	
2995	4th	DCN			Beaumont, TX	1904	
2996	4th	DCN			Beaumont, TX	1904	
2997	2nd	DCN	L.C. Cron		Piqua, OH	1904	*
2998	2nd	DCN		2	Topeka, KS	1904	*
2999	X 1st	DCN			Omaha, NE	1904	
3000	2nd	DCN			Marysville, CA	1904	
3001	2nd	DCN			Youngstown, OH	1904	
3002	2nd	DCN		3	Grand Rapids, MI	1904	
3003	2nd	DCN			Riverside, CA	1904	
3004	3rd	DCN			East St. Louis, IL	1904	
3005	3rd	DCN			Evansville, IN	1904	
3006	X 1st	DCN		2	Eureka, CA	1904	*
3007	1st	DCN			Montgomery, AL	1904	*
3008	4th	DCN			Dallas, TX	1904	
3009	1st	DCN			Portland, OR	1904	
3010	1st	DCN			Cincinnati, OH	1904	
3011	1st	DCN			Cincinnati, OH	1904	

Ahrens Continental

Reg. #	Class	Style	Name	Co.	City & State	Date	
101	3rd	DCN			East St. Louis, IL	1905	
102	3rd	DCN		2	Indianapolis, IN	11- -1905	
103	2nd	DCN			Cincinnati, OH	1905	
104	1st	DCN			Charleston, WV	1905	
105	1st	DCN		9	Louisville, KY	1905	
106	3rd	DCN	H.C. Sponhauer		Galion, OH	1906	*
107	1st	DCN		1	Toledo, OH	1906	
108	3rd	DCN	G.F.E. Kanfleiter		Terre Haute, IN	1906	
109	3rd	DCN	Edwin J. Bidaman		Terre Haute, IN	1906	
110	X 1st	DCN	John Schafer Jr.	1	Superior, WI	1906	*
111	2nd	DCN		16	Louisville, KY	1906	
112	2nd	DCN		4	Seattle, WA	1906	
113	3rd	DCN	North Montgomery	5	Montgomery, AL	7-23-1906	
114	1st	DCN		1	Denver, CO	1906	
115	2nd	DCN		3	Dallas, TX	10- -1906	
116	2nd	DCN		18	Buffalo, NY	1906	
117	1st	DCN		10	Grand Rapids, MI	1906	
118	2nd	DCN	Geo. R. Harrison		San Diego, CA	12-11-1906	
119	X 1st	DCN		2	Kansas City, MO	1907	
120	2nd	DCN		59	Chicago, IL	1907	
121	2nd	DCN		64	Chicago, IL	1907	
122					Unknown		
123	3rd	DCN			Everett, WA	1907	*
124	2nd	DCN		15	Louisville, KY	1907	
125	3rd	DCN			Canal Dover, OH	1907	*
126	DX 1st	Pump	S.S. Alabama		Chicago, IL	1907	
127					Unknown		
128	4th	DCN		21	Los Angeles, CA	1907	
129	2nd	DCN		3	Toledo, OH	1907	
130	2nd	DCN		7	Toledo, OH	1907	
131	2nd	DCN		2	Valdez, AK	10-1-1907	*
132					Unknown		
133	2nd	DCN		38	Chicago, IL	1907	
134	2nd	DCN		10	Chicago, IL	1907	
135	2nd	DCN		39	Chicago, IL	1907	*
136	2nd	DCN		120	Chicago, IL	1907	
137	3rd	DCN		3	Seattle, WA	1907	
138					Unknown		
139	3rd	DCN		1	Van Wert, OH	10-1-1907	*
140					Unknown		
141	1st	DCN			Omaha, NE	1908	
142	1st	DCN		1	St. Paul, MN	1908	
143	1st	DCN		16	St. Paul, MN	1908	
144	1st	DCN			Houston, TX	1908	
145					Unknown		
146					Unknown		
147					Unknown		
148	1st	DCN			Cincinnati, OH	1908	
149					Unknown		
150					Unknown		
151					Unknown		
152					Unknown		
153					Unknown		
154					Unknown		
155	2nd	DCN		13	Detroit, MI	1908	*
156	2nd	DCN		34	San Francisco, CA	1908	
157	2nd	DCN			Toledo, OH	1908	
158	3rd	DCN			St. Paul, MN	1908	
159					Unknown		
160	3rd	DCN			Merced, CA	1908	
161					Unknown		
162	1st	DCN		4	Baltimore, MD	1909	
163	3rd	DCN			New Orleans, LA	1909	
164	3rd	DCN			New Orleans, LA	1909	
165	3rd	DCN		2	Denver, CO	1909	
166		DCN			Columbus, OH	1909	
167					Unknown		
168					Unknown		
169	3rd	DCN		41	San Francisco, CA	2- -1910	
170	1st	DCN	Columbia	4	Allentown, PA	1909	*
171	2nd	DCN			Seattle, WA	3-7-1910	
172	1st	DCN	Wm. A. Hancock	4	Atlanta, GA	6- -1910	
173	3rd	DCN		6	Los Angeles, CA	1910	
174		DCN			Key West, FL	1910	
175					Unknown		
176	3rd	DCN			New Orleans, LA	6- -1910	
177	3rd	DCN			New Orleans, LA	6- -1910	

Reg. #	Class	Style	Name	Co.	City & State	Date	
178		DCN			Spokane, WA	1910	
179	1st	DCN	E.H. Crump	5	Memphis, TN	7-4-1910	*
180	2nd	DCN			Memphis, TN	7-4-1910	

Ahrens-Fox Continental

Reg. #	Class	Style	Name	Co.	City & State	Date	
181					Unknown		
182					Unknown		
183					Unknown		
184	3rd	DCN	Friendship	1	Woodbury, NJ	1911	*
185	2nd	DCN		32	Chicago, IL	1911	
186					Unknown		
187	DX 1st	DCN		1	Beaumont, TX	11- -1911	
188	2nd	DCN		1	Washington, DC	10-31-1911	
189	2nd	DCN			Indianapolis, IN	1911	
190	2nd	DCN		122	Chicago, IL	1912	
191	2nd	DCN		43	Chicago, IL	1912	
192	2nd	DCN		106	Chicago, IL	1912	
193	2nd	DCN		123	Chicago, IL	1912	
194					Unknown		
195					Unknown		
196					Unknown		
197	1st	DCN		23	Rochester, NY	1913	
198	2nd	DCN		3	Paterson, NJ	9-16-1913	
199	2nd	DSF		2	Somerville, MA	1913	*
200	4th	DCN		39	Philadelphia, PA	1914	
201	4th	DCN		54	Philadelphia, PA	1914	
202	4th	DCN		35	Philadelphia, PA	1914	
203	4th	DCN		19	Philadelphia, PA	1914	
204	1st	DSF			Hamtramck, MI	1915	
205	2nd	DSF	American	1	Lansford, PA	1916	*

Ahrens Continental (Numbers Unknown)

Reg. #	Class	Style	Name	Co.	City & State	Date	
AAA	2nd	DCN		2	Tacoma, WA	1906	*
BBB	3rd	DCN			Columbus, OH	10- -1906	
CCC	1st	DCN		2	Roanoke, VA	1907	
DDD	1st	DCN		1	San Jose, CA	1907	
EEE	3rd	DCN			East St. Louis, IL	1907	
FFF	2nd	DCN		1	Berkeley, CA	1907	
GGG	2nd	DCN		1	Covington, KY	1908	
HHH	2nd	DCN		4	Dayton, OH	1908	
III	2nd	DCN		6	Dayton, OH	1908	
JJJ	2nd	DCN		13	Dayton, OH	1908	
KKK	2nd	DCN		5	Berkeley, CA	1908	
LLL	2nd	DCN		3	Louisville, KY	1908	
MMM	4th	DCN		30	Cleveland, OH	1908	
NNN	3rd	DCN		1	Charelston, WV	1908	
OOO		DCN			Wheeling, WV		
PPP		DCN			Wheeling, WV		
QQQ	2nd	DCN			Janesville, WI	1909	
RRR	1st	DCN		2	Baltimore, MD	1909	
SSS		DCN			Chicago Heights, IL		
TTT	1st	DCN		23	Baltimore, MD	1909	
UUU	1st	DCN		38	Baltimore, MD	1909	
VVV	2nd	DCN		1	Milwaukee, WI	2- -1910	
WWW	1st	DCN		13	Denver, CO	1910	
XXX		DCN			Clarksburg, WV	1910	
YYY	2nd	DCN	Herbert P. Keller	7	St. Paul, MN	11- -1910	
ZZZ	2nd	DCN		14	St. Paul, MN	11- -1910	
aaa	2nd	DCN		3	St. Paul, MN	1911	
bbb		DCN			Springfield, MO		
ccc		DCN			Springfield, MO		
ddd	1st	DCN			Denver, CO	11- -1912	

Alphabetical Index of Engines

City & State	Fire Co.	Make	Reg. No.	Year
Aberdeen, MS		American	2695	1899
Aberdeen, WA		American	2892	1902
Adrian, MI	(rblt 1873 Silsby)	Ahrens Mfg.	446	1885
Akron, OH		Lane & Bodley	12	1866
Akron, OH	1	Ahrens Mfg.	465	1886
Akron, OH		American	2766	1900
Allentown, PA	4	Ahrens F.E.	170	1909
Alton, IL		C. Ahrens	9	1871
Anacortes, WA	1	American	2103	1892
Arcanum, OH		Ahrens Mfg.	425	1885
Ardmore, OK		American	2696	1900
Ashtabula, OH		Ahrens Mfg.	455	1886
Atchinson, KS		American	2739	1900
Atlanta, GA	4	Ahrens F.E.	172	1910
Auburn, IN		Ahrens Mfg.	416	1885
Aurora, IN		C. Ahrens	166	1876
Aurora, IN	1	Ahrens Mfg.	394	1885
Babahoyo, Ecuador		American	2809	1901
Bakersfield, CA		American	2767	1900
Baltimore, MD	4	Ahrens F.E.	162	1909
Baltimore, MD	2	Ahrens F.E.	SSS	1909
Baltimore, MD	23	Ahrens F.E.	TTT	1909
Baltimore, MD	38	Ahrens F.E.	UUU	1909
Bardstown, KY		Ahrens Mfg.	453	1885
Battle Creek, MI		Lane & Bodley	6	1863
Battle Creek, MI		Ahrens Mfg.	290	1881
Battle Creek, MI		American	2838	1902
Beaumont, TX		ALF	2995	1904
Beaumont, TX		ALF	2996	1904
Beaumont, TX		Ahrens-Fox	187	1911
Belle Plaine, IA	1	Ahrens Mfg.	641	1891
Belleville, IL		C. Ahrens	9	1876
Belleville, IL		Ahrens Mfg.	304	1888
Berkeley, CA	1	Ahrens F.E.	FFF	1907
Berkeley, CA	5	Ahrens F.E.	KKK	1908
Birmingham, AL		Ahrens Mfg.	459	1886
Birmingham, AL		Ahrens Mfg.	520	1887
Birmingham, AL		American	2260	1894
Birmingham, AL		American	2785	1900
Birmingham, AL		American	2908	1902
Bloomington, IL		American	2753	1900
Bloomington, IL		American	2909	1902
Boise, ID		American	2910	1902
Boston, MA		Latta	5	1855
Bryan, OH		American	2723	??
Buffalo, NY	18	Ahrens F.E.	116	1906
Cambridge City, IN		C. Ahrens	163	1876
Camp Carlin, MO	U.S. Army	Ahrens Mfg.	242	1880
Canal Dover, OH		Ahrens F.E.	125	1907
Canton, OH		Ahrens Mfg.	549	1888
Cardington, OH		Ahrens Mfg.	348	1903
Carlisle, KY		ALF	2963	1904
Catlettsburg, KY		Ahrens Mfg.	417	1884
Cavite, Philipines		American	2890	1902
Cedarburg, WI		Ahrens Mfg.	412	1889
Celina, OH		Ahrens Mfg.	390	1884
Centralia, WA		Ahrens Mfg.	625	1890
Charleston, WV		Ahrens F.E.	104	1905
Charleston, WV		Ahrens F.E.	NNN	1908
Chattanooga, TN	1	C. Ahrens	146	1875
Chattanooga, TN	1	Ahrens Mfg.	570	1888
Chattanooga, TN	2	Ahrens Mfg.	418	1885
Chattanooga, TN		Ahrens Mfg.	461	1886
Chattanooga, TN	3	Ahrens Mfg.	558	1888
Chattanooga, TN	3	American	2446	1896
Chattanooga, TN	4	American	2104	1892
Checotah, OK	Lafayette Bros.	C. Ahrens	149	1899
Cheyenne, WY	U.S. Army	Ahrens Mfg.	242	1880
Chicago, IL	1	American	2545	1897
Chicago, IL	2,38,113	Ahrens Mfg.	264	1880
Chicago, IL	2	American	2668	1899
Chicago, IL	3,95,117	Ahrens Mfg.	374	1883
Chicago, IL	5,9	Ahrens Mfg.	254	1880
Chicago, IL	5	American	2858	1902
Chicago, IL	7,2,56,52,86	Ahrens Mfg.	248	1880
Chicago, IL	8,26,79	Ahrens Mfg.	298	1881
Chicago, IL	8,20	American	2667	1899

City & State	Fire Co.	Make	Reg. No.	Year
Chicago, IL	10,21	Ahrens Mfg.	666	1891
Chicago, IL	10,21	American	2835	1902
Chicago, IL	10,90	Ahrens F.E.	134	1907
Chicago, IL	11	Ahrens Mfg.	306	1881
Chicago, IL	11	American	2857	1902
Chicago, IL	12,24	American	2670	1899
Chicago, IL	13,59	Ahrens Mfg.	268	1881
Chicago, IL	13,53	American	2756	1900
Chicago, IL	14,50	Ahrens Mfg.	308	1881
Chicago, IL	15,78,48,95,109	Ahrens Mfg.	334	1882
Chicago, IL	15,39	Ahrens Mfg.	366	1883
Chicago, IL	15,20,65	Ahrens Mfg.	655	1891
Chicago, IL	15,36	American	2610	1898
Chicago, IL	16,27	ALF	2966	1904
Chicago, IL	17,33	American	2175/2226	1893
Chicago, IL	17	American	2612	1898
Chicago, IL	18,57	American	2672	1899
Chicago, IL	19,44	American	2611	1898
Chicago, IL	22,76	Ahrens Mfg.	523	1887
Chicago, IL	23,77	Ahrens Mfg.	429	1885
Chicago, IL	24,54	Ahrens Mfg.	634	1890
Chicago, IL	24,81	American	2127	1892
Chicago, IL	25	C. Ahrens	31	1873
Chicago, IL	25,115	American	2466	1896
Chicago, IL	26,49,68	Ahrens Mfg.	342	1882
Chicago, IL	28,114	Ahrens Mfg.	656	1891
Chicago, IL	28,80,73	American	2384	1895
Chicago, IL	29,116	Ahrens Mfg.	427	1885
Chicago, IL	30,43	Ahrens Mfg.	260	1880
Chicago, IL	30,112	American	2467	1896
Chicago, IL	31,100	Ahrens Mfg.	262	1880
Chicago, IL	31,65	Ahrens Mfg.	300	1881
Chicago, IL	31,74	Ahrens Mfg.	407	1884
Chicago, IL	32,3	Ahrens Mfg.	296	1881
Chicago, IL	32,67	Ahrens Mfg.	367	1883
Chicago, IL	32,34	ALF	2964	1904
Chicago, IL	32,75	Ahrens-Fox	185	1911
Chicago, IL	34,17	American	2859	1902
Chicago, IL	37 (boat)	American	2592	1898
Chicago, IL	37 (boat)	American	2593	1898
Chicago, IL	37 (boat)	American	2594	1898
Chicago, IL	38,35,109	American	2176/2227	1893
Chicago, IL	38	Ahrens F.E.	133	1907
Chicago, IL	39,56	Ahrens F.E.	135	1907
Chicago, IL	40,4	Ahrens Mfg.	501	1886
Chicago, IL	40,29	ALF	2965	1904
Chicago, IL	42	American	2546	1897
Chicago, IL	43	Ahrens-Fox	191	1912
Chicago, IL	44,75	Ahrens Mfg.	567	1888
Chicago, IL	46	American	2754	1900
Chicago, IL	47	Ahrens Mfg.	619	1890
Chicago, IL	49,91	Ahrens Mfg.	491	1886
Chicago, IL	51,85	American	2128	1892
Chicago, IL	53,69	Ahrens Mfg.	568	1888
Chicago, IL	54,67	Ahrens Mfg.	408	1884
Chicago, IL	55,89	Ahrens Mfg.	658	1891
Chicago, IL	56	Ahrens Mfg.	336	1882
Chicago, IL	57,96	Ahrens Mfg.	409	1884
Chicago, IL	59	Ahrens F.E.	120	1907
Chicago, IL	60	American	2171/2222	1893
Chicago, IL	62,99	Ahrens Mfg.	440	1885
Chicago, IL	63	American	2150/2208	1893
Chicago, IL	64	Ahrens Mfg.	601	1889
Chicago, IL	64	Ahrens F.E.	121	1907
Chicago, IL	66	Ahrens Mfg.	657	1891
Chicago, IL	69	American	2147	1893
Chicago, IL	70,112	American	2142	1893
Chicago, IL	72	American	2234	1893
Chicago, IL	73,61	Ahrens Mfg.	603	1890
Chicago, IL	73,111,113	American	2144/2225	1893
Chicago, IL	74	American	2148	1893
Chicago, IL	75,108	American	2277	1894
Chicago, IL	76,88	American	2279	1894
Chicago, IL	77	American	2286	1894
Chicago, IL	78,105	Ahrens Mfg.	524	1887
Chicago, IL	79	American	2149	1893
Chicago, IL	81	American	2447	1896
Chicago, IL	82,119	American	2468	1896
Chicago, IL	83,97	American	2469	1896

City & State	Fire Co.	Make	Reg. No.	Year
Chicago, IL	85	Ahrens Mfg.	576	1889
Chicago, IL	86	American	2288	1894
Chicago, IL	87	Ahrens Mfg.	649	1891
Chicago, IL	88,108,72	American	2145/2224	1893
Chicago, IL	89	American	2150/2208	1893
Chicago, IL	90	American	2755	1900
Chicago, IL	91,84,87	Ahrens Mfg.	441	1885
Chicago, IL	93	American	2547	1897
Chicago, IL	94,36	Ahrens Mfg.	439	1885
Chicago, IL	94	American	2413	1895
Chicago, IL	97,101	American	2109	1892
Chicago, IL	105,27	Ahrens Mfg.	521	1887
Chicago, IL	106	Ahrens-Fox	193	1912
Chicago, IL	120	Ahrens F.E.	136	1907
Chicago, IL	122	Ahrens-Fox	190	1912
Chicago, IL	123	Ahrens-Fox	193	1912
Chicago, IL	Columbian Expo	American	2142	1892
Chicago, IL	Columbian Expo	American	2143/2209	1892
Chicago, IL	Columbian Expo	American	2144/2225	1892
Chicago, IL	Columbian Expo	American	2145/2224	1893
Chicago, IL	Columbian Expo	American	2147	1893
Chicago, IL	Columbian Expo	American	2148	1893
Chicago, IL	Columbian Expo	American	2149	1893
Chicago, IL	Columbian Expo	American	2150/2208	1893
Chicago, IL	Goodrich Trans	American	2528	1897
Chicago, IL	N. American	American	2589	1898
Chicago, IL	S.S. Alabama	Ahrens F.E.	126	1907
Chicago Heights, IL		Ahrens F.E.	RRR	??
Chicora, MS	Robinson Lumber	Ahrens Mfg.	338	1911
Chillicothe, OH	1	Ahrens Mfg.	232	1880
Cincinnati, OH		Latta	2	1853
Cincinnati, OH		Lane & Bodley	8	1864
Cincinnati, OH		Ahrens Mfg.	631	1890
Cincinnati, OH		American	2516	1897
Cincinnati, OH		American	2713	1900
Cincinnati, OH		American	2816	1902
Cincinnati, OH		American	2817	1902
Cincinnati, OH		American	2818	1902
Cincinnati, OH		ALF	3010	1904
Cincinnati, OH		ALF	3011	1904
Cincinnati, OH		Ahrens F.E.	103	1905
Cincinnati, OH		Ahrens F.E.	148	1908
Cincinnati, OH	1	Latta	11	1856
Cincinnati, OH	1	C. Ahrens	3	1877
Cincinnati, OH	1,17	Ahrens Mfg.	322	1883
Cincinnati, OH	2	Latta	27	1861
Cincinnati, OH	2,4	Ahrens Mfg.	316	1882
Cincinnati, OH	3	Latta	3	1854
Cincinnati, OH	3,10,21,27	Ahrens Mfg.	228	1879
Cincinnati, OH	3,25,3,12	Ahrens Mfg.	280	1881
Cincinnati, OH	3,25	Ahrens Mfg.	534	1888
Cincinnati, OH	4	Latta	7	1854
Cincinnati, OH	4,13	Ahrens Mfg.	198	1878
Cincinnati, OH	4	Lane & Bodley	21	1867
Cincinnati, OH	5	Latta	28	1861
Cincinnati, OH	5,17	C. Ahrens	11	1871
Cincinnati, OH	5	C. Ahrens	36	1874
Cincinnati, OH	5	Ahrens Mfg.	621	1891
Cincinnati, OH	6	Latta	29	1861
Cincinnati, OH	6	C. Ahrens	7	1870
Cincinnati, OH	7	Latta	30	1861
Cincinnati, OH	7	C. Ahrens	43	1875
Cincinnati, OH	8	Latta	8	1855
Cincinnati, OH	8	C. Ahrens	4	1870
Cincinnati, OH	8,14	Ahrens Mfg.	314	1882
Cincinnati, OH	8,32	Ahrens Mfg.	443	1885
Cincinnati, OH	9,3	C. Ahrens	27	1873
Cincinnati, OH	9,2	C. Ahrens	34	1874
Cincinnati, OH	9	Ahrens Mfg.	672	1892
Cincinnati, OH	10	Latta	9	1855
Cincinnati, OH	10,2	C. Ahrens	1	1869
Cincinnati, OH	10	American	2248	1893
Cincinnati, OH	12	Latta	10	1855
Cincinnati, OH	12,20,12	C. Ahrens	17	1871
Cincinnati, OH	13,1	C. Ahrens	3	1869
Cincinnati, OH	13,18	Ahrens Mfg.	503	1887
Cincinnati, OH	14	Latta	20	1858
Cincinnati, OH	16,12	C. Ahrens	6	1870
Cincinnati, OH	16	Ahrens Mfg.	410	1884

City & State	Fire Co.	Make	Reg. No.	Year
Cincinnati, OH	17,5,7,14,13	C. Ahrens	10	1871
Cincinnati, OH	18,9	C. Ahrens	13	1871
Cincinnati, OH	18	C. Ahrens	172	1877
Cincinnati, OH	19	C. Ahrens	14	1871
Cincinnati, OH	19	Ahrens Mfg.	186	1877
Cincinnati, OH	21	Ahrens Mfg.	573	1889
Cincinnati, OH	22,20	Ahrens Mfg.	550	1889
Cincinnati, OH	22	American	2517	1897
Cincinnati, OH	23	Ahrens Mfg.	551	1889
Cincinnati, OH	24	Ahrens Mfg.	604	1890
Cincinnati, OH	26	Ahrens Mfg.	552	1889
Cincinnati, OH	29,6,4	Ahrens Mfg.	571	1889
Cincinnati, OH	Fleischman Co.	C. Ahrens	3	1887
Claremore, OK		ALF	2959	1904
Clarksburg, WV		Ahrens-Fox	XXX	1910
Clarksville, TN	1	American	2146/2235	1893
Cleveland, OH		Lane & Bodley	5	1863
Cleveland, OH	10,22	C. Ahrens	168	1877
Cleveland, OH	8,20	Ahrens Mfg.	200	1878
Cleveland, OH	30	Ahrens F.E.	MMM	1908
Cochran, IN		C. Ahrens	45	1898
Colorado Springs, CO	1	American	2606	1898
Columbia, TN		Ahrens Mfg.	190	1878
Columbus, OH		Latta	13	1856
Columbus, OH	1	Ahrens Mfg.	607	1890
Columbus, OH	2	Ahrens Mfg.	574	1889
Columbus, OH	2	Ahrens Mfg.	611	1890
Columbus, OH	3	Ahrens Mfg.	608	1890
Columbus, OH	4	Ahrens Mfg.	596	1889
Columbus, OH	4	Ahrens Mfg.	614	1890
Columbus, OH	7	Ahrens Mfg.	613	1890
Columbus, OH	8	Ahrens Mfg.	610	1890
Columbus, OH		American	2124/2258	1894
Columbus, OH		American	2130	1895
Columbus, OH		American	2460	1896
Columbus, OH		American	2490	1897
Columbus, OH		American	2491	1897
Columbus, OH		American	2527	1897
Columbus, OH		Ahrens F.E.	166	1909
Columbus, OH		Ahrens F.E.	BBB	1906
Columbus, OH	Seagrave & Co.	Ahrens Mfg.	302	1902
Coney Island, NY		American	2937	1903
Council Bluffs, IA		Ahrens Mfg.	212	1880
Covington, KY		C. Ahrens	169	1876
Covington, KY		C. Ahrens	171	1877
Covington, KY		Ahrens Mfg.	346	1882
Covington, KY	1	Ahrens F.E.	GGG	1908
Crystal City, MO	Glass Works	C. Ahrens	34	1881
Cynthiana, KY	1	C. Ahrens	159	1876
Dallas, TX	1,6	Ahrens Mfg.	392	1884
Dallas, TX	2	Ahrens Mfg.	643	1891
Dallas, TX	3	Ahrens Mfg.	644	1891
Dallas, TX	3	Ahrens F.E.	115	1906
Dallas, TX	4	American	2274	1893
Dallas, TX	4	American	2274	1893
Dallas, TX		American	2903	1902
Dallas, TX		American	2904	1902
Dallas, TX		ALF	3008	1904
Dayton, OH	1	Lane & Bodley	1	1863
Dayton, OH	2	Lane & Bodley	2	1863
Dayton, OH	3	Lane & Bodley	3	1863
Dayton, OH	3	Ahrens Mfg.	445	1885
Dayton, OH	4	Ahrens Mfg.	589	1889
Dayton, OH	4	Ahrens F.E.	HHH	1908
Dayton, OH	6	Ahrens F.E.	III	1908
Dayton, OH	13	Ahrens F.E.	JJJ	1908
Dayton, OH		American	2746	1900
Decatur, IL		Ahrens Mfg.	647	1891
Decatur, IL		American	2723	1900
Delphos, OH	1	Ahrens Mfg.	272	1881
Delphos, OH	1	Ahrens Mfg.	376	1883
Denver, CO	1	Ahrens F.E.	114	1906
Denver, CO	2	Ahrens F.E.	165	1909
Denver, CO	4	Ahrens Mfg.	628	1890
Denver, CO		American	2706	1900
Denver, CO		American	2883	1902
Denver, CO		American	2893	1902
Denver, CO		American	2914	1902
Denver, CO	13	Ahrens-Fox	WWW	1910

City & State	Fire Co.	Make	Reg. No.	Year
Denver, CO		Ahrens-Fox	ddd	1912
Deshler, OH		Ahrens Mfg.	487	1887
Detroit, MI	6,7,15	Ahrens Mfg.	623	1890
Detroit, MI	13,29	Ahrens F.E.	155	1908
Detroit, MI	14,12,38	American	2362	1896
Detroit, MI	15,22	Ahrens Mfg.	538	1888
Detroit, MI	18,1,8,9,17,21	Ahrens Mfg.	518	1887
Dubuque, IA	4	Ahrens Mfg.	640	1891
Duluth, MN	1	Ahrens Mfg.	626	1890
Duluth, MN		American	2705	1900
Durban, South Africa		American	2939	1903
East St. Louis, IL		ALF	3008	1904
East St. Louis, IL		Ahrens F.E.	101	1905
East St. Louis, IL		Ahrens F.E.	EEE	1907
Eaton, OH	1	C. Ahrens	18	1872
Eaton, OH		Ahrens Mfg.	272	1884
El Paso, TX		American	2789	1901
Elwood, IN		Ahrens Mfg.	625	1890
Erie, PA		American	2261	1894
Eureka, CA	2	ALF	3006	1904
Evanston, IL		American	2249	1893
Evanston, IL		American	2150/2208	1894
Evansville, IN	2	Latta	12	1856
Evansville, IN	4	American	2101	1892
Evansville, IN	5	American	2102	1892
Evansville, IN		American	2946	1903
Evansville, IN		American	2882	??
Evansville, IN		ALF	3005	1904
Everett, WA		Ahrens F.E.	123	1907
Fond-du-Lac, WI		Ahrens Mfg.	505	1918
Fort Brady, MI		American	2937	1903
Fort Duchesne, UT	U.S. Army	Ahrens Mfg.	242	1897
Fort Smith, AR		American	2741	1900
Fort Thomas, KY		American	2943	1903
Fort Worth, TX		American	2368	1895
Fountain City, WI	1	American	2172	1893
Frankfort, KY		Ahrens Mfg.	292	1881
Franklin, OH		Ahrens Mfg.	178	1877
Franklin, OH		Ahrens Mfg.	461	1886
Fresno, CA	1	Ahrens Mfg.	526	1887
Fresno, CA	5	ALF	2985	1904
Galion, OH		C. Ahrens	25	1873
Galion, OH		Ahrens F.E.	106	1906
Gallipolis, OH		Ahrens Mfg.	184	1877
Galveston, TX		Lane & Bodley	15	1866
Galveston, TX	6	Ahrens Mfg.	174	1877
Galveston, TX		ALF	2979	1904
Georgetown, KY		C. Ahrens	170	1877
Germantown, OH		Ahrens Mfg.	475	1886
Glenville, OH		American	2768	1900
Grand Rapids, MI		Lane & Bodley	10	1865
Grand Rapids, MI	3	American	2438	1896
Grand Rapids, MI	3	ALF	3002	1904
Grand Rapids, MI	4	American	2458	1896
Grand Rapids, MI	6,7	Ahrens Mfg.	632	1890
Grand Rapids, MI	10	Ahrens F.E.	117	1906
Greencastle, IN		Ahrens Mfg.	302	1882
Greensburg, IN	1	C. Ahrens	39	1874
Guayaquil, Ecuador		American	2659	1899
Hamilton, IL		C. Ahrens	157	1894
Hamilton, OH		Lane & Bodley	7	1864
Hamilton, OH	1	C. Ahrens	32	1873
Hamilton, OH	3	C. Ahrens	2	1869
Hammond, IN		American	2687	1899
Hamtramck, MI		Ahrens-Fox	204	1915
Harrisburg, PA	2	Latta	4	1854
Harrison, OH		Ahrens Mfg.	388	1883
Hillsboro, OH		C. Ahrens	151	1875
Honolulu, HI		American	2780	1900
Hot Springs, AR		Ahrens Mfg.	256	1880
Houston, TX	7	American	2798	1901
Houston, TX		Ahrens F.E.	144	1908
Huntington, WV		Ahrens Mfg.	328	1882
Indianapolis, IN	2	Latta	26	1860
Indianapolis, IN	2	C. Ahrens	160	1876
Indianapolis, IN	2	Ahrens F.E.	102	1905
Indianapolis, IN	3	Ahrens Mfg.	669	1892
Indianapolis, IN	4	C. Ahrens	164	1876
Indianapolis, IN	5	C. Ahrens	37	1874

City & State	Fire Co.	Make	Reg. No.	Year
Indianapolis, IN	5	Ahrens Mfg.	670	1892
Indianapolis, IN	6	C. Ahrens	41	1874
Indianapolis, IN	6,7,9	Ahrens Mfg.	530	1888
Indianapolis, IN	7	C. Ahrens	155	1876
Indianapolis, IN	7,8	Ahrens Mfg.	671	1892
Indianapolis, IN		American	2420	1896
Indianapolis, IN		American	2782	1901
Indianapolis, IN		American	2783	1901
Indianapolis, IN		American	2925	1903
Indianapolis, IN		Ahrens-Fox	189	1911
Jackson, MI		American	2773	1900
Jackson, MS		American	2762	1900
Jacksonville, FL		Ahrens Mfg.	459	c-1912
Jacksonville, IL		ALF	2955	1904
Janesville, WI		Ahrens F.E.	QQQ	1909
Jefferson City, MO	Penetentiary	Ahrens Mfg.	393	1884
Kalamazoo, MI		American	2338	1895
Kalamazoo, MI		American	2882	1902
Kansas City, KS	Swift & Co.	Ahrens Mfg.	212	1893
Kansas City, MO	2	Ahrens F.E.	119	1907
Kansas City, MO	6,25	Ahrens Mfg.	234	1889
Kewaunee, WI		American	2510	1897
Key West, FL		Ahrens F.E.	174	1910
Lake, IL	5	Ahrens Mfg.	312	1882
Lake Providence, LA	1	Ahrens Mfg.	204	1879
Lake View, IL		Ahrens Mfg.	527	1888
Lansford, PA	1	Ahrens-Fox	205	1916
Lawrenceburg, IN	1	Ahrens Mfg.	348	1883
Lawrenceburg, IN	1	C. Ahrens	158	1901
Lawrenceburg, IN	2	Ahrens Mfg.	340	1883
Leavenworth, KS		Lane & Bodley	13	1866
Leavenworth, KS	3	C. Ahrens	157	1875
Lebanon, OH		Ahrens Mfg.	423	1887
Leipsic, OH		American	2364	1895
Lena, IL	1	Ahrens Mfg.	244	1881
Los Angeles, CA	3,5,9	American	2638	1899
Los Angeles, CA	4,8	Ahrens Mfg.	493	1887
Los Angeles, CA	4,24	American	2639	1899
Los Angeles, CA	5,2,39	Ahrens Mfg.	543	1888
Los Angeles, CA	6,20,6,25	Ahrens Mfg.	544	1888
Los Angeles, CA	6,20	Ahrens F.E.	173	1910
Los Angeles, CA	7,14	Ahrens Mfg.	537	1888
Los Angeles, CA	11,29,2,15	American	2640	1899
Los Angeles, CA	21,8	Ahrens F.E.	128	1907
Louisville, KY	1	Latta	17	1857
Louisville, KY	1,3	C. Ahrens	165	1877
Louisville, KY	2	Latta	21	1858
Louisville, KY	2,3,16	Ahrens Mfg.	276	1881
Louisville, KY	2,7	ALF	2992	1904
Louisville, KY	3	Latta	22	1858
Louisville, KY	3,12	Ahrens Mfg.	270	1880
Louisville, KY	3	Ahrens F.E.	LLL	1908
Louisville, KY	4	Latta	23	1859
Louisville, KY	4	Ahrens Mfg.	278	1881
Louisville, KY	4	ALF	2993	1904
Louisville, KY	5	Latta	24	1859
Louisville, KY	5,18	Ahrens Mfg.	250	1880
Louisville, KY	5	American	2405	1895
Louisville, KY	6	Lane & Bodley	9	1865
Louisville, KY	6,16,2	Ahrens Mfg.	220	1879
Louisville, KY	7	C. Ahrens	20	1872
Louisville, KY	7	Ahrens Mfg.	646	1891
Louisville, KY	7	American	2819	1902
Louisville, KY	8	C. Ahrens	22	1872
Louisville, KY	8	American	2820	1902
Louisville, KY	9	C. Ahrens	24	1872
Louisville, KY	9	Ahrens Mfg.	535	1888
Louisville, KY	9	Ahrens F.E.	105	1905
Louisville, KY	10	C. Ahrens	33	1874
Louisville, KY	10	Ahrens Mfg.	400	1884
Louisville, KY	11,20	Ahrens Mfg.	344	1882
Louisville, KY	12,2,1	Ahrens Mfg.	384	1883
Louisville, KY	12	American	2406	1895
Louisville, KY	13	Ahrens Mfg.	397	1884
Louisville, KY	14	Ahrens Mfg.	629	1891
Louisville, KY	15	Ahrens F.E.	124	1907
Louisville, KY	16	Ahrens F.E.	111	1906
Louisville, KY	19	American	2414	1896
Louisville, KY	(Not sold)	American	2124	1892

City & State	Fire Co.	Make	Reg. No.	Year
Loveland, OH		American	2252	1894
Madison, IN	1	C. Ahrens	5	1870
Madison, IN	2	C. Ahrens	12	1871
Madison, IN	2	American	2658	1899
Madison, IN	3	Ahrens Mfg.	660	1891
Manila, Philipines		American	2258	1894
Manila, Philipines		American	2259	1894
Manila, Philipines		American	2828	1902
Manila, Philipines		American	2829	1902
Manila, Philipines		American	2931	1903
Manila, Philipines		American	2932	1903
Manila, Philipines		American	2933	1903
Marysville, CA		ALF	3000	1904
Marysville, OH		Ahrens Mfg.	180	1879
Matanzas, Cuba		American	2807	1901
McKinney, TX		American	2383	1895
McKinney, TX		American	2385	1895
Memphis, TN		Latta	31	1861
Memphis, TN		Lane & Bodley	14	1866
Memphis, TN		Ahrens Mfg.	479	1900
Memphis, TN		Ahrens F.E.	180	1910
Memphis, TN	1	Ahrens Mfg.	318	1882
Memphis, TN	2,4	Ahrens Mfg.	320	1882
Memphis, TN	3	C. Ahrens	40	1874
Memphis, TN	4,6	Ahrens Mfg.	414	1884
Memphis, TN	5	Ahrens Mfg.	395	1884
Memphis, TN	5	Ahrens F.E.	179	1910
Memphis, TN	6,2	Ahrens Mfg.	405	1884
Memphis, TN	7	C. Ahrens	155	1888
Memphis, TN	8	C. Ahrens	41	1888
Memphis, TN	Merchants Cotton	Ahrens Mfg.	479	1887
Merced, CA		Ahrens F.E.	160	1908
Miamisburg, OH		Ahrens Mfg.	470	1886
Michigamme, MI		American	2722	1900
Midway, KY		C. Ahrens	170	1893
Midway, KY		C. Ahrens	11	1894
Millersburg, KY		American	2945	1903
Milwaukee, WI	1,8,16	Ahrens Mfg.	224	1879
Milwaukee, WI	1,2,13	Ahrens Mfg.	635	1891
Milwaukee, WI	1,21,18,7	American	2123	1892
Milwaukee, WI	1	Ahrens F.E.	VVV	1910
Milwaukee, WI	2,21,6	American	2905	1902
Milwaukee, WI	3,8,22	Ahrens Mfg.	401	1884
Milwaukee, WI	3,5	ALF	2991	1904
Milwaukee, WI	4,6,9,27,6	Ahrens Mfg.	246	1880
Milwaukee, WI	4,18	Ahrens Mfg.	372	1883
Milwaukee, WI	4	ALF	2989	1904
Milwaukee, WI	5,25	Ahrens Mfg.	591	1889
Milwaukee, WI	6	American	2173/2214	1893
Milwaukee, WI	7,24	Ahrens Mfg.	505	1887
Milwaukee, WI	7,10,16	American	2907	1902
Milwaukee, WI	9,10,14	Ahrens Mfg.	396	1884
Milwaukee, WI	9,16	Ahrens Mfg.	451	1886
Milwaukee, WI	9	American	2906	1902
Milwaukee, WI	10,12,7	ALF	2990	1904
Milwaukee, WI	11,19,11	Ahrens Mfg.	617	1890
Milwaukee, WI	12	Ahrens Mfg.	546	1888
Milwaukee, WI	13,26	Ahrens Mfg.	547	1888
Milwaukee, WI	14,28,30	Ahrens Mfg.	587	1889
Milwaukee, WI	17,18,8,25,31	American	2177/2228	1893
Milwaukee, WI	19,8	American	2737	1900
Milwaukee, WI	20,28	American	2738	1900
Milwaukee, WI	21,28	American	2280	1894
Milwaukee, WI	22	American	2445	1896
Milwaukee, WI	(Not sold)	American	2171	1893
Milwaukee, WI	Fireboat	American	2861	1902
Milwaukee, WI	Fireboat	American	2862	1902
Milwaukee, WI	Fireboat	American	2863	1902
Minneapolis, MN	3,9,4,15,18	Ahrens Mfg.	352	1883
Minneapolis, MN	3,6,5,14,8	Ahrens Mfg.	500	1886
Minneapolis, MN	6,12,19	Ahrens Mfg.	437	1885
Minneapolis, MN	8,22	Ahrens Mfg.	638	1890
Minneapolis, MN	9,14,7	Ahrens Mfg.	637	1890
Minneapolis, MN	11,22	Ahrens Mfg.	499	1886
Mobile, AL		Ahrens Mfg.	310	1888
Montgomery, AL	1	Ahrens Mfg.	226	1879
Montgomery, AL		ALF	3007	1904
Montgomery, AL	5,6	Ahrens F.E.	113	1906
Montreal, Quebec	Exposition	Ahrens Mfg.	180	1877

City & State	Fire Co.	Make	Reg. No.	Year
Montreal, Quebec	Exposition	Ahrens Mfg.	182	1877
Montreal, Quebec	Exposition	Ahrens Mfg.	196	1877
Morgan City, LA		American	2770	1900
Morrow, OH		Ahrens Mfg.	512	1888
Morrow, OH		Ahrens Mfg.	555	1889
Moss Point, MS	Dantzler Lumber	Ahrens Mfg.	310	1891
Mount Sterling, KY	2	Ahrens Mfg.	580	1889
Nashville, TN	1	Lane & Bodley	4	1863
Nashville, TN	1	Ahrens Mfg.	190	1895
Nashville, TN	2	C. Ahrens	38	1874
Nashville, TN	3	C. Ahrens	158	1875
Nashville, TN	3	American	2765	1900
Nashville, TN	4	C. Ahrens	173	1877
Nashville, TN	5	Ahrens Mfg.	380	1883
Nashville, TN	6	Ahrens Mfg.	382	1883
Nashville, TN		American	2108	1892
Nashville, TN		American	2181	??
Nashville, TN		American	2511	1897
Newark, OH		Lane & Bodley	22	1867
Newark, OH		Ahrens Mfg.	216	1879
New London, WI		American	2287	1894
New Orleans, LA		Latta	6	1855
New Orleans, LA	1	Ahrens Mfg.	324	1882
New Orleans, LA	1	Ahrens Mfg.	338	1882
New Orleans, LA	2	Ahrens Mfg.	192	1878
New Orleans, LA	3	Ahrens Mfg.	294	1881
New Orleans, LA	5	Ahrens Mfg.	238	1880
New Orleans, LA	6	Ahrens Mfg.	431	1885
New Orleans, LA	7	Ahrens Mfg.	230	1880
New Orleans, LA	9	Ahrens Mfg.	258	1880
New Orleans, LA	10	Ahrens Mfg.	222	1879
New Orleans, LA	11	Ahrens Mfg.	206	1879
New Orleans, LA	13	Ahrens Mfg.	288	1881
New Orleans, LA	14	Ahrens Mfg.	326	1882
New Orleans, LA	18	Ahrens Mfg.	214	1879
New Orleans, LA	21	Ahrens Mfg.	240	1880
New Orleans, LA	23	Ahrens Mfg.	370	1883
New Orleans, LA	Hose	Ahrens Mfg.	413	1884
New Orleans, LA		American	2423	1896
New Orleans, LA		American	2428	1896
New Orleans, LA		American	2430	1896
New Orleans, LA		American	2493	1897
New Orleans, LA		American	2502	1897
New Orleans, LA		American	2503	1897
New Orleans, LA		American	2504	1897
New Orleans, LA		American	2519	1897
New Orleans, LA		American	2520	1897
New Orleans, LA		American	2522	1897
New Orleans, LA		American	2523	1897
New Orleans, LA		American	2524	1897
New Orleans, LA		American	2526	1897
New Orleans, LA		American	2554	1897
New Orleans, LA		American	2555	1897
New Orleans, LA		American	2561	1897
New Orleans, LA		American	2569	1898
New Orleans, LA		American	2570	1898
New Orleans, LA		American	2577	1898
New Orleans, LA		American	2578	1898
New Orleans, LA		Ahrens F.E.	163	1909
New Orleans, LA		Ahrens F.E.	164	1909
New Orleans, LA		Ahrens F.E.	176	1910
New Orleans, LA		Ahrens F.E.	177	1910
New York, NY	13-1,34	Ahrens Mfg.	252	1880
New York, NY	45	Ahrens Mfg.	354	1883
New York, NY	42	Ahrens Mfg.	356	1883
New York, NY	13-2,28	Ahrens Mfg.	358	1883
New York, NY	20,11,45,208	Ahrens Mfg.	360	1883
Niles Center, OH		American	2869	1902
North Amherst, OH		Ahrens Mfg.	541	1888
North Lewisburg, OH		American	2942	1903
North Vernon, IN		Ahrens Mfg.	477	1886
North Yakima, WA		Ahrens Mfg.	577	??
Oakland, CA	5,16	Ahrens Mfg.	411	1885
Oakland, CA	5	American	2715	1900
Oklahoma City, OK		C. Ahrens	164	1906
Omaha, NE		Lane & Bodley	17	1866
Omaha, NE	4	Ahrens Mfg.	210	1879
Omaha, NE		ALF	2999	1904
Oshkosh, WI		Lane & Bodley	19	1866

City & State	Fire Co.	Make	Reg. No.	Year
Oshkosh, WI		Ahrens Mfg.	274	1881
Ottawa, IL		American	2143/2209	1894
Ottawa, KS		Ahrens Mfg.	386	1883
Owensboro, KY	1	C. Ahrens	26	1873
Oxford, OH		C. Ahrens	45	1875
Oxford, OH		American	2386	1895
Paducah, KY		Ahrens Mfg.	284	1881
Paris, KY		C. Ahrens	188	1878
Parkersburg, WV		Ahrens Mfg.	182	1878
Paterson, NJ	3,10	Ahrens-Fox	198	1913
Pekin, IL		Ahrens Mfg.	405	1884
Pendleton, OR		Ahrens Mfg.	432	1885
Pensacola, FL	2	Ahrens Mfg.	310	1881
Peoria, IL		Ahrens Mfg.	364	1883
Philadelphia, PA	Hose 1	Shawk	1	1855
Philadelphia, PA	19	Ahrens-Fox	203	1914
Philadelphia, PA	35	Ahrens-Fox	202	1914
Philadelphia, PA	39	Ahrens-Fox	200	1914
Philadelphia, PA	54	Ahrens-Fox	201	1914
Phoenix, AZ	1	Ahrens Mfg.	506	1887
Piqua, OH	1	American	2141	1892
Piqua, OH		ALF	2997	1904
Portland, IN		Ahrens Mfg.	479	1887
Portland, OR		American	2791	1901
Portland, OR		American	2824	1902
Portland, OR		ALF	3009	1904
Portsmouth, OH		Ahrens Mfg.	362	1883
Portsmouth, OH		American	2948	1903
Pueblo, CO	1	Ahrens Mfg.	583	1889
Quincy, IL		Lane & Bodley	16	1866
Quincy, IL	1	Ahrens Mfg.	616	1890
Quincy, IL	3	C. Ahrens	35	1874
Quincy, IL		American	2842	1902
Racine, WI		Lane & Bodley	18	1866
Racine, WI		Ahrens Mfg.	412	1885
Reading, PA	12	American	2122	1893
Reading, PA	Hose 1	American	2203	1893
Reno, NV		American	2823	1902
Reno, NV		ALF	2969	1904
Ripley, OH		Ahrens Mfg.	481	1887
Riverside, CA		ALF	3003	1904
Roanoke, VA	2	Ahrens F.E.	CCC	1907
Rochester, NY	23	Ahrens-Fox	197	1913
Rockford, IL	3	American	2105	1892
Rockford, IL		American	2319	1894
Rockford, IL		American	2656	1899
Rushville, IN	1	Ahrens Mfg.	282	1881
Rushville, IN		American	2250	1893
Sacramento, CA		American	2928	1903
St. Joseph, MO		American	2262	1894
St. Louis, MO	1	Latta	14	1856
St. Louis, MO	1	C. Ahrens	147	1875
St. Louis, MO	1	American	2285	1894
St. Louis, MO	2	Shawk	2	1855
St. Louis, MO	2	C. Ahrens	8	1870
St. Louis, MO	2	American	2867	1902
St. Louis, MO	3	Latta	15	1857
St. Louis, MO	3	C. Ahrens	152	1875
St. Louis, MO	4	Latta	16	1857
St. Louis, MO	4	C. Ahrens	149	1875
St. Louis, MO	4,29	Ahrens Mfg.	599	1890
St. Louis, MO	5	Shawk	3	1856
St. Louis, MO	5	C. Ahrens	156	1876
St. Louis, MO	5	ALF	2961	1904
St. Louis, MO	6	Latta	18	1857
St. Louis, MO	6	C. Ahrens	148	1875
St. Louis, MO	6	American	2584	1898
St. Louis, MO	7	Latta	19	1858
St. Louis, MO	7	C. Ahrens	150	1875
St. Louis, MO	7	American	2868	1902
St. Louis, MO	8	Lane & Bodley	11	1865
St. Louis, MO	8	C. Ahrens	15	1871
St. Louis, MO	8	American	2337	1895
St. Louis, MO	9	Lane & Bodley	23	1867
St. Louis, MO	9,12	C. Ahrens	19	1872
St. Louis, MO	9	C. Ahrens	154	1875
St. Louis, MO	10	Lane & Bodley	24	1867
St. Louis, MO	10,14	C. Ahrens	23	1872
St. Louis, MO	10	C. Ahrens	162	1876

City & State	Fire Co.	Make	Reg. No.	Year
St. Louis, MO	11	C. Ahrens	16	1871
St. Louis, MO	11	Ahrens Mfg.	398	1884
St. Louis, MO	11,43	American	2125	1892
St. Louis, MO	11	American	2562	1897
St. Louis, MO	12	Ahrens Mfg.	330	1882
St. Louis, MO	12	American	2563	1897
St. Louis, MO	13	C. Ahrens	21	1872
St. Louis, MO	13	American	2880	1902
St. Louis, MO	14	American	2747	1900
St. Louis, MO	15	C. Ahrens	28	1873
St. Louis, MO	15,9	Ahrens Mfg.	651	1891
St. Louis, MO	15	American	2585	1898
St. Louis, MO	16	C. Ahrens	29	1873
St. Louis, MO	16,23	American	2181/2247	1893
St. Louis, MO	17	C. Ahrens	30	1873
St. Louis, MO	18	C. Ahrens	44	1875
St. Louis, MO	18	American	2799	1901
St. Louis, MO	19	Ahrens Mfg.	196	1878
St. Louis, MO	20	Ahrens Mfg.	208	1878
St. Louis, MO	20	American	2421	1896
St. Louis, MO	21	Ahrens Mfg.	236	1880
St. Louis, MO	21	Ahrens Mfg.	508	1887
St. Louis, MO	22,3	Ahrens Mfg.	399	1884
St. Louis, MO	22	Ahrens Mfg.	600	1890
St. Louis, MO	23,34	Ahrens Mfg.	402	1884
St. Louis, MO	24,44	Ahrens Mfg.	419	1885
St. Louis, MO	24	American	2881	1902
St. Louis, MO	25	Ahrens Mfg.	420	1885
St. Louis, MO	25	ALF	2962	1904
St. Louis, MO	26,14,45	Ahrens Mfg.	467	1886
St. Louis, MO	26,35	Ahrens Mfg.	532	1888
St. Louis, MO	27	Ahrens Mfg.	502	1887
St. Louis, MO	27,10,8	Ahrens Mfg.	564	1888
St. Louis, MO	28	Ahrens Mfg.	562	1888
St. Louis, MO	29	Ahrens Mfg.	565	1888
St. Louis, MO	30	Ahrens Mfg.	653	1891
St. Louis, MO	31,46	American	2126	1892
St. Louis, MO	31	American	2321	1895
St. Louis, MO	32,26	American	2121	1892
St. Louis, MO	32	American	2492	1896
St. Louis, MO	33	American	2360	1895
St. Louis, MO	37	American	2253	1894
St. Louis, MO	38	American	2694	1899
St. Louis, MO	40	ALF	2960	1904
St. Louis, MO	41	ALF	2994	1904
St. Louis, MO	42	American	2320	1895
St. Louis, MO	47	American	2693	1899
St. Louis, MO		American	2174/2246	1893
St. Louis, MO		American	2381	1895
St. Louis, MO		American	2417	1896
St. Louis, MO	Exposition	American	2902	1902
St. Paris, OH		Ahrens Mfg.	415	1885
St. Paul, MN		Lane & Bodley	20	1866
St. Paul, MN	1	Ahrens F.E.	142	1908
St. Paul, MN	2	Ahrens Mfg.	391	1884
St. Paul, MN	3,22,1	Ahrens Mfg.	483	1886
St. Paul, MN	3	Ahrens-Fox	aaa	1910
St. Paul, MN	7	Ahrens-Fox	YYY	1910
St. Paul, MN	8,4	Ahrens Mfg.	448	1885
St. Paul, MN	9,16,20	Ahrens Mfg.	529	1887
St. Paul, MN	10	Ahrens Mfg.	509	1887
St. Paul, MN	14	Ahrens-Fox	ZZZ	1910
St. Paul, MN	15	Ahrens Mfg.	511	1887
St. Paul, MN	16	Ahrens F.E.	143	1908
St. Paul, MN		ALF	2978	1904
St. Paul, MN		Ahrens F.E.	158	1908
St. Regis Falls, NY	Bklyn Cooperage	American	2103	1908
Salt Lake City, UT		American	2844	1902
San Antonio, TX	1,3	Ahrens Mfg.	449	1886
San Antonio, TX	2	American	2106	1892
San Antonio, TX	1	American	2107	1892
San Antonio, TX	4	American	2129	1892
San Diego, CA	2	Ahrens Mfg.	561	1888
San Diego, CA		Ahrens F.E.	118	1906
San Francisco, CA		American	2399	1895
San Francisco, CA		American	2404	1895
San Francisco, CA		American	2623	1899
San Francisco, CA		American	2624	1899
San Francisco, CA		American	2652	1899

City & State	Fire Co.	Make	Reg. No.	Year
San Francisco, CA		American	2653	1899
San Francisco, CA		American	2675	1899
San Francisco, CA		American	2676	1899
San Francisco, CA		American	2786	1901
San Francisco, CA	34	Ahrens F.E.	156	1908
San Francisco, CA	41	Ahrens F.E.	169	1910
San Jose, CA	1	American	2804	1901
San Jose, CA	1	Ahrens F.E.	DDD	1907
San Jose, CA	1	American	2676	1914
San Salvador, El Salvador		American	2938	1903
Seattle, WA	1,4	Ahrens Mfg.	597	1889
Seattle, WA	2,3,1	Ahrens Mfg.	598	1889
Seattle, WA	2,14	American	2686	1899
Seattle, WA		American	2854	1902
Seattle, WA		American	2856	1902
Seattle, WA	3	Ahrens F.E.	137	1907
Seattle, WA	4	Ahrens F.E.	112	1906
Seattle, WA		Ahrens F.E.	171	1910
Seymour, IN	2	Ahrens Mfg.	433	1885
Sheboygan, WI		Ahrens Mfg.	350	1883
Shelbyville, IN	3	C. Ahrens	42	1875
Sidney, OH	1	Ahrens Mfg.	378	1883
Somerville, MA	2	Ahrens-Fox	199	1913
Spencerville, OH		American	2387	1895
Spokane, WA	4	American	2777	1900
Spokane, WA		Ahrens F.E.	178	1910
Springfield, IL	3	Ahrens Mfg.	622	1890
Springfield, OH	2,8	American	2839	1903
Springfield, OH	4	American	2841	1903
Springfield, MO	1	American	2911	1902
Springfield, MO	2	American	2912	1902
Springfield, MO		Ahrens-Fox	bbb	??
Springfield, MO		Ahrens-Fox	ccc	??
Sullivan, IN		Ahrens Mfg.	435	1885
Superior, WI	1,2	Ahrens Mfg.	648	1891
Superior, WI	1	Ahrens F.E.	110	1906
Tacoma, WA	4,5	Ahrens Mfg.	577	1889
Tacoma, WA	2	Ahrens F.E.	AAA	1906
Taunton, MA		Ahrens-Fox	199	??
Terre Haute, IN		Ahrens F.E.	108	1906
Terre Haute, IN		Ahrens F.E.	109	1906
Toledo, OH		Latta	25	1860
Toledo, OH	1	Ahrens F.E.	107	1906
Toledo, OH	2	Ahrens Mfg.	593	1889
Toledo, OH	3	Ahrens F.E.	129	1907
Toledo, OH	5	Ahrens Mfg.	594	1889
Toledo, OH	7	Ahrens F.E.	130	1907
Toledo, OH		Ahrens F.E.	157	1908
Topeka, KS	1	Ahrens Mfg.	194	1877
Topeka, KS	2	ALF	2998	1904
Tower, MN	2	Ahrens Mfg.	668	1892
Traverse City, MI		American	2378	1895
Traverse City, MI		American	2806	1901
Uniontown, PA		Ahrens Mfg.	176	1877
U.S. Army		Ahrens Mfg.	242	1880
Valdez, AK		Ahrens F.E.	131	1907
Vancouver, WA		ALF	2986	1904
Van Wert, OH	1	C. Ahrens	167	1877
Van Wert, OH	1	Ahrens F.E.	139	1907
Versailles, KY		Ahrens Mfg.	540	1888
Vicksburg, MS	2	Ahrens Mfg.	422	1885
Vicksburg, MS	3	Ahrens Mfg.	332	1882
Vincennes, IN		Ahrens Mfg.	234	1880
Walla Walla, WA		ALF	2968	1904
Washington, DC	1,14	Ahrens-Fox	188	1911
Washington, IN	1	Ahrens Mfg.	202	1878
Watertown, WI	1	C. Ahrens	161	1876
Waukesha, WI	Indus'l School	Ahrens Mfg.	286	1882
Wausau, WI		Ahrens Mfg.	266	1880
Waverly, OH		C. Ahrens	153	1875
Waynesville, OH		Ahrens Mfg.	457	1886
West Belleville, IL		Ahrens Mfg.	304	1881
West Bend, WI		American	2870	1902
Westwood, OH	1	Ahrens Mfg.	661	1891
Wheeling, WV	4	American	2110	1892
Wheeling, WV	2	American	2130	1892
Wheeling, WV	8	American	2339	1895
Wheeling, WV		American	2945	1903
Wheeling, WV		Ahrens F.E.	OOO	??

City & State	Fire Co.	Make	Reg. No.	Year
Wheeling, WV		Ahrens F.E.	PPP	??
Wichita, KS	1	Ahrens Mfg.	602	1890
Wichita, KS	2	American	2821	1902
Wilkes-Barre, PA	4	Ahrens Mfg.	582	1889
Wilmington, OH		Ahrens Mfg.	405	1884
Wilmington, OH	1	Ahrens Mfg.	605	1890
Wilmington, OH		Ahrens Mfg.	348	1902
Woodbury, NJ	1	Ahrens-Fox	184	1911
Xenia, IL		American	2866	1902
Youngstown, OH		ALF	3001	1904

Index of Ahrens Patents

Date	Patent #	Inventor(s)	Description
9-20-1853	10041	Abel Shawk	improved Buchanan boiler
6-6-1854	11025	A.B. Latta	steam generator
4-10-1855	12682	A.B. Latta	steam generator
5-22-1855	12912	A.B. Latta	fire engine chassis
5-27-1856	14963	A.B. Latta	safety valve
7-8-1856	15297	A.B. Latta	iron wheels for fire engines
10-20-1857	18460	A.B. Latta	steam generator
4-12-1859	23589	A.B. Latta	steam generator
1-31-1871	111,298	Chris Ahrens/Frank Kamman	Upright coil boiler
4-10-1877	189,414	Chris Ahrens	Upright pump
10-23-1883	287,211	Chris Ahrens	improved coil boiler
1-22-1884	292,396	Chris Ahrens	pump valve
8-11-1885	324,207	Chris Ahrens	platform springs
4-18-1893	495,868	Chris Ahrens	tubular boiler
5-1-1894	519,013	Chris Ahrens	flue boiler
5-12-1896	559,881	Chris Ahrens/Chas. H. Fox	automatic relief valve
9-29-1896	568,438	Charles H. Fox	steam boiler
10-20-1896	569,966	Charles H. Fox	steam boiler
5-3-1898	603,482	Charles H. Fox	vertical water-tube boiler
5-3-1898	603,502	Charles H. Fox	sectional water-tube boiler
9-26-1899	633,644	Charles H. Fox	stationary heater
2-4-1902	692,436	Charles H. Fox	steam generator
5-20-1902	700,320	Charles H. Fox	steam generator
10-25-1902	485,112	Hiram H. Nieman	flue protector
4-7-1903	724,564	Charles H. Fox	steam generator
9-12-1905	799,196	William H. Shafer	helical water-tube boiler
9-12-1905	799,197	William H. Shafer	inclined water-tube boiler
9-12-1905	799,198	William H. Shafer	radial water-tube boiler
11-7-1905	803,889	Charles H. Fox	throttle valve
11-7-1905	805,568	William H. Shafer	Continental pump barrel
5-29-1906	821,674	William H. Shafer	Continental pump
11-12-1907	870,646	William H. Shafer	water-tower fire engine
11-19-1907	871,147	William H. Shafer	chemical fire extinguisher
12-24-1907	874,546	William H. Shafer	aerial ladder
1-14-1908	876,404	William H. Shafer	traction fire engine
1-28-1908	877,601	William H. Shafer	inclined radial water-tube boiler
3-3-1908	880,513	Charles H. Fox	advertisement
4-14-1908	884,769	William H. Shafer	force pump
3-29-1910	953,166	Charles H. Fox	steam boiler

Siblings and Children of Chris Ahrens

Friedrich Ahrens
b. Brettbeck, Hanover
1800
d. Brettbeck 1869

Anna (Hartmann) Ahrens
b. Brettbeck, Hanover
1804
d. Brettbeck 1866

Children

Hermann Ahrens
b. Brettbeck
2-22-1829
Emigrated 1847
m. Rebecca Eggers
d. Cincinnati
8-17-1886

Frederick Ahrens
b. Brettbeck
12-1830
Emigrated 1849
d. Cincinnati
4-17-1868

Elizabeth Ahrens
b. Brettbeck
12-16-1833
Emigrated 1852
m. Mr. Willman
d. Cincinnati
9-2-1859

Chris Ahrens
b. Brettbeck
12-7-1834
Emigrated 1853
m. Mary Seidenburg
12-11-1862
d. Cincinnati
6-24-1919
wife b. 9-22-1843
wife d. 1-10-1923

Anna Ahrens
b. Brettbeck
(1836?)
m. Mr. Albers
d. Brettbeck

Mary Ahrens
b. Brettbeck
8-12-1839
Emigrated 1858
m. Mr. Embhoff
d. Cincinnati
9-13-1911

Peter Ahrens
b. Brettbeck
4-14-1841
Emigrated 1861
never married
d. Cincinnati
12-18-1908

Children of Chris Ahrens

Geo. Fred Ahrens
b. Cincinnati
9-1-1863
d. Cincinnati
8-16-1953

Lettie Ahrens
b. Cincinnati
7-24-1865
m. Geo. W. Krapp
9-14-1887
d. Cincinnati
2-12-1956
hus b. 5-2-1863
hus d. 1-13-1953

Lillie Ahrens
b. Cincinnati
7-7-1870
m. Chas H Fox
11-2-1892
d. Cincinnati
9-17-1904
hus b. 12-16-1860
hus d. 1-26-1950

John P. Ahrens
b. Cincinnati
8-11-1872
m. Louise Krug
d. Cincinnati
6-11-1943
wife b. 3-30-1874
wife d. 1-28-1950

Christine Ahrens
b. Cincinnati
2-24-1878
never married
d. Cincinnati
7-15-1947

Clara Ahrens
b. Cincinnati
12-15-1879
m. Robert Abbihl
d. Cincinnati
11-11-1966

Florence Ahrens
b. Cincinnati
9-13-1885
m. Albert H. Meyer
d. Cincinnati
2-2-1946

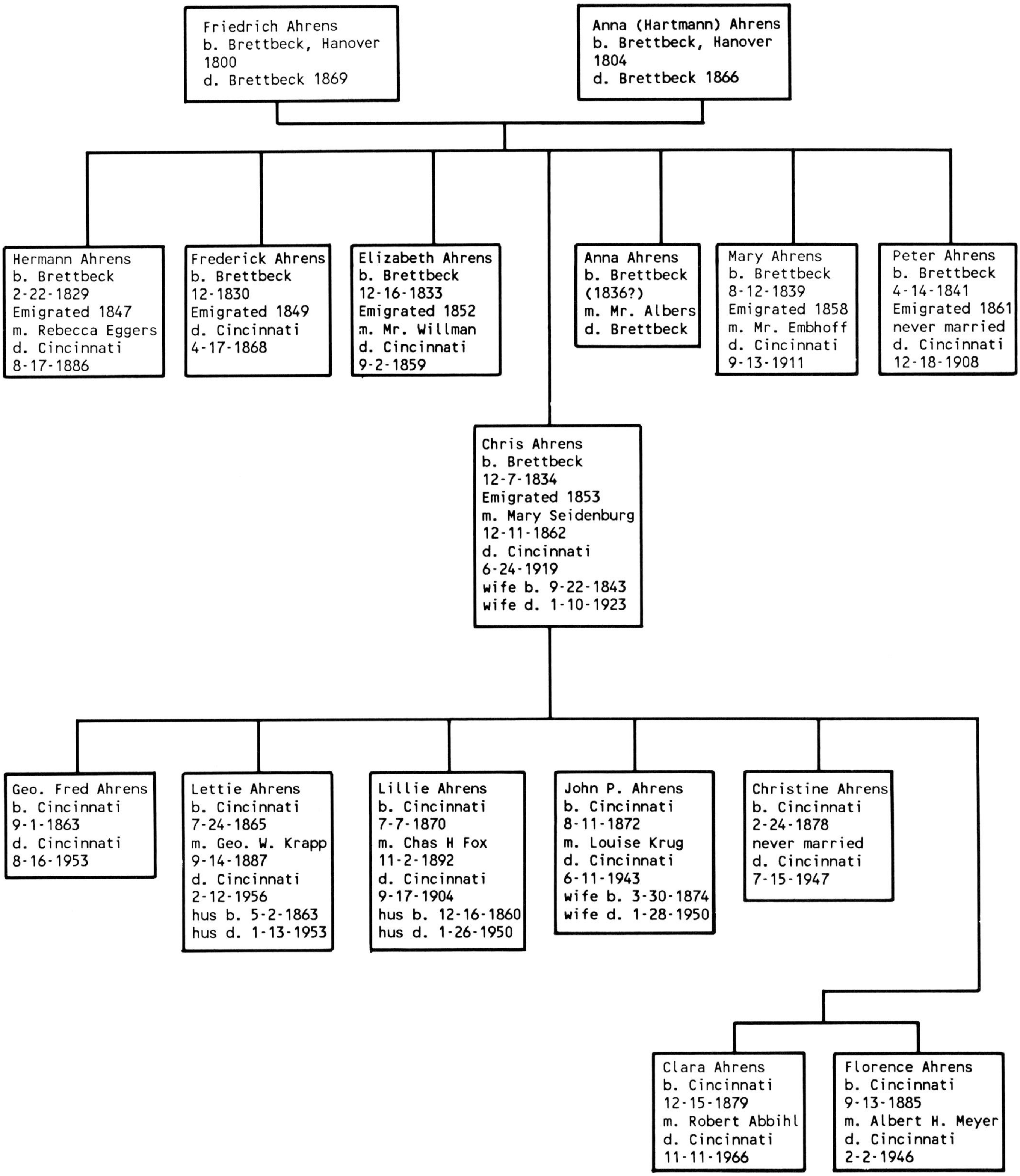